Solder Joint Reliability of BGA, CSP, Flip Chip, and Fine Pitch SMT Assemblies

Electronic Packaging and Interconnection Series

Charles M. Harper, Series Advisor

ALVINO • *Plastics for Electronics*

CLASSON • *Surface Mount Technology for Concurrent Engineering and Manufacturing*

GINSBERG, SCHNOOR • *Multichip Module and Related Technologies*

HARPER • *Electronic Packaging and Interconnection Handbook*

HARPER, MILLER • *Electronic Packaging, Microelectronics, and Interconnection Dictionary*

HARPER, SAMPSON • *Electronic Materials and Processes Handbook*, 2/e

LAU • *Ball Grid Array Technology*

LAU • *Flip Chip Technologies*

LICARI • *Multichip Module Design, Fabrication, and Testing*

HWANG • *Modern Solder Technology for Competitive Electronics Manufacturing*

Related Books of Interest

BOSWELL • *Subcontracting Electronics*

BOSWELL, WICKAM • *Surface Mount Guidelines for Process Control, Quality, and Reliability*

BYERS • *Printed Circuit Board Design with Microcomputers*

CAPILLO • *Surface Mount Technology*

CHEN • *Computer Engineering Handbook*

COOMBS • *Electronic Instrument Handbook*, 2/e

COOMBS • *Printed Circuits Handbook*, 4/e

DI GIACOMO • *Digital Bus Handbook*

DI GIACOMO • *VLSI Handbook*

FINK, CHRISTIANSEN • *Electronics Engineers' Handbook*, 3/e

GINSBERG • *Printed Circuits Design*

JURAN, GRYNA • *Juran's Quality Control Handbook*

JURGEN • *Automotive Electronics Handbook*

MANKO • *Solders and Soldering*, 3/e

RAO • *Multilevel Interconnect Technology*

SZE • *VLSI Technology*

VAN ZANT • *Microchip Fabrication*

To order or receive additional information on these or any other McGraw-Hill titles, please call 1-800-822-8158 in the United States. In other countries, contact your local McGraw-Hill representative. **KEY = WM16XXA**

Solder Joint Reliability of BGA, CSP, Flip Chip, and Fine Pitch SMT Assemblies

John H. Lau
Express Packaging Systems, Inc.

Yi-Hsin Pao
Ford Motor Company

McGraw-Hill
New York San Francisco Washington, D.C. Auckland Bogotá
Caracas Lisbon London Madrid Mexico City Milan
Montreal New Delhi San Juan Singapore
Sydney Tokyo Toronto

Library of Congress Cataloging-in-Publication Data

Lau, John H.
 Solder joint reliability of BGA, CSP, flip chip, and fine pitch SMT assemblies /
John H. Lau, Yi-Hsin Pao.
     p.   cm. — (Electronic packaging and interconnection series)
 Includes bibliographical references and index.
 ISBN 0-07-036648-9 (hc)
 1. Microelectronic packaging—Reliability.   2. Solder and
soldering—Testing.   3. Multichip modules (Microelectronics)—
Testing.   I. Pao, Yi-Hsin.   II. Title.   III. Series.
 TK7874.L317   1997
 621.381'046—dc20   96-31361
                                                                          CIP

McGraw-Hill

A Division of The **McGraw·Hill** Companies

ISBN 0-07-036648-9

*The sponsoring editor for this book was Stephen S. Chapman, the
editing supervisor was Christina M. Palaia, and the production
supervisor was Pamela A. Pelton. It was set in Century Schoolbook by
North Market Street Graphics.*

Printed and bound by R. R. Donnelley & Sons Company.

To our parents for their encouragement

To our wives Teresa Lau and Yin-Fen Pao for their support

To our daughters Judy Lau, Janet Pao, and Tiffany Pao for their understanding

To the ones who published peer-reviewed papers for sharing their knowledge

Contents

Foreword

Terry T. M. Gou

The invention of the bipolar junction transistor and the junction field-effect transistor by Bardeen, Brattain, and Shockley at Bell Laboratories in 1947 foreshadowed the development of generations of computers yet to come. The invention of the silicon integrated circuit (IC) by Jack Kilby of Texas Instruments in 1958 and by Robert Noyce and Gordon Moore of Fairchild Semiconductor in 1959 excited the development of generations of (small, medium, large, very large, ultra large, and other yet to come) scale integrations. Fifty years ago, the average family probably owned five active electronic devices. Today, the average family owns several million transistors, and it is probable that by the end of the century, each family will own more than one billion transistors.

There was no semiconductor industry until the early 1950s, and last year the total worldwide merchant semiconductor usage was $144 billion. ICs were first marketed in 1961, and last year about $110 billion worth of IC chips were sold. These results have produced a global revolution of major proportions that affects the lives of people everywhere.

The IC chip is not an isolated island. It must communicate with other IC chips in a circuit through an input/output (I/O) system of interconnects. Furthermore, the IC chip and its embedded circuitry are delicate, requiring the package to both carry and protect it. Consequently, the major functions of the electronics package are: (1) to provide a path for the electrical current that powers the circuits on the IC chip, (2) to distribute the signals onto and off of the IC chip, (3) to remove the heat generated by the circuits, and (4) to support and protect the IC chip from hostile environments.

As the trend toward higher I/O, higher performance, and higher printed circuit board (PCB) manufacturing yield continues, ball grid array (BGA), chip-scale package (CSP), and flip chip become the packages of choice. For these packages the solder joints are the only mechanical means of attaching them to the PCB. Unfortunately, for most of the practicing engineers and managers, solder joint reliability is the least understood in electronics packaging and interconnect systems.

Thus, there was an urgent need, both in industry and in universities, to create a comprehensive book on the current state of knowledge of reliability testing, failure analysis, and design for solder joint reliability. This book would be written so that anyone could quickly learn about the basics and problem-solving methods, understand the trade-offs, and make system-level decisions.

To meet this need, John H. Lau and Yi-Hsin Pao have collected much useful information from the latest technical publications. Together they have produced *Solder Joint Reliability of BGA, CSP, Flip Chip, and Fine Pitch SMT Assemblies,* an excellent book for both industry and university use. It is equally appropriate as an introduction to solder joint reliability for those just entering the field and as an up-to-date reference for those already engaged in interconnection design and development.

This book deals with the subject of solder joint reliability along three parts: reliability testing and statistical data analysis, failure analysis and root cause of solders, and design for reliability of BGA, CSP, flip chip, and fine pitch SMT assemblies under thermal, mechanical, and shock and vibration conditions. I join the authors in hoping that this book will help focus the attention of practicing engineers and scientists, as well as faculty and students, on the complex solder joint reliability challenges that must be solved. I also hope that it will serve a future generation of engineers, scientists, and students who will continue to further the science and engineering of solder joint reliability of interconnect systems. Since they are armed with their predecessors' knowledge and the creativity of their minds, there is no doubt these wishes will come true. It is an exciting time for solder!

Terry T. M. Gou
Chairman and C.E.O.
Hon Hai Precision Industry Co., Ltd.

Foreword

by Walter L. Winterbottom

Design is an interdisciplinary activity that relies on the expertise of the engineering profession and is buttressed by the methodology and innovations developed within the fields of science. The integration of science and engineering has produced remarkable changes in the commercial products of today as illustrated by the progression from an understanding of the solid-state physics of the semiconductor to the development of the transistor and to the applications in computers and microprocessors. Similarly in the transportation industry, electronics and the microprocessor are responsible for major changes in the control systems, functional capability, and performance of every vehicle.

The methodology of design is being adjusted in order to meet the needs of society and to address the conservation and environmental requirements of a full-life cycle, "concept-to-grave" product responsibility. In responding to customer needs, the designer's goals are reliable and cost-effective products. To achieve such products there is a need for a level of design optimization that is only now becoming possible through the use of tools developed in the science of materials and mechanics.

The extended-life goals of the automakers have imposed extreme demands on electronic packaging reliability. Currently, the emphasis is focused on implementing extensive testing programs to assure reliability growth during the product development process. Although this iterative, build-test-fix process has a long history of application in new product development, concerns are now beginning to surface that world-class, cost-effective, and fast-to-market needs may not be attainable with such an approach. Hardware buildup and testing are costly and time-consuming, and severely limit the number of design alternatives that can be explored in a shortened development cycle to promote truly optimal and innovative products. In order to achieve the ambitious reliability goals, the design procedure will have to be modified to become preventive, with increased analytical effort earlier in the process.

The authors have been promoting the advantages of physically based models to provide interconnect design information useful in the early stages of the design process where the uncertainties of material selection, manufacturing variability quality, and the spectrum of possible service conditions can be addressed prior to hardware build and test. They have demonstrated the need to consider the full system complexity of packaging structures and for accurate material properties and constitutive equations in order to adequately model the solder joint deformation.

This book on solder joint reliability and interconnect design illustrates a design-for-reliability methodology. The approach is mechanistic and is used to address the reliability issues of the most recent packaging interconnect designs required in packaging miniaturization. The reliability risks posed by harsh environments on advanced packaging designs of solder-bumped flip chip-on-board (FCOB), ball grid array (BGA), chip-scale package (CSP), and other surface-mount assemblies are explored. The authors, in describing their work and the contributions of others, present a design approach that must ultimately replace the build-test-fix process if the efficiencies and potential cost benefits of advanced electronics are to be fully realized.

Walter L. Winterbottom, Ph.D.
Formerly, Manager of the Material Systems Reliability Department
Ford Research Laboratory, Dearborn, Michigan

Preface

Twelve years ago, when the principal author (Lau) was a newcomer in the electronics industry, he stated that from a mechanics point of view, solders were very poor at resisting thermal fatigue and should not be used as joining materials for interconnections. Since then, after assembling many electronics components on printed circuit boards (PCB), he realizes that he was absolutely wrong about solders as joining materials and greatly appreciates the "convenience" solders provide. Solders have given us a remarkable flexibility to interconnect electronics components. The unique properties of solders have facilitated PCB assembly choices that have fueled creative advance packaging developments, e.g., solder-bumped flip chip technology, solder ball grid array (BGA) technology, and solder chip-scale packaging (CSP). For these technologies, solder is the electrical and mechanical "glue," and, thus, solder joint reliability is one of the most critical issues in the development of these technologies. In this book, the solder joint reliability of various BGA, CSP, flip chip, and fine pitch surface-mount technology (SMT) assemblies will be presented.

A variety of environmental stress factors (e.g., temperature and humidity, shock and vibration, and thermal/power cycles) may lead to solder joint failure, and the most commonly observed failure modes in practice are overload and fatigue. *Overload failure* occurs whenever the stress in the solder joint, brought about by the imposed stress factors, is greater than the strength or fracture toughness of the solder alloy. An example would be excessive bending and twisting of a PCB with flip chips, CSP, or BGA soldered to its surface. On the other hand, *fatigue failure* takes place via the initiation and slow propagation of a crack until it becomes unstable. The stress factors that typically cause failure by fatigue are far below the overload failure levels. Examples of fatigue failure include the vibration of a PCB with flip chips, CSP, or BGA soldered to its surface and the power cycling of a flip chip, CSP, or BGA which is soldered to the surface of a PCB. In this book the solder joint reliability of BGA, CSP, flip chip, and fine pitch SMT assemblies under thermal, mechanical bending, twisting and shearing, and shock and vibration conditions will be thoroughly discussed.

Reliability engineering of solder joints consists of three major tasks, namely, design for reliability (DFR), reliability testing and data analysis, and failure analysis. Usually, it starts off with a design (or a few designs) and demonstrates that it (or one of the few) is electrically, thermally, and mechanically reliable by analysis, e.g., the finite element method. Then it builds the reliable one with a certain sample size and tests it under certain conditions. The objective of the test is to find failures (the more the better) and to choose a failure distribution (e.g., Weibull, lognormal, or exponential) to fit the failure data (to determine the parameters of the distribution). The distribution then becomes the life distribution of the solder joints. Failure analysis is then done on the failed sample in order to understand why it failed. This information is very useful for the next round of DFR. The first part of this book presents the fundamental theories and principles of reliability testing and data analysis (Chap. 2), failure modes and analysis methods of SMT solder joints (Chap. 3), and DFR of solder joints (Chap. 4).

Chapter 5, 6, 7, and 8, respectively, present the solder joint reliability of BGA, flip chip, CSP, and fine pitch assemblies under thermal, mechanical, and shock and vibration conditions. Details on no-clean solder reflow processes of BGA, flip chip, and fine pitch surface-mount components on PCB are also discussed.

For whom is this book intended? Undoubtedly, it will be of interest to three groups of specialists: (1) those who are active or intend to become active in research and development of BGA, CSP, flip chip, and fine pitch technologies; (2) those who have encountered practical solder joint reliability problems and wish to understand and learn other methods of solving such problems; and (3) those who must choose a reliable, creative, high-performance, robust, and cost-effective packaging technique for their interconnect system. This book also can be used as a text for college and graduate students who may become our future leaders, scientists, and engineers in the electronics industry.

We hope this book will serve as a valuable source of reference to all those faced with the challenging problems created by the ever-expanding use of solder BGA, solder CSP, and solder-bumped flip chip technologies in electronics packaging and interconnection. We also hope that it will aid in stimulating further research and development on solder materials, testing, and analytical methods, and more sound use of solders in new packaging applications.

The organizations that learn how to design and manufacture reliable solder joints in their BGA, CSP, and flip chip systems have the potential to make major advances in electronics packaging and to gain great benefits in cost, performance, quality, size, and weight. It is our hope that the information presented in this book may assist in removing road blocks, avoiding unnecessary false starts, and accelerate the applications of these technologies. Reliability of the BGA, CSP, and flip chip solder joints is limited only by the ingenuity and imagination of engineers, managers, and researchers.

John H. Lau, Ph.D., PE, IEEE Fellow
Express Packaging Systems, Inc.
Palo Alto, California

Yi-Hsin Pao, Ph.D.
Research Laboratory
Ford Motor Company
Dearborn, Michigan

Acknowledgments

Development and preparation of *Solder Joint Reliability of BGA, CSP, Flip Chip, and Fine Pitch SMT Assemblies* was facilitated by the efforts of a number of dedicated people at McGraw-Hill and North Market Street Graphics. We would like to thank them all, with special mention to Christina Palaia of North Market Street Graphics and Donna Namorato and David Fogarty of McGraw-Hill for their unswerving support and advocacy. Our special thanks to Steve Chapman (Editor of Electronics and Optical Engineering) who made our dream of this book come true by effectively sponsoring the project and solving many problems that arose during the book's preparation. It has been a great pleasure and fruitful experience to work with them in transferring our messy manuscript into a very attractive printed book.

The material in this book has clearly been derived from many sources including individuals, companies, and organizations, and we have attempted to acknowledge, in the appropriate parts of the book, the assistance that we have been given. It would be quite impossible for us to express our thanks to everyone concerned for their cooperation in producing this book, but we would like to extend due gratitude. Especially, we want to thank several professional societies and publishers for permitting us to reproduce some of their illustrations and information in this book. For example: the American Society of Mechanical Engineers (ASME) Conferences, Proceedings, and Transactions (*Journal of Electronic Packaging*); the Institute of Electrical and Electronic Engineers (IEEE) Conferences, Proceedings, and Transactions (*Components, Packaging, and Manufacturing Technology*); the International Society of Hybrid Microelectronics (ISHM) and the International Electronic Packaging Society (IEPS) Conferences, Proceedings, and Transactions (*Microcircuits & Electronic Packaging*); American Society of Metals (ASM) Conferences, Proceedings, and books (e.g., *Electronic Materials Handbook,* vol. 1, *Packaging*); the Surface Mount Technology Association (SMTA) Conferences, Proceedings, and Journals (e.g., *Journal of Surface Mount Technology*); the National Electronic Packaging Conferences (NEPCON) and Proceedings; the *IBM Journal of Research and Development; Electronic Packaging & Production; Circuits Assembly; Surface Mount Technology; Connection Technology; Solid State Technology; Circuit World; Hybrid Circuits;* and *Soldering and Surface Mount Technology.*

Each chapter of the book was reviewed by at least three individuals who are experts in solder joint reliability, BGA, CSP, flip chip, fine pitch SMT, and related areas. These reviewers are Dr. Kuan-Luen Chen, Express Packaging Systems; Dr. Yung-Shih Chen, Express Packaging Systems; Dr. Tai-Yu Chou, Express Packaging Systems; Dr. Jimmy M. Hu, Ford Motor Company; Dr. Edward Jih, Ford Motor Company; Dr. Wen-Je Jung, Ford Motor Company; Dr. Wei H. Koh, Express Packaging Systems; and Dr. Frank Wu, Express Packaging Systems. We want to thank them for their many helpful comments and constructive suggestions that added significantly to this book. Their depth of knowledge and dedication have been demonstrated throughout the process of reviewing this book.

The first author (Lau) wants to thank his former employer, Hewlett-Packard Company, for providing him with an excellent working environment that has nurtured him as a human being, fulfilled his job satisfaction, and enhanced his professional reputation. He also wants to thank his eminent colleagues (the enumeration of whom would not be practical here) at Hewlett-Packard Company, Express Packaging Systems, and throughout the electronics industry for their useful help, strong support, and stimulating discussions. Working and socializing with them has been a privilege and an adventure. He learned a lot about life and electronics packaging from them.

The second author (Pao) would like to thank his friends and colleagues, Ed Jih, Wayne Jung, Venu Siddapureddy, Ratan Govila, Scott Badgley, Xu Song, Ron Cooper, Chuck Larner, Jimmy Hu, Brad Boswell, and others at Ford Research Laboratory and Automotive Component Division for their long-term friendship and teamwork that have made part of this book possible. It has been a privilege and great pleasure to work with this strong team that possesses a variety of expertise and skills in such an interdisciplinary research area as electronics packaging. Also, he would like to express his sincere appreciation to Ford Motor Company and the management team of Ford Research Laboratory, Bill Powers, Norm Gjostein, Charles Wu, Walt Winterbottom, and Ken Huebner, for their strong and continued support in the research of electronics packaging reliability.

Lastly, the first author (Lau) wants to thank his daughter, Judy, and his wife, Teresa, and the second author (Pao) would like to thank his wife, Yin-Fen, and daughters, Janet and Tiffany, for their love, consideration, and patience in allowing them to work on many weekends for this book. Their simple belief that the authors are making a small contribution to the electronics industry was a strong motivating factor, and to their families the authors have dedicated their efforts on this book. The authors also would like to dedicate this book to their parents.

John H. Lau, Ph.D., PE, IEEE Fellow
Palo Alto, California

Yi-Hsin Pao, Ph.D.
Dearborn, Michigan

Chapter

1

Introduction

1.1 Integrated Circuit (IC) Trends

Figures 1.1 and 1.2 show the trends of IC density (number of transistors per chip) and chip size, respectively. It can be seen that there are more and more transistors on the chip (increasing at an average rate of 1.4/year), and the chip sizes are getting bigger and bigger (at an average rate of 1.13/year). One of the key reasons is because of the CMOS (complementary metal-oxide-semiconductor) process, which has very fine feature sizes and high yield. Right now, 0.35-μm drawing gate length (0.29 μm effective) semiconductor ICs are in volume production. Before the turn of the century, the 0.2-μm CMOS technology is forecast by the SIA (Semiconductor Industry Association) to be in production.

In most electronics products, there are four major IC devices, namely, the microprocessor, the ASIC (application-specific integrated circuit), the cache memory, and the main memory. For example, a personal computer usually has one microprocessor; a few cache memories, e.g., fast SRAM (static random access memory); a few ASICs, e.g., video, sound, data path, high-speed memory controller, NuBus controller, and I/O controller; and many system memories, such as ROM (read only memory) which contains permanent code used by software applications and DRAM (dynamic random access memory) to store the information while the power is turned on.

The microprocessor is the brain of a computing system. Some well-known ones are: Intel's CISC (complex instruction-set computing)-based microprocessor family (e.g., the Pentium and Pentium-Pro); IBM, Motorola, and Apple's RISC (reduced instruction-set computing)-based PowerPC microprocessors; Hewlett-Packard's RISC-based PA8000; Digital Equipment Corporation's RISC-based Alpha chipset; Silicon Graph-

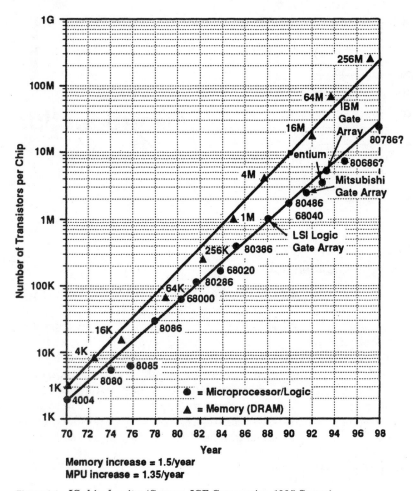

Figure 1.1   IC chip density. (*Source: ICE Corporation 1995 Status*)

ics' RISC-based MIPS; and Sun Microsystems' RISC-based UltraSparc. Both RISC-based and CISC-based microprocessors are expected to require over 1000 package pin counts and to perform over a 400-MHz on-chip clock frequency (Fig. 1.3).

The SRAMs for cache memories are expected to perform at high speeds similar to that of the microprocessors to prevent system data bottlenecks. Even the ASICs are expected to run faster than 200-MHz on-chip clock frequencies and have up to 900 package pin counts. (The ASICs of some telecommunications products need more than 1000 pin counts.) Packaging technology will be hard-pressed to meet all of these future requirements.

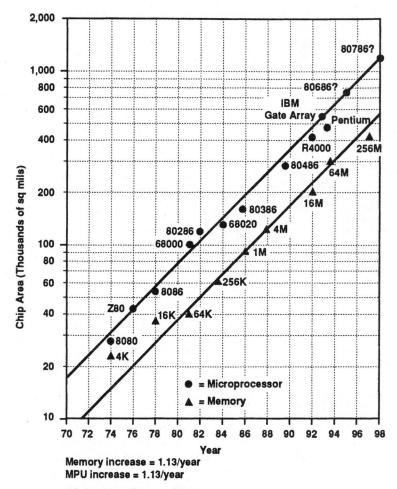

Figure 1.2   IC chip size. (*Source: ICE Corporation 1995 Status*)

1.2   Packaging Technology Update

Different packaging technologies are required for different semiconductor IC devices and applications.[1-8] Figure 1.3 shows some of the well-known packaging technologies for four of the most common IC devices.

1.2.1   Solder-bumped flip chip technologies

For high-speed, single-chip microprocessors and ASICs, the complex IC designs require very high I/O packages. For these types of ICs and subsystems, area-arrayed flip chip technologies provide a viable answer to

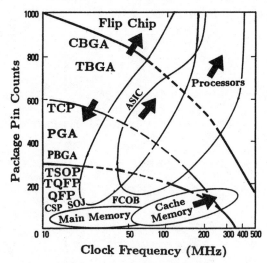

Figure 1.3 Device and packaging trends.

the needs (Fig. 1.3). The design, material, and process of flip chip technologies[1,2] are beyond the scope of this book; however, the solder joint reliability of various flip chip technologies under thermal, mechanical, and vibration conditions will be discussed in Chap. 6.

1.2.2 Ball grid array (BGA) technologies

There are many different kinds of BGAs.[3] Depending on their substrates, there are ceramic BGA (CBGA), tape-automated bonding BGA (TBGA), plastic BGA (PBGA), metal BGA (MBGA), dimple BGA (DBGA), etc. For high-I/O and -performance ASICs and microprocessors, CBGA, TBGA, MBGA, and DBGA could meet the high pin counts and clock frequency requirements (Fig. 1.3). The PBGA is cost-effective between 300 and 600 package pin counts. The solder joint reliability of PBGA, CBGA, TBGA, and DBGA surface-mount technology (SMT) assemblies under thermal, mechanical, and vibration conditions will be discussed in Chap. 5.

1.2.3 TSOP and PQFP

Up to 208-pin (0.5-mm pitch and 28-mm body size), 240-pin (0.5-mm pitch, 32-mm body size), and 304-pin (0.5-mm pitch and 40-mm body size), plastic quad flat packs (PQFPs)[4] are the most cost-effective packages for SMT. They have been used extensively for ASICs and low-speed microprocessors. Sometimes they are used for housing a few cache memories. Currently, the price for PQFP packages is less than one penny a pin. For example, the listing price for the 208-pin PQFP is $1.80.

Thin small outline package (TSOP)[4] is a very low profile plastic package which is specifically designed for SRAM, DRAM, and flash memory (which retains information even when the power is turned off) devices for space-limited applications. Right now, the price for TSOP is about one penny a pin. The solder joint reliability of the PQFP and TSOP SMT assemblies under different environmental stress factors will be discussed in Chap. 8.

1.2.4 Chip-scale package (CSP)

One of the most cost-effective packaging technologies is solder-bumped flip chip on board (FCOB), also called direct chip attach (DCA).[1,2] However, because of the infrastructure in supplying the solder-bumped known good die (KGD) and the corresponding fine-line and -spacing printed circuit board (PCB), most in the industry are still working on these issues. In the meantime, a class of new technology called chip-scale packaging (CSP) has surfaced (Table 1.1). It can be used for SRAMs, DRAMs, flash memories (Figs. 1.3 and 1.4), and not-so-high-pin-count ASICs and microprocessors. The advantages of CSP versus DCA are shown in Table 1.2. It can be seen that the advantages of CSP are that it is easy to test-at-speed and burn-in for known good die (KGD), to handle, to assemble, to rework, to standardize, to protect the die, to deal with die shrink or expand, and it is subject to less infrastructure constraints. On the other hand, the advantages of DCA are that it has better electrical and thermal performance, and less weight, size, and cost.

It can be seen from Table 1.1 that there are at least 18 different kinds and five different groups of CSPs. One example of each

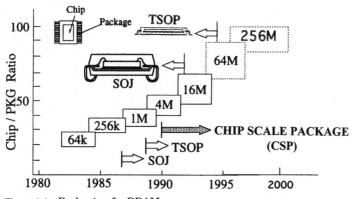

Figure 1.4   Packaging for DRAMs.

TABLE 1.1 Chip-Scale Packages (CSP)

CSP types	Company names	Chip-level interconnect	Interposer	Chip-to-interposer interconnects	Interposer-to-next-level interconnects	Reference no.
Lead on chip (LOC)	Fujitsu	Wire bonded*	Lead frame	Wire bonded	Lead (chip face-down)	9
	Hitachi Cable	Wire bonded	Lead frame	Wire bonded	Lead (chip face-up)	10
	LG Semicon	Wire bonded	Lead frame	Wire bonded	Lead (chip face-up)	11
	Samsung	Wire bonded	Lead frame	Wire bonded	Lead (chip face-up)	12
	TI	Wire bonded	Lead frame	Wire bonded	Lead (chip face-up)	13
	Toshiba	Wire bonded	Lead frame	Wire bonded	Lead (chip face-up)	14, 15
Rigid substrate	Matsushita	Au stud	Ceramic (2 ~ 4 layers)	Ag-Pd paste with underfill	Land grid array (LGA)	16
	IBM	Solder bump	Ceramic (multilayers)	C4$^+$	Composite solder balls	17
	Motorola	Solder bump	FR-4 or BT (2 layers)	C4 with underfill	Eutectic solder balls	18
	Toshiba	Au bump	Ceramic (2 layers)	Au-Au solid phase diffusion with underfill	LGA	19
Flexible substrate	GE	Ti/Cu-Cu	Cu/PI flex	Laser drilled vias plated with Cu	Eutectic solder balls	20
	NEC	Au bump	Cu/PI flex	Thermosonic Au-Cu bonded	Eutectic solder balls	21
	Nitto Denko	None	Cu/PI with Au bump	Au bump on tape to Al (350°C)	Eutectic solder balls	22
	Tessera	Wire bonded	Cu/PI with Ni/Au bump	Wire bonded	Eutectic solder balls	23
Wafer-level process	ChipScale	Ti/W/Au	Silicon post	Au beam lead	Lead (chip face-up)	24
	ShellCase	None	Ni/Au/solder-plated glass plates	Extended chip pad to the Ni/Au/solder-plated metal	Ni/Au/solder lead (chip face-down)	25
Wafer-level process with redistribution	Mitsubishi	Metallization	PI/metal/solder	Redistributed with metal & solder	Solder balls	26
	Sandia	Ti/W and Cu	2 layers of PI & Cu/Ni	Redistributed with metal	Solder balls	27

* Wire bonded: thermosonic Au-Al bonded.
† C4: Controlled-collapse chip connection.

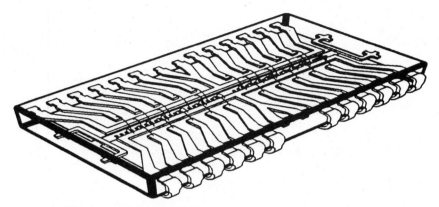

Figure 1.5   TI's chip-scale package (CSP).

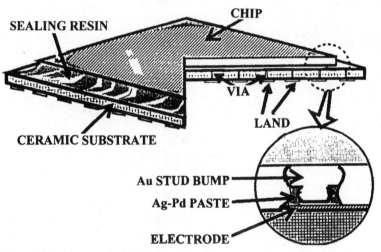

SEALING RESIN

CHIP

VIA

LAND

CERAMIC SUBSTRATE

Au STUD BUMP

Ag-Pd PASTE

ELECTRODE

Figure 1.6   Matsushita's chip-scale package (CSP).

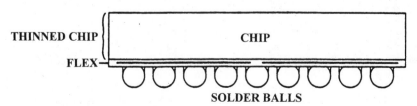

THINNED CHIP

CHIP

FLEX

SOLDER BALLS

Figure 1.7   GE's chip-scale package (CSP).

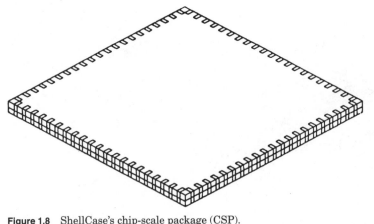

Figure 1.8 ShellCase's chip-scale package (CSP).

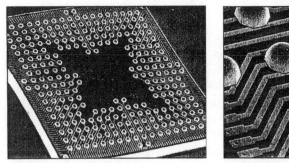

Figure 1.9 Sandia's chip-scale package (CSP). LSI Logic ASIC LCA100106, 275 pads, 0.020-in pitch.

group is shown in Figs. 1.5 through 1.9, respectively, for the CSPs of TI, Matsushita, GE, ShellCase, and Sandia. The solder joint reliability of seven other CSP assemblies (Mitsubishi, Motorola, Tessera, NEC, LG Semicon, Nitto Denko, and Toshiba) will be discussed in Chap. 7.

TABLE 1.2   Advantages: CSP versus DCA

CSP	DCA
* Test-at-speed for KGD	* Electrical performance
* Burn-in for KGD	* Thermal performance
* Handling	* Weight
* Assembly	* Size
* Rework	* Cost
* Die shrink or expand	
* Standards	
* Die protection	
* Infrastructure	

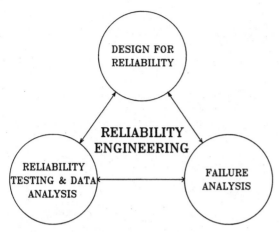

Figure 1.10 Reliability engineering.

1.3 Reliability Engineering

Figure 1.10 shows the concept of reliability engineering. It consists of three major tasks, namely, design for reliability (DFR), reliability testing and data analysis, and failure analysis. Usually, it starts off with a design (or a few designs) and demonstrates that it (or one of the few) is reliable by analysis, e.g., by the finite element method. Then it builds the reliable one with a certain sample size and tests it under certain conditions. The objective of the test is to find failures (the more the better) and choose a failure distribution (such as Weibull) to fit the failure data. The distribution then becomes the life distribution of the solder joints. Failure analysis should then be done on the failed sample in order to understand why it fails. This information is very useful for the next round of DFR. In this book, the reliability testing and data analysis are presented in Chap. 2. Chapter 3 discusses different failure modes and analysis methods of SMT solder joints. The DFR of solder joints is elucidated in Chap. 4.

References

1. Lau, J. H., *Flip Chip Technologies*, McGraw-Hill, New York, 1996.
2. Lau, J. H., *Chip On Board Technologies for Multichip Modules*, Van Nostrand Reinhold, New York, 1994.
3. Lau, J. H., *Ball Grid Array Technology*, McGraw-Hill, New York, 1995.
4. Lau, J. H., *Handbook of Fine Pitch Surface Mount Technology*, Van Nostrand Reinhold, New York, 1993.
5. Lau, J. H., *Handbook of Tape Automated Bonding*, Van Nostrand Reinhold, New York, 1992.
6. Lau, J. H., *Thermal Stress and Strain in Microelectronic Packaging*, Van Nostrand Reinhold, New York, 1993.

7. Frear, D., H. Morgan, S. Burchett, and J. H. Lau, *The Mechanics of Solder Alloy Interconnects,* Van Nostrand Reinhold, New York, 1993.
8. Lau, J. H., *Solder Joint Reliability: Theory and Applications,* Van Nostrand Reinhold, New York, 1991.
9. Kasai, J., M. Sato, T. Fujisawa, T. Uno, M. Waki, K. Hayashida, and T. Kawahara, "Low Cost Chip Scale Package for Memory Products," *Proceedings of SMI Conference,* August 1995, pp. 6–17.
10. Murakami, G., "Rationale for Chip Scale Packaging (CSP) Rather Than Multichip Modules (MCM)," *Proceedings of SMI Conference,* August 1995, pp. 1–5.
11. Cha, K., Y. Kim, T. Kang, D. Kang, and S. Back, "Ultra-Thin and Crack-Free Bottom Leaded Plastic (BLP) Package Design," *Proceedings of IEEE Electronic Components & Technology Conference,* May 1995, pp. 224–228.
12. Lee, S., J. Lee, S. Oh, and H. Chung, "Passivation Cracking Mechanism in High Density Memory Devices Assembled in SOJ Packages Adopting LOC Die Attach Technique," *Proceedings of IEEE Electronic Components & Technology Conference,* May 1995, pp. 455–462.
13. Amagai, M., "The Effect of Stress Intensity of Package Cracking in Lead-On-Chip (LOC) Packages," *Proceedings of IEEE Japan International Electronics Manufacturing Technology Symposium,* December 1995, pp. 415–420.
14. Okugawa, Y., T. Yoshida, T. Suzuki, and H. Nakayoshi, "New Tape LOC Adhesive Tapes," *Proceedings of IEEE Electronic Components & Technology Conference,* May 1994, pp. 570–574.
15. Nakayoshi, H., N. Izawa, T. Ishikawa, and T. Suzuki, "Memory Package with LOC Structure Using New Adhesive Material," *Proceedings of IEEE Electronic Components & Technology Conference,* May 1994, pp. 575–579.
16. Kunitomo, Y., "Practical Chip Size Package Realized by Ceramic LGA Substrate and SBB Technology," *Proceedings of SMI Conference,* August 1995, pp. 18–25.
17. Master, R., R. Jackson, S. Ray, and A. Ingraham, "Ceramic Mini-Ball Grid Array Package for High Speed Device," *Proceedings of IEEE Electronic Components & Technology Conference,* May 1995, pp. 46–50.
18. Lall, P., G. Gold, B. Miles, K. Banerji, P. Thompson, C. Koehler, and I. Adhihetty, "Reliability Characterization of the SLICC Package," *Proceedings of IEEE Electronic Components & Technology Conference,* May 1996, pp. 1202–1210.
19. Forman, G., R. Fillion, R. Kole, R. Wojnarowski, and J. Rose, "Development of GE's Plastic Thin-Zero Outline Package (TZOP) Technology," *Proceedings of IEEE Electronic Components & Technology Conference,* May 1995, pp. 664–668.
20. Matsuda, S., K. Kata, and E. Hagimoto, "Simple-Structure, Generally Applicable Chip-Scale Package," *Proceedings of IEEE Electronic Components & Technology Conference,* May 1995, pp. 218–223.
21. Tanigawa, S., K. Igarashi, M. Nagasawa, and N. Yoshio, "The Resin Molded Chip Size Package (MCSP)," *Proceedings of IEEE Japan International Electronics Manufacturing Technology Symposium,* December 1995, pp. 410–415.
22. Koyama, T., K. Abe, N. Sakaguchi, and S. Wakabayashi, "Reliability of mBGA Mounted on a Printed Circuit Board," *Proceedings of SMI Conference,* August 1995, pp. 43–56.
23. Young, J., "Chip Scale Packaging Provides Known Good Die," *Proceedings of SMI Conference,* August 1995, pp. 52–59.
24. Badihi, A., and E. Por, "ShellCase—A True Miniature Integrated Circuit Package," *Proceedings of Flip Chip, BGA, TAB, and Advanced Packaging Symposium,* February 1995, pp. 244–252.
25. Yasunaga, M., S. Baba, M. Matsuo, H. Matsushima, S. Nako, and T. Tachikawa, "Chip Scale Package (CSP) A Lightly Dressed LSI Chip," *Proceedings of IEEE Japan International Electronics Manufacturing Technology Symposium,* December 1994, pp. 169–176.
26. Chanchani, R., K. Treece, and P. Dressendorfer, "mini Ball Grid Array (mBGA) Technology," *Proceedings of NEPCPN West,* February 1995, pp. 938–945.
27. Iwasaki, H., "CSTP: Chip Scale Thin Package," *Proceedings of SEMICON Japan,* November 1994, pp. 488–495.

Reliability Testing and Data Analysis

In the past three decades reliability has been formulated as the science of estimating and predicting the life distribution of products by the application of theories of probability and statistics.[1-25] Evidence of the intimate relationship among reliability, probability, and statistics is demonstrated in numerous articles written on statistical methods in reliability. See, for example, *IEEE Transactions on Reliability, ASME Transactions on Reliability, Journal of the American Statistical Association, Technometrics,* or *Annals of Mathematical Statistics.* On reviewing some of the articles, you will soon find that reliability has become the playground of mathematicians. Many methods based on sound statistical theories have been proposed for the analysis and solution of reliability problems. However, most of these methods are still either in an ivory tower or not practical enough for routine use. In this book we will focus only on a very simple method, which has been used extensively in the automobile, aerospace, and electronics industries. With this method, reliability of a product is modeled by choosing a probability function (failure model) and then is analyzed by Johnson's theory of order statistics.[1-3]

There are many different kinds of reliability testing. However, the most common tests for electronics packaging are the thermal cycling tests; mechanical bending, twisting, and shearing tests; and vibration tests. The applications of these tests to BGA, flip chip, CSP, and fine pitch SMT assemblies are shown in Chaps. 5, 6, 7, and 8, respectively.

2.1 Basic Statistics and Probability

The definitions of probability density function, cumulative distribution function, mean, median, and standard deviation will be very briefly mentioned in this section.

2.1.1 Probability density functions

Table 2.1 shows 100 measurements of the resistance (ohms) of a resistor. The resistance is assumed to be a continuous random variable. These results are reproduced in Table 2.2 and Fig. 2.1. In this histogram (Fig. 2.1) the base of each vertical rectangle (cell) represents the interval of resistance (class boundary), and the height represents the number of observed frequencies within that class boundary. Figure 2.2 shows the frequency bar chart, which uses bars centered on the midpoint of the intervals. The heights of the bars are proportional to the frequencies in the respective intervals. Figure 2.3 shows the frequency polygon, which consists of a series of straight lines joining small circles. These circles are plotted at interval midpoints with a height proportional to the frequencies in the respective intervals. If a much larger number of measurements were taken, a smooth curve could be drawn through the peaks of the midpoint of the vertical rectangle. This curve would then represent probability, that is, frequency of occurrence. The curve would thus be a density function of the resistor's resistance.

In general, the probability density function (PDF) $f(x)$ (probability of obtaining the value x) where x denotes a continuous random variable, is defined as

$$f(x) \geq 0 \qquad \text{where } -\infty \leq x \leq \infty \qquad (2.1)$$

$$\int_{-\infty}^{+\infty} f(x)\, dx = 1 \qquad (2.2)$$

2.1.2 Cumulative distribution function

Plotting the frequency histogram (Fig. 2.1), frequency bar chart (Fig. 2.2), and frequency polygon (Fig. 2.3) are not the only ways of presenting the measurement data (Table 2.1). It is sometimes advantageous to tabulate the frequencies of values less than the respective class bound-

TABLE 2.1   Resistance (Ohms) Measurements of a Resistor

1.003	1.009	1.002	1.018	1.015	1.001	0.991	1.004	0.935	1.001
0.943	1.016	1.048	1.018	1.016	1.002	1.002	0.993	0.993	1.022
0.995	1.052	1.017	0.982	1.017	1.050	0.992	0.954	1.013	0.996
0.994	1.047	0.965	0.981	0.993	0.980	1.003	1.004	0.971	0.963
1.004	0.991	0.990	1.028	1.022	0.983	1.021	0.987	1.006	0.987
1.046	0.978	0.991	1.053	0.993	1.020	1.065	1.022	1.061	1.001
0.988	1.016	1.017	0.996	1.003	0.990	1.021	0.976	1.023	1.024
1.025	1.007	1.003	1.002	1.054	1.010	1.041	1.012	1.042	1.013
1.015	1.026	1.027	1.005	1.008	0.990	1.011	0.979	1.005	1.014
1.038	1.037	1.036	1.032	1.031	1.030	1.036	1.030	1.033	1.039

TABLE 2.2   Rearrange the Data from Table 2.1
for Histogram Plot

Class boundaries	Class midpoints	Observed frequency
0.9295–0.9395	0.9345	1
0.9395–0.9495	0.9445	1
0.9495–0.9595	0.9545	1
0.9595–0.9695	0.9645	2
0.9695–0.9795	0.9745	4
0.9795–0.9895	0.9845	7
0.9895–0.9995	0.9945	15
0.9995–1.0095	1.0045	20
1.0095–1.0195	1.0145	16
1.0195–1.0295	1.0245	12
1.0295–1.0395	1.0345	10
1.0395–1.0495	1.0445	5
1.0495–1.0595	1.0545	4
1.0595–1.0695	1.0645	2

aries as shown in Table 2.3, which is plotted in Fig. 2.4. This type of
graph is called an *ogive*. Again, if a much larger number of measurements was taken, a smooth curve could be obtained. This curve would
then represent probability of events defined in terms of its corresponding random variable. The curve would thus be a distribution function
of the resistor's resistance. It tells us how the values of the random
variable (resistance) are distributed and it is a cumulated distribution
function since it gives the distribution of values in cumulative form.

In general, the cumulative distribution function (CDF) $F(x)$ where x
denotes a continuous random variable is defined as (Fig. 2.4)

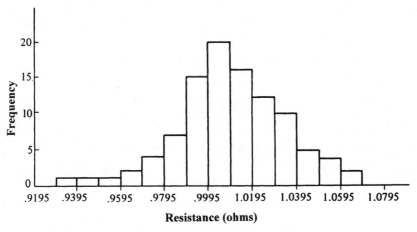

Figure 2.1   Histogram for the resistance measurements.

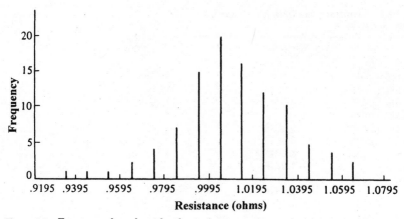

Figure 2.2   Frequency bar chart for the resistance measurements.

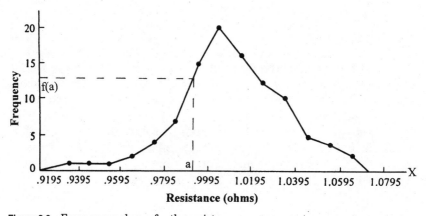

Figure 2.3   Frequency polygon for the resistance measurements.

$$F(x \leq a) = F(a) = \int_{-\infty}^{a} f(x)\, dx \qquad (2.3)$$

or
$$f(x) = \frac{dF}{dx} \qquad (2.4)$$

2.1.3 Measures of central tendency of a population

The expected value (mean) of a random variable (x) with its PDF $f(x)$ is

$$\mu = E(x) = \int_{-\infty}^{+\infty} x f(x)\, dx \qquad (2.5)$$

where μ is the population mean.

TABLE 2.3   Rearrange the Data from Table 2.2 for Ogive Plot

Resistance	Number of resistors having less than given resistance
0.9295	0
0.9395	1
0.9495	2
0.9595	3
0.9695	5
0.9795	9
0.9895	16
0.9995	31
1.0095	51
1.0195	67
1.0295	79
1.0395	89
1.0495	94
1.0595	98
1.0695	100

The median or the x_{50} value of a random variable (x) with its PDF $f(x)$ is

$$0.5 = \int_{-\infty}^{x_{50}} f(x)\, dx \qquad (2.6)$$

That is, the population median is the value of x at which the CDF $F(x)$ has a value of 0.5.

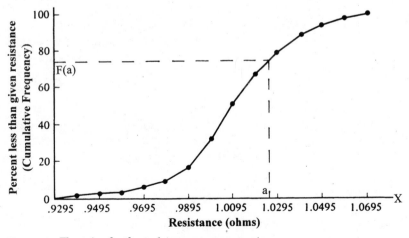

Figure 2.4   The ogive for the resistance measurements.

2.1.4   Measures of variability of a population

A population can generally be described in terms of two parameters: central tendency (generally, mean) and variation about the mean (generally, standard deviation or square root of the variance).

$$\sigma^2 = E[(x - \mu)^2] = \int_{-\infty}^{+\infty} (x - \mu)^2 f(x)\, dx \qquad (2.7)$$

where σ is the population standard deviation and σ^2 is the variance.

2.2   Life Distribution

One of the central concepts in reliability is the cumulative distribution function (CDF). Life distribution is a theoretical population model used to describe the lifetime of a component/product/system and is defined as the CDF $F(x)$ for the population. Consequently, life distribution has two useful interpretations: (1) $F(x)$ is the probability a random unit drawn from the population fails by x hours, and (2) $F(x)$ is the fraction of all units in the population which fails by x hours.

2.3   Definition of Reliability

Reliability of a component/product/system $R(t)$ can be defined as the probability that the component/product/system will perform its intended function for a specified period of time, under a given operation condition, without failure. Numerically, reliability is the percent of survivors, i.e.,

$$R(x) = 1 - F(x) \qquad (2.8)$$

where $R(x)$ is the reliability (survival) function and has two useful interpretations: (1) $R(x)$ is the probability a random unit drawn from the population will still be operating after x hours, and (2) $R(x)$ is the fraction of all units in the population that will survive at least x hours.

2.4   Failure (Hazard) Rate

The instantaneous failure rate (or, simply, the failure rate) $h(x)$ represents the instantaneous rate of failure for units of a population that have survived to time x and is defined as follows:

$$h(x) = \frac{f(x)}{R(x)} \qquad (2.9)$$

Equation (2.9) is typically represented by the bathtub curve shown in Fig. 2.5. It can be seen that for the early failure period, it has a decreasing $h(x)$; for the steady-state (stable) period, it has a constant $h(x)$; and for the wear-out failure period, it has an increasing $h(x)$. In general, most of the useful life of a device should take place in the steady-state region of the curve. One of the units of failure rate is parts per million per thousand hours (ppm/K) or FIT (fails in time). One ppm/K or FIT means one failure is expected out of one million devices operating for 1000 hours.

The cumulative failure rate $H(x)$ prior to time x is defined as

$$H(x) = -\ln R(x) \tag{2.10}$$

The average failure rate $\text{AFR}(x_1, x_2)$ between time x_1 and x_2 is defined as

$$\text{AFR}(x_1, x_2) = \frac{\ln R(x_1) - \ln R(x_2)}{x_2 - x_1} = \frac{H(x_1) - H(x_2)}{x_2 - x_1} \tag{2.11}$$

2.5 Mean Time to Failure

Mean time to failure (MTTF) is defined as the expected time to (first) failure. It is computed with the probability (of failure) density function $f(x)$:

$$\text{MTTF} = \int_0^\infty x f(x)\, dx \tag{2.12}$$

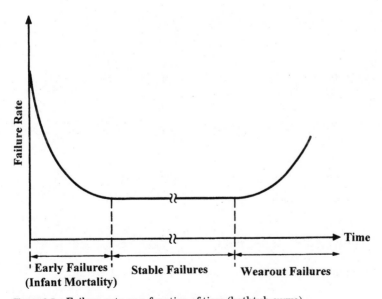

Figure 2.5   Failure rate as a function of time (bathtub curve).

The application of the previous equations is shown in the following example. Say the life distribution of an IC component has been experimentally determined as

$$F(x) = 1 - (1 + 0.002x)^{-1} \qquad (2.13)$$

Determine the following:

1. The probability a new unit will fail by 2000 h
2. The probability a new unit will fail by 4000 h
3. The probability a new unit will fail between 2000 and 4000 h
4. The proportion surviving past 9000 h
5. The proportion surviving past 2000 h
6. The failure rate at 100 h
7. The failure rate at 1000 h
8. The failure rate at 10,000 h
9. The average failure rate between 2000 and 9000 h

solution

$$F(x) = 1 - (1 + 0.002x)^{-1} \qquad (2.14)$$

$$R(x) = 1 - F(x) = (1 + 0.002x)^{-1} \qquad (2.15)$$

$$f(x) = \frac{dF}{dx} = 0.002(1 + 0.002x)^{-2} \qquad (2.16)$$

$$h(x) = \frac{f(x)}{R(x)} = 0.002(1 + 0.002x)^{-1} \qquad (2.17)$$

1. $F(2000) = 1 - (1 + 0.002 \times 2000)^{-1} = 0.8$
2. $F(4000) = 0.889$
3. $F(4000) - F(2000) = 0.889 - 0.8 = 0.089$
4. $R(9000) = (1 + 0.002 \times 9000)^{-1} = 0.053$
5. $R(2000) = 1 - 0.8 = 0.2$
6. $h(100) = 0.002(1 + 0.002 \times 100)^{-1} = 0.0017$ failures/h
7. $h(1000) = 0.0007$ failures/h $= 0.0007 \times 10^9$ FIT
8. $h(10,000) = 0.0001$ failure/h
9. $AFR(2000,9000) = (\ln 0.2 - \ln 0.053)/(9000 - 2000) = 0.0002$ failures/h

2.6 Exponential Life Distribution

The exponential PDF, CDF, $R(x)$, and $h(x)$ are given below and shown in Figs. 2.6 through 2.9, respectively.

Exponential probability density function (PDF)

$$f(x) = \lambda e^{-\lambda x} \qquad (2.18)$$

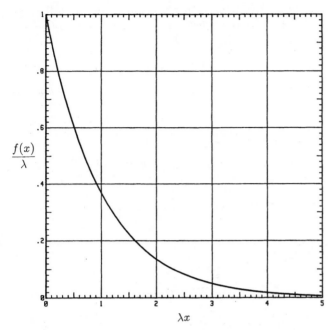

Figure 2.6 Exponential probability density function (PDF).

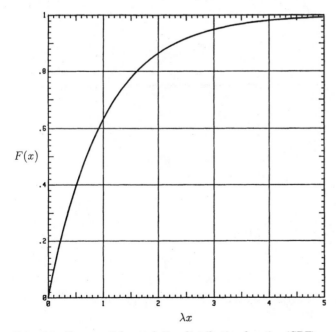

Figure 2.7 Exponential cumulative distribution function (CDF).

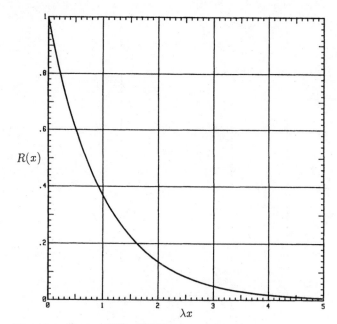

Figure 2.8   Exponential reliability function.

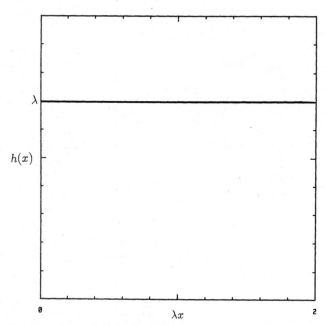

Figure 2.9   Exponential failure rate.

Exponential cumulative distribution function (CDF)

$$F(x) = 1 - e^{-\lambda x} \tag{2.19}$$

Exponential reliability function

$$R(x) = 1 - F(x) = e^{-\lambda x} \tag{2.20}$$

Exponential failure rate

$$h(x) = \frac{f(x)}{R(x)} = \lambda \tag{2.21}$$

where
x = random variable
λ = failure rate
Mean = $1/\lambda$ (2.22)
Variance = $1/\lambda^2$

The exponential distribution is a one-parameter (λ) function which can be applied to the life of systems, assemblies, and other situations where failures occur by chance alone and do not depend on time in service. Examples are vacuum-tube-failure life, expected cost to detect bad equipment during reliability testing, expected life of indicator tubes used in radar sets, life to failure of lightbulbs, electric generators, and automobile transmissions. A more precise example follows.

Example *Given:* The life distribution of an IC component is ($\lambda = 0.001$/day)

$$F(x) = 1 - e^{-0.001x} \tag{2.23}$$

Determine the following:

1. The probability that a component will last more than 1200 days
2. The probability that a component will last more than 1200 days given that it has already served 1000 days

solution

1. $R(1200) = 1 - F(1200) = e^{-0.001 \times 1200} = 0.301$
2. $P\{x > 200\} = 1 - F(200) = e^{-0.001 \times 200} = 0.819$

2.7 Normal and Lognormal Life Distributions

The normal PDF and CDF are given by

Normal probability density function (PDF)

$$f(x) = \frac{1}{\sigma\sqrt{2\pi}} \exp\left[-\frac{1}{2}\left(\frac{x - \mu}{\sigma}\right)^2 \right] \tag{2.24}$$

Normal cumulative distribution function (CDF)

$$F(x) = \frac{1}{\sigma\sqrt{2\pi}} \int_{-\infty}^{x} \exp\left[-\frac{1}{2}\left(\frac{\zeta - \mu}{\sigma} \right)^2 \right] d\zeta \qquad (2.25)$$

where   Random variable $= x$
Mean $= \mu$ (2.26)
Variance $= \sigma^2$
Standard deviation $= \sigma$

The normal distribution is a two-parameter (μ, σ) function, and Fig. 2.10 shows the plots of Eq. (2.24). It can be seen that the distribution is symmetrical about its center μ, and σ is a scale parameter that shows how close to the center the area under the curve is packed. It is well known that the integral in Eq. (2.25) cannot be integrated in closed form. Thus, an infinite number of tables would be required for infinite pairs of these two parameters. Fortunately, a simple transformation combining the two parameters into a new parameter,

$$z = \frac{x - \mu}{\sigma} \qquad (2.27)$$

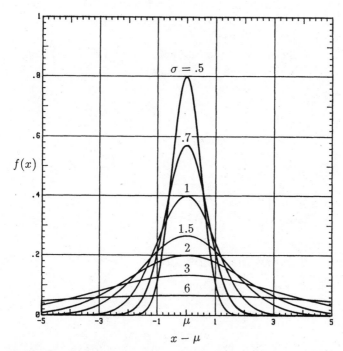

Figure 2.10 Normal probability density function (PDF).

will lead us to the standard normal PDF and CDF:

Standard normal PDF

$$f(z) = \frac{1}{\sqrt{2\pi}} \exp\left[-z^2/2\right]$$ (2.28)

where $f(z)$ has a normal distribution with mean = 0 and variance = 1.

Standard normal CDF

$$F(x) = \frac{1}{\sqrt{2\pi}} \int_{-\infty}^{(x-\mu)/\sigma} e^{-z^2/2} \, dz$$ (2.29)

Figures 2.11 and 2.12 show the plots of Eqs. (2.28) and (2.29) for the standard normal PDF and CDF, respectively. It is worthwhile to point out that Eq. (2.29) can be approximated with great accuracy by the following expression:

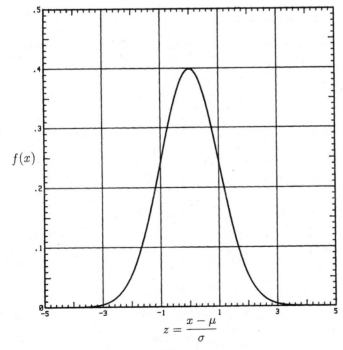

Figure 2.11 Standard normal PDF.

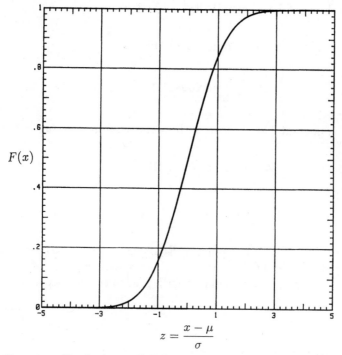

$$z = \frac{x - \mu}{\sigma}$$

Figure 2.12   Standard normal CDF.

$$F(x) \approx 1 - \frac{1}{2}\,(1 + c_1 z + c_2 z^2 + c_3 z^3 + c_4 z^4)^{-4} + \epsilon(z)$$

$$|\,\epsilon(z)\,| \leq 2.5 \times 10^{-4}$$

$$c_1 = 0.196854$$

$$c_2 = 0.115194$$

$$c_3 = 0.000344$$

$$c_4 = 0.019527$$

(2.30)

$$F(x) \approx 1 - \frac{1}{2}\,(1 + d_1 z + d_2 z^2 + d_3 z^3 + d_4 z^4 + d_5 z^5 + d_6 z^6)^{-16} + \epsilon(z)$$

$$|\,\epsilon(z)\,| \leq 1.5 \times 10^{-7}$$

$$d_1 = 0.0498673470$$

$$d_2 = 0.0211410061$$

(2.31)

$$d_3 = 0.0032776263$$

$$d_3 = 0.0000380036$$

$$d_5 = 0.0000488906$$

$$d_6 = 0.0000053830$$

Equations (2.30) and (2.31) are plotted in Fig. 2.12. It can be seen that the differences are invisible.

The lognormal PDF and CDF are given by

Lognormal probability density function (PDF)

$$f(x) = \frac{1}{\sigma\sqrt{2\pi}x} \exp\left[-\frac{1}{2\sigma^2}(\ln x - \ln T_{50})^2 \right] \qquad (2.32)$$

Lognormal cumulative distribution function (CDF)

$$F(x) = \frac{1}{\sigma\sqrt{2\pi}} \int_{-\infty}^{x} \exp\left[-\frac{1}{2\sigma^2}(\ln \zeta - \ln T_{50})^2 \right] d\zeta \qquad (2.33)$$

where
$$\mu = \ln T_{50}$$
$$\sigma = \text{Shape parameter}$$
$$T_{50} = \text{Median lifetime or 50\% failure point} \qquad (2.34)$$
$$\text{Mean} = T_{50}e^{\sigma^2/2}$$
$$\text{Variance} = T_{50}e^{\sigma^2}(e^{\sigma^2} - 1)$$

The lognormal distribution is a two-parameter function (μ, σ), and Fig. 2.13 shows the plots of Eq. (2.32). It can be seen that the shape of this distribution is determined by the parameter σ. Again, Eq. (2.33) cannot be integrated explicitly. Let

$$z = \frac{\ln x - \ln T_{50}}{\sigma} \qquad (2.35)$$

Then we have the standard lognormal PDF and CDF:

Standard lognormal PDF

$$f(z) = \frac{1}{\sqrt{2\pi}} \exp[-z^2/2] \qquad (2.36)$$

Standard lognormal CDF

$$F(x) = \frac{1}{\sqrt{2\pi}} \int_{-\infty}^{(\ln x - \ln T_{50})/\sigma} e^{-z^2/2}\, dz \qquad (2.37)$$

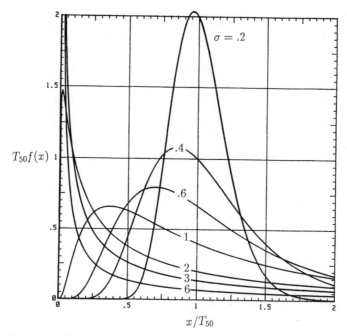

Figure 2.13 Lognormal probability density function (PDF).

Again, by using the approximate expressions, Eqs. (2.30) and (2.31), we can determine the standard lognormal CDF, $R(x)$, and $h(x)$ and plot them in Figs. 2.14 through 2.17 for engineering-practice convenience. The application of lognormal distribution is for life phenomena, situations where occurrences are concentrated at the tail end of the range, and where differences in observations are of a large order of magnitude. Examples are downtime of a large number of electrical systems, solder joint reliability, concentration of chemical process residues, or automotive mileage accumulation by different customers. A more precise example follows.

Example *Given:* The life distribution of an IC component is ($T_{50} = 5000$ and $\sigma = 0.7$)

$$F(x) = \frac{1}{\sqrt{2\pi}} \int_{-\infty}^{(\ln x - \ln T_{50})/\sigma} e^{-z^2/2}\, dz$$

What percent of failures are expected at 2000 hours of operation?
solution

$$z = \frac{\ln x - \ln T_{50}}{\sigma} = \frac{\ln 2000 - \ln 5000}{0.7} = -1.309$$

Substituting $z = -1.309$ in Eq. (2.30), we have $F(2000) = 0.95$, i.e., 95 percent of the IC components are expected to fail at 2000 hours of operation.

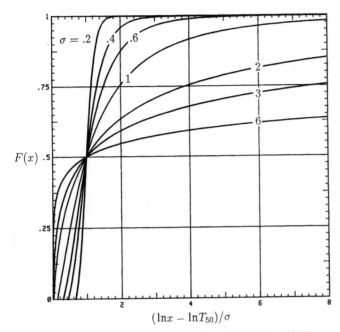

Figure 2.14   Lognormal cumulative distribution function (CDF).

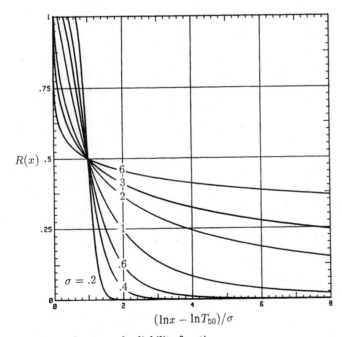

Figure 2.15   Lognormal reliability function.

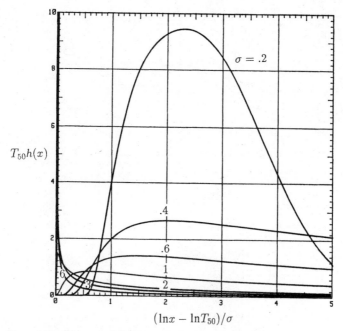

Figure 2.16 Lognormal failure rate.

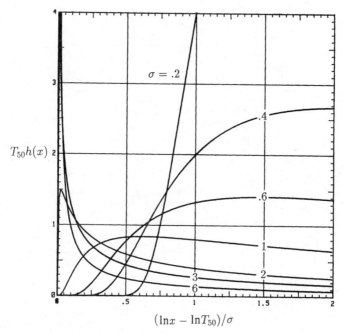

Figure 2.17 Lognormal failure rate.

2.8   Weibull Life Distribution

The Weibull PDF, CDF, $R(x)$, and $h(x)$ of a random variable x are shown, respectively, in Figs. 2.18 through 2.21, and given by[12,13]

Weibull probability density function (PDF)

$$f(x) = (\beta/x)(x/\theta)^\beta \, e^{-(x/\theta)^\beta} \qquad (2.38)$$

Weibull cumulative distribution function (CDF)

$$F(x) = 1 - e^{-(x/\theta)^\beta} \qquad (2.39)$$

Weibull reliability function

$$R(x) = 1 - F(x) = e^{-(x/\theta)^\beta} \qquad (2.40)$$

Weibull failure rate

$$h(x) = \frac{f(x)}{R(x)} = (\beta/\theta)(x/\theta)^{\beta-1} \qquad (2.41)$$

where
$\quad x$ = random variable (e.g., life)
$\quad \beta$ = shape parameter (Weibull slope)
$\quad \theta$ = characteristic life ($F(\theta)$ = 62.3 percent)   (2.42)
$\quad$ Mean = $\theta\Gamma(1 + 1/\beta)$
$\quad$ Variance = $\theta^2\Gamma(1 + 2/\beta) - [\theta\Gamma(1 + 1/\beta)]^2$

The Weibull distribution is a two-parameter function (θ, β) and can be used for life phenomena, situations where the percent occurrences (say, failure rates) may decrease, increase, or remain constant with increase in the characteristic measured, for devices at debut, wear-out, and chance failure stages of the product's life. Examples are life of electronics components, solder joint fatigue, corrosion life, and wear-out life. Precise examples for solder joint reliability will be discussed in Chaps. 5 through 8. The Weibull distribution is very general and can be degenerated to the following special cases.

1. For $\beta \leq 1$, the Weibull PDF approaches to infinity as time approaches zero and moves toward zero rapidly as time increases. The failure rate behaves the same way.
2. For $\beta = 1$, the Weibull PDF degenerates to the exponential PDF and has a constant failure rate.
3. For $\beta = 2$, the Weibull PDF degenerates to the Rayleigh PDF and has a linearly increasing failure rate.

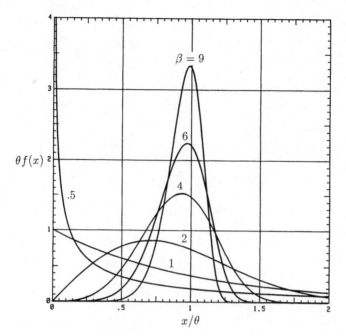

Figure 2.18   Weibull probability density function (PDF).

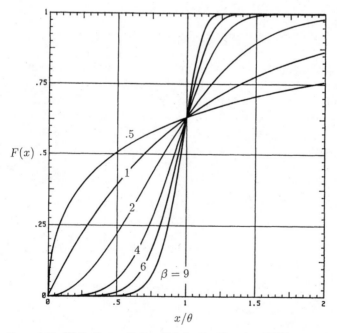

Figure 2.19   Weibull cumulative distribution function (CDF).

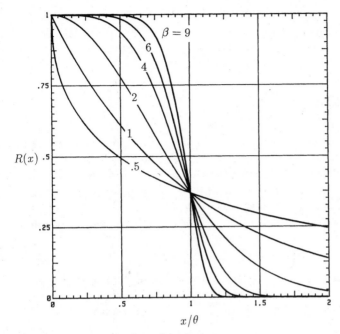

Figure 2.20 Weibull reliability function.

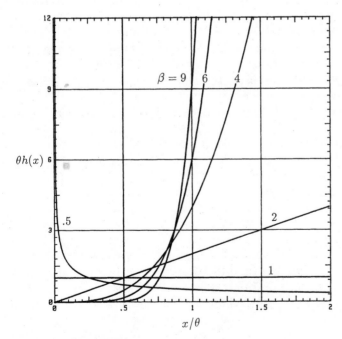

Figure 2.21 Weibull failure rate.

4. For $3 \leq \beta \leq 4$, the Weibull PDF approaches to the normal PDF and has a rapidly increasing failure rate. In fact, for $\beta = 4$ the Weibull and the normal PDFs are almost indistinguishable.

2.9 Reliability Testing

As mentioned in the previous sections, the reliability function, failure rate, and mean time to failure can be determined as soon as the life distribution is estimated by reliability tests. Some common tests for electronics packaging assemblies are thermal cycling, shock and vibration, and mechanical shear, pull, bend, and twist. In order to achieve a good reliability test, the program must be well designed. Specifically, among many others, the following three factors should be considered:

1. Since the objective for doing the reliability test is to estimate the probability of failure, the number of items (sample size) to be tested should be such that the final data are statistically significant. On the other hand, the reliability test should not be designed in such a complicated fashion that the failure probability cannot be estimated by the available statistical methods.

2. It is generally easy to say that the reliability test conditions should be close to the normal operating conditions. However, with today's environments (e.g., short time to market and requirements for highly reliable components and assemblies), we are unable to obtain enough test data to estimate the failure probability if the test is run at the normal operating conditions. Thus, most of the reliability tests are run at higher magnitudes to force the components and assemblies to fail (accelerated tests). In that case we have enough test data to fit the life distribution (to determine its parameters) with practical test times and reasonable sample sizes. It should be noted that the failure mode(s) under the accelerated test conditions must be close to that under the anticipated normal operating conditions. Thus, the time-, rate-, frequency-, and temperature-dependent physical and mechanical properties of the components and assemblies must be understood to design the accelerated tests.

3. To justify the accelerated test (with increased intensity and realistic sample sizes and test times) we must establish the acceleration models. These models transfer the failure probability, reliability function, failure rate, and mean time to failure from a test condition to a normal operating condition. In establishing the accelerated model(s), the assemblies' materials (e.g., solder, plastic, ceramic, copper, silicon), loadings (e.g., stress, strain, temperature, voltage, humidity, current density), and failure modes (e.g., fatigue, corrosion, electromigration) must be considered.

2.10   Parameter Estimation

After a failure model (such as exponential, lognormal, or Weibull) has been chosen for the statistical analysis, the next step is to determine the parameter(s) in the model by fitting a large number of the life-test data into a frequency histogram, frequency bar chart, or frequency polygon. However, life testing of this sort is often very time-consuming and costly, and is limited to a small sample size. Hence, the shape of the histogram, bar chart, or polygon varies considerably with changes in class interval. Therefore, instead of fitting the probability density function (PDF) it is more accurate to fit the cumulative distribution function (CDF) with the life-test data. This involves plotting the life-test data on the abscissa against the ranking of these life-test data on the ordinate.

In general, successive cycle (or time)-to-failure data are noted during the life test and the failures occur in order. The rankings of these life-test data can be determined by the theory of order statistics. The equation obtained by Johnson[1–3] is summarized as follows:

$$1 - (1 - z)^n - nz(1 - z)^{n-1} - \frac{n(n-1)}{2!} z^2 (1 - z)^{n-2} - \cdots$$

$$- \frac{n(n-1)\cdots(n-j+1)}{(j-1)!} z^{j-1}(1-z)^{n-j+1} = G \quad (2.43)$$

where j is the failure order number, n is the number of samples, G is the required ranking, and z is the percent rank of the jth value in n. Thus, for a given set of values of G, n, and j, the value of z ($0 < z < 1$) can be determined with a calculator or computer from Eq. (2.43).

For example, given the sample size $n = 10$, and the required rankings $G = 95$ percent and $G = 7$ percent, determine the percent rank (z) of the jth values in n. Table 2.4 shows the 95 percent and 7 percent ranks of the jth value in $n = 10$. The meaning of the numbers in Table 2.4 can be explained as follows: the 95 percent rank of the lowest of 10 measurements is 25.89 percent. This means that in 95 percent of the cases the lowest of 10 would represent as much as 25.89 percent of the population. In only 5 percent of the cases would the lowest of 10 represent even more than 25.89 percent of population. Similarly, for the 7 percent rank the lowest of ten measurements is 0.72 percent. This means that only in 7 percent of the cases the lowest of 10 measurements would represent less than 0.72 percent of the population; in 93 percent of cases the lowest of 10 would represent more than 0.72 percent of the population.

If only the median ranks are of interest, then use the following approximate formula:

$$\text{Median or 50\% rank} = \frac{j - 0.3}{n + 0.4} \quad (2.44)$$

TABLE 2.4   95 Percent and 7 Percent
Failure Ranks ($n = 10$)

j	95% rank	7% rank
1	0.25887	0.00723
2	0.39416	0.04441
3	0.50690	0.09989
4	0.60662	0.16687
5	0.69646	0.24280
6	0.77756	0.32673
7	0.84997	0.41870
8	0.91274	0.51975
9	0.96323	0.63275
10	0.99488	0.76650

Since the Weibull cumulative distribution function is the most popular failure model for solder joint reliability, we will use it to show how its parameters are determined. Recall the Weibull CFD

$$F(x) = 1 - e^{-(x/\theta)^\beta} \qquad (2.45)$$

Applying the principles of least squares and regression analysis by linearizing transformation to the Weibull data, it can be shown that[7,9,11,17]

$$\beta = \frac{n \sum_{i=1}^{n} X_i Y_i - \sum_{i=1}^{n} X_i \sum_{i=1}^{n} Y_i}{n \sum_{i=1}^{n} (X_i^2) - \left(\sum_{i=1}^{n} X_i\right)^2} \qquad (2.46)$$

$$C = \frac{\sum_{i=1}^{n} (X_i^2) \sum_{i=1}^{n} Y_i - \sum_{i=1}^{n} X_i \sum_{i=1}^{n} X_i Y_i}{n \sum_{i=1}^{n} (X_i)^2 - \left(\sum_{i=1}^{n} X_i\right)^2} \qquad (2.47)$$

where $\theta = e^{-C/\beta}$ $\qquad\qquad\qquad$ (2.48)

$X = \ln(x)$ $\qquad\qquad\qquad\qquad$ (2.49)

$$Y = \ln \ln \frac{1}{1 - F(x)} = \beta X + C \qquad (2.50)$$

For example, the test data on the failure of 30 electronic components are given in the first column of Table 2.5. The second column is for the failure order (j). The third column is for the median ranks ($n = 30$) $F(x)$, and the successive columns are the transformations as indicated in column headings. Thus, we have

TABLE 2.5 Failure Data of 30 Electronic Components

x (hours)	Failure count j	Median rank $F(x) = (j - .3)/(n + .4)$	$X = \ln x$	$Y = \ln \ln \dfrac{1}{1 - F(x)}$	$X^2 = (\ln x)^2$	$XY = \ln x \ln \ln \dfrac{1}{1 - F(x)}$
79	1	0.0231	4.37	−3.73	19.10	−16.30
98	2	0.0559	4.58	−2.86	20.98	−13.10
124	3	0.0888	4.82	−2.36	23.23	−11.38
128	4	0.1217	4.85	−2.04	23.52	−9.89
150	5	0.1546	5.01	−1.78	25.10	−8.92
160	6	0.1875	5.08	−1.57	25.81	−7.99
175	7	0.2204	5.16	−1.39	26.63	−7.17
180	8	0.2533	5.19	−1.23	26.96	−6.38
190	9	0.2862	5.25	−1.09	27.56	−5.72
198	10	0.3191	5.29	−0.95	27.98	−5.03
210	11	0.3519	5.35	−0.83	28.62	−4.45
215	12	0.3848	5.37	−0.72	28.84	−3.87
240	13	0.4177	5.48	−0.61	30.03	−3.34
268	14	0.4506	5.59	−0.51	31.25	−2.85
282	15	0.4835	5.64	−0.42	31.81	−2.37
300	16	0.5164	5.70	−0.32	32.49	−1.82
320	17	0.5403	5.77	−0.23	33.29	−1.33
360	18	0.5822	5.89	−0.14	34.69	−0.82
390	19	0.6151	5.97	−0.05	35.64	−0.30
420	20	0.6480	6.04	0.04	36.48	0.24
450	21	0.6808	6.11	0.13	37.33	0.79
520	22	0.7137	6.25	0.22	39.06	1.38
550	23	0.7466	6.31	0.31	39.82	1.96
590	24	0.7795	6.38	0.41	40.70	2.62
650	25	0.8124	6.48	0.51	41.99	3.30
690	26	0.8453	6.54	0.63	42.77	4.12
750	27	0.8782	6.62	0.75	43.82	4.97
840	28	0.9111	6.73	0.88	45.29	5.92
890	29	0.9440	6.79	1.06	46.10	7.20
930	30	0.9768	6.84	1.33	46.79	9.10
			$\sum X = 171.45$	$\sum Y = -16.56$	$\sum X^2 = 993.68$	$\sum XY = -71.43$

$$\beta = \frac{n\left(\sum XY\right) - \left(\sum X\right)\left(\sum Y\right)}{n\left(\sum X^2\right) - \left(\sum X\right)^2}$$

$$= \frac{(30)(-71.43) - (171.45)(-16.56)}{(30)(993.68) - 171.45^2} \tag{2.51}$$

$$= 1.68$$

$$C = \frac{\left(\sum X^2\right)\left(\sum Y\right) - \left(\sum X\right)\left(\sum XY\right)}{n\left(\sum X^2\right) - \left(\sum X\right)^2}$$

$$= \frac{(993.68)(-16.56) - (171.45)(-71.43)}{(30)(993.68) - 171.45^2} \tag{2.52}$$

$$= -10.13$$

$$\theta = e^{-C/\beta}$$

$$= e^{10.13/1.68} \tag{2.53}$$

$$= 423 \text{ hours}$$

With $\beta = 1.68$ and $\theta = 423$, Eq. (2.45) can be plotted in Fig. 2.22 and is given by

$$F(x) = 1 - e^{-(x/423)^{1.68}} \tag{2.54}$$

2.11   Acceleration Factor and Models

As mentioned earlier, an acceleration factor is needed to bridge the stress gap between the test condition and operating condition. Let

$$x_o = \text{AF } x_t \tag{2.55}$$

where AF = linear acceleration factor
x_o = a random variable (time to failure at operating condition)
x_t = a random variable (time to failure at test condition)

Then, we have

$$f_o(x_o) = \frac{1}{\text{AF}} f_t\left(\frac{x_o}{\text{AF}}\right) \tag{2.56}$$

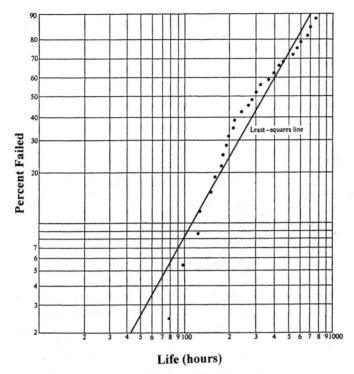

Life (hours)

Figure 2.22 Weibull life distribution of the electronic components.

$$F_o(x_o) = F_t\left(\frac{x_o}{\text{AF}}\right) \tag{2.57}$$

$$R_o(x_o) = 1 - F_t\left(\frac{x_o}{\text{AF}}\right) \tag{2.58}$$

$$h_o(x_o) = \frac{1}{\text{AF}}\, h_t\left(\frac{x_o}{\text{AF}}\right) \tag{2.59}$$

where $f_o(x_o)$, $F_o(x_o)$, $R_o(x_o)$, and $h_o(x_o)$ are the PDF, CDF, reliability function, and failure rate at operating condition. For example, with the Weibull distribution, we have

$$F_o(x_o) = 1 - e^{-(x_o/\text{AF}\theta)^\beta} \tag{2.60}$$

$$R_o(x_o) = e^{-(x_o/\text{AF}\theta)^\beta} \tag{2.61}$$

$$h_o(x_o) = \text{AF}^{-\beta}(\beta/\theta)(x_o/\theta)^{\beta-1} \tag{2.62}$$

With the exponential distribution, we have

$$F_o(x_o) = 1 - e^{-(\lambda/\mathrm{AF})X_o} \tag{2.63}$$

$$R_o(x_o) = e^{-(\lambda/\mathrm{AF})X_o} \tag{2.64}$$

$$h_o(x_o) = \frac{\lambda}{\mathrm{AF}} \tag{2.65}$$

How do you determine the AF? The answer is that you need an acceleration model. How do you choose an acceleration model? The answer is failure mechanisms! Failure mechanisms are the most important factors in choosing an acceleration model for determining the AF. For more information please read Chap. 3. For example, if the key factor for failure is temperature then the Arrhenius model can be used with great success.

$$\mathrm{AF} = e^{(\Delta H/k)[1/T_o - 1/T_t]} \tag{2.66}$$

Where ΔH is the unknown constant, k is Boltzmann's constant (8.617×10^{-5} in eV/K or 1.380×10^{-16} in erg/K), T_t is the temperature at test condition, and T_o is the temperature at operating condition. These temperatures, measured in kelvins, are obtained by adding 273.16 to the temperatures in degrees Celsius. The unknown constant ΔH can be estimated from two cells of experimental test data consisting of time to failure of units tested at temperature T_t and time to failure of units tested at temperature T_o.

For solder joint failure under thermal fatigue (temperature and frequency are key factors), the most widely used model is the modified Coffin-Manson equation and is given in the following forms.[26-38]

$$\mathrm{AF} = \left(\frac{f_o}{f_t}\right)^m \left(\frac{\Delta T_t}{\Delta T_o}\right)^n \frac{\Phi_o}{\Phi_t} \tag{2.67}$$

$$\mathrm{AF} = \left(\frac{f_o}{f_t}\right)^m \left(\frac{\Delta T_t}{\Delta T_o}\right)^n e^{1414(1/T_o - 1/T_t)} \tag{2.68}$$

$$\mathrm{AF} = \left(\frac{f_o}{f_t}\right)^m \left(\frac{\Delta \gamma_t}{\Delta \gamma_o}\right)^n e^{1414(1/T_o - 1/T_t)} \tag{2.69}$$

In Eqs. (2.67) to (2.69), $m = \frac{1}{3}$, $n = 1.9 \sim 2.0$, f_t is the temperature cycling frequency at test condition, f_o is the temperature cycling frequency at operating condition, T_t is the temperature (in kelvins) at test condition, and T_o is the temperature (in kelvins) at operating condition. In Eq. (2.67), Φ_t is the isothermal fatigue test lifetime at the peak temperature (in degrees Celsius) attained during the testing cycle and Φ_o

is the isothermal fatigue test lifetime at the peak temperature (in degrees Celsius) attained during the operating cycle. In Eqs. (2.67) and (2.68), ΔT_t is the temperature range (in degrees Celsius) at test condition and ΔT_o is the temperature range (in degrees Celsius) at operating condition. In Eq. (2.69), $\Delta \gamma_t$ is the shear strain range at test condition and $\Delta \gamma_o$ is the shear strain range at operating condition.

Among these three equations, Eq. (2.68) is the easiest to use. In order to use Eq. (2.67) you must run at least two more isothermal fatigue tests for Φ_t and Φ_o. (Sometimes we can obtain these values from the published data; however, the chances may not be great of finding your particular solder and temperature conditions.) In order to use Eq. (2.69) you must run a finite element analysis to determine the shear strain range for the given solder joint's material, geometry, and temperature boundary conditions. Sometimes we can obtain these values by means of the concept of distance to neutral point (DNP) and pure shear theory as long as the predicted life is above the reliability requirement, since it overpredicts the shear strain range by neglecting the contribution of resistance from the noncorner interconnects and the warpage of the assembly. A critical discussion of Eqs. (2.67) through (2.69) is given by Solomon.[26] A reliable AF for solders remains to be determined.

The use of the preceding equations is demonstrated in the following example with the Weibull distribution. Figure 2.23 shows a board layout for the solder joint reliability thermal cycling test of the 208-pin fine pitch (0.5-mm) plastic quad flat pack (PQFP). There are 25 placement sites for the PQFPs. A total of four boards (100 PQFPs or 20,800 solder joints) were tested.

These four test boards were grouped and cabled together in such a way that the resistance of all the solder joints could be measured. The boards were tested in an air-to-air thermal cycling chamber. They were subjected to the temperature cycling profile shown in Fig. 2.24. It can be seen that the cycle period was one hour with a maximum temperature of 125°C and a minimum temperature of −40°C. A data acquisition system continuously monitored the electrical resistance of the PQFP solder joints and logged the failure times. The number of cycles to failure N_f (the random variable) was defined as the first solder joint failure in any 208-pin PQFP.

Applying the principles of least squares and ranking to the thermal cycling test results, it was shown by Lau et al.[28] that the best-fit Weibull parameters are $\beta = 3.66$ and $\theta = 6357$. The life distribution (CDF) and failure rate h of the PQFP solder joints under test condition, 1 cycle/h or $f_t = 1$ cycle/h and from −40°C to +125°C or $\Delta T_t = 165°C$, are shown in Figs. 2.25 and 2.26, respectively.

Let us say that the operating condition is one cycle per day ($f_o = 1$ cycle/day = 1/24 h) and the operating temperature range is from 0 to

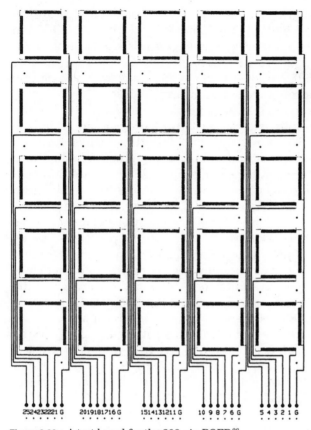

Figure 2.23 A test board for the 208-pin PQFP.[28]

$85°C$ ($\Delta T_o = 85°C$). Also, from the literature[36] we find that Φ_o ($85°C$)/Φ_t ($125°C$) is approximately equal to 1.7. Then we have

$$\text{AF} = \left(\frac{1}{24}\right)^{1/3} \left(\frac{165}{85}\right)^2 (1.7) = 2.22 \qquad (2.70)$$

Once the acceleration factor has been determined, the life distribution, reliability function, and failure rate under operating conditions can be determined by Eqs. (2.60) through (2.62), respectively, and are given by

$$F_o(x_o) = 1 - e^{-(x_o/14113)^{3.66}} \qquad (2.71)$$

$$R_o(x_o) = e^{-(x_o/14113)^{3.66}} \qquad (2.72)$$

$$h_o(x_o) = 0.0000311 \left(\frac{x_o}{6357}\right)^{2.66} \qquad (2.73)$$

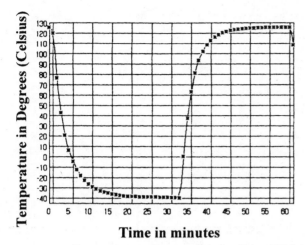

Figure 2.24 Temperature cycling condition, −40 to 125°C, one cycle per hour.[28]

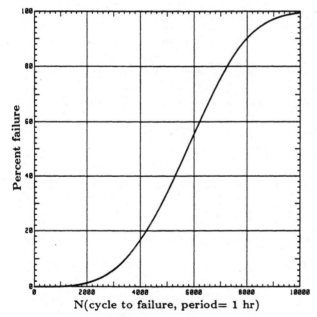

Figure 2.25 Life distribution of the 208-pin solder joints under test condition, −40 to 125°C, one cycle per hour.[28]

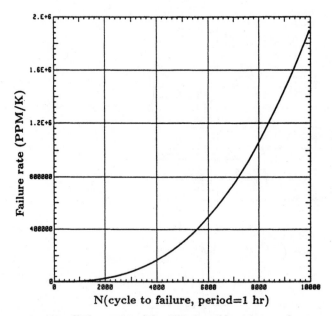

Figure 2.26 Failure rate of the 208-pin solder joints under test condition, –40 to 125°C, one cycle per hour.[28]

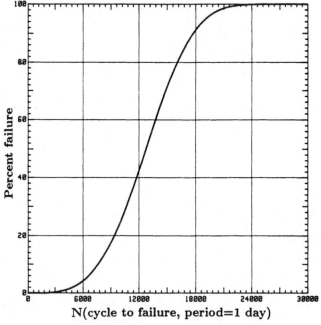

Figure 2.27 Life distribution of the 208-pin solder joints under operating condition, 0 to 85°C, one cycle per day.[28]

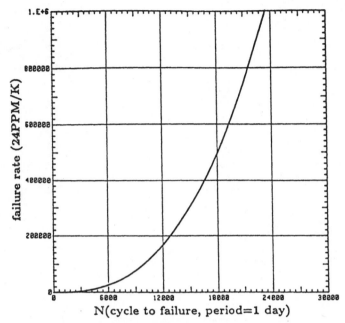

Figure 2.28   Failure rate of the 208-pin solder joints under operating condition, 0 to 85°C, one cycle per day.[28]

The life distribution and failure rate of the 208-pin PQFP solder joints under operating conditions are shown in Figs. 2.27 and 2.28, respectively. It can be seen that the 1 percent cumulative failure point occurred at approximately 104,000 hours, or 11.8 years, and the 50 percent cumulative failure point occurred at about 304,560 hours, or 34.7 years. These values are beyond most of the design lifetimes. It should be noted that the present results are for the solder joints subjected only to the thermal cycling condition. Other environmental stress factors such as humidity, shock and vibration, and corrosion were not considered.

Acknowledgments

The principal author (Lau) would like to thank G. Dody, W. Chen, M. McShane, D. Rice, S. Erasmus, and W. Adamjee for their contribution on Ref. 28. It has been a great pleasure and fruitful experience to work with them. Most of the materials in this chapter are reproduced from Ref. 39. The kind permission of NEPCON is greatly appreciated.

References

1. Johnson, L. G., *The Statistical Treatment of Fatigue Experiments,* Elsevier Publishing Company, Amsterdam, 1964.

2. Johnson, L. G., *Theory and Technique of Variation Research,* Elsevier Publishing Company, Amsterdam, 1964.
3. Johnson, L. G., "The Median Ranks of Sample Values in Their Population with an Application to Certain Fatigue Studies," *Industry Mathematics,* **2:**1–9, 1951.
4. Shooman, M. L., *Probabilistic Reliability: An Engineering Approach,* McGraw-Hill, New York, 1968.
5. Billinton, R., and R. N. Allan, *Reliability Evaluation of Engineering Systems Concepts and Techniques,* Plenum Press, New York, 1983.
6. Endrenyi, J., *Reliability Modeling in Electric Power Systems,* John Wiley & Sons, New York, 1978.
7. Dhillon, B. S., and C. Singh, *Engineering Reliability, New Techniques and Applications,* John Wiley & Sons, New York, 1981.
8. Ang, A. H., and W. H. Tang, *Probability Concepts in Engineering Planning and Design,* John Wiley & Sons, New York, 1984.
9. Lipson, C., and N. J. Sheth, *Statistical Design and Analysis of Engineering Experiments,* McGraw-Hill, New York, 1973.
10. Serfling, R. Y., *Approximation Theorems of Mathematical Statistics,* John Wiley & Sons, New York, 1980.
11. Lawless, J. F., *Statistical Methods and Methods for Lifetime Data,* John Wiley & Sons, New York, 1982.
12. Weibull, W., "A Statistical Distribution Function of Wide Applicability," *ASME Transactions, Journal of Applied Mechanics,* **18:**293–297, 1951.
13. Weibull, W., *Fatigue Testing and Analysis of Results,* The MacMillan Company, New York, 1961.
14. Barlow, R. E., and F. Proschan, *Mathematical Theory of Reliability,* John Wiley & Sons, New York, 1965.
15. Henley, E. J., and H. Kumamoto, *Reliability Engineering and Risk Assessment,* Prentice-Hall, New Jersey, 1981.
16. Henley, E. J., and H. Kumamoto, *Designing for Reliability and Safety Control,* Prentice-Hall, New Jersey, 1981.
17. Little, R. E., and E. H. Jebe, *Statistical Design of Fatigue Experiments,* Applied Science Publishers, London, 1975.
18. Little, R. E., and E. H. Jebe, *Manual on Statistical Planning and Analysis for Fatigue Experiments,* STP 588, ASTM, Philadelphia, 1975.
19. Mann, N. R., R. E. Schafer, and N. D. Singpurwalla, *Methods for Statistical Analysis of Reliability and Life Data,* John Wiley & Sons, New York, 1974.
20. Young, L., and J. C. Ekvall, *Reliability of Fatigue Testing, Statistical Analysis of Fatigue Data,* STP 744, ASTM, Philadelphia, 1981, pp. 55–74.
21. Grant, E. L., and R. S. Leavenworth, *Statistical Quality Control,* McGraw-Hill, New York, 1980.
22. Barlow, R. E., and F. Proschan, *Statistical Theory of Reliability and Life Testing,* Holt, Reinhart and Winston, New York, 1975.
23. Tobias, P. A., and D. Trindade, *Applied Reliability,* Van Nostrand Reinhold, New York, 1986.
24. Gibra, I. N., *Probability and Statistical Inference for Scientists and Engineers,* Prentice-Hall, New Jersey, 1973.
25. Mood, A. M., F. A. Graybill, and D. C. Boss, *Introduction to the Theory of Statistics,* 3d ed., McGraw-Hill, New York, 1974.
26. Solomon, H. D., "The Solder Joint Fatigue Life Acceleration Factor," *ASME Transactions, Journal of Electronic Packaging,* **113:**186–190, June 1991.
27. Solomon, H. D., "Fatigue of 60/40 Solder," *IEEE Transactions on Components, Hybrids, and Manufacturing Technology,* vol. CHMT-9: 423–432, December 1986.
28. Lau, J. H., G. Dody, W. Chen, M. McShane, D. Rice, S. Erasmus, and W. Adamjee, "Experimental and Analytical Studies of 208-pin Fine Pitch Quad Flat Pack Solder-Joint Reliability," *Circuit World,* **18:**13–19, January 1992.
29. Lau, J. H., S. Golwalkar, D. Rice, S. Erasmus, and R. Foehringer, "Experimental and Analytical Studies of 28-pin Thin Small Outline Package Solder-Joint Reliability," *ASME Transactions, Journal of Electronic Packaging,* **114:**169–176, June 1992.

30. Lau, J. H., S. Golwalkar, P. Boysan, R. Surratt, D. Rice, R. Forhringer, and S. Erasmus, "Solder Joint Reliability of a Thin Small Outline Package (TSOP)," *Circuit World*, **20:**12–19, November 1993.
31. Lau, J. H., Y.-H. Pao, C. Larner, S. Twerefour, R. Govila, D. Gilbert, S. Erasmus, and S. Dolot, "Reliability of 0.4 mm Pitch, 256-pin Plastic Quad Flat Pack No-Clean and Water-Clean Solder Joints," *Soldering & Surface Mount Technology*, no. 16:42–50, February 1994.
32. Lau, J. H., S. Golwalkar, and S. Erasmus, "Advantages and Disadvantages of Thin Small Outline Packages (TSOP) with Copper Gull-Wing Leads," *ASME Transactions, Journal of Electronic Packaging*, **116:**234–237, September 1994.
33. Lau, J. H., G. Harkins, D. Rice, J. Kral, and B. Wells, "Experimental and Statistical Analyses of Surface-Mount Technology PLCC Solder-Joint Reliability," *IEEE Transactions of Reliability*, **37:**524–530, December 1988.
34. Lau, J. H., L. Powers, J. Baker, D. Rice, and W. Shaw, "Solder Joint Reliability of Fine Pitch Surface Mount Technology Assemblies," *IEEE Transactions on Components, Hybrids, and Manufacturing Technology*, **13:**534–544, September 1990.
35. Norris, K. C., and A. H. Landzberg, "Reliability of Controlled Collapse Interconnections," *IBM Journal of Research and Development*, **13:**266–271, May 1969.
36. Lau, J. H., *Solder Joint Reliability: Theory and Applications*, Van Nostrand Reinhold, New York, 1991.
37. Master, R. N., M. S. Cole, and G. B. Martin, "Ceramic Column Grid Array for Flip Chip Applications," *IEEE Proceedings of Electronic Components & Technology Conference*, May 1995, pp. 925–929.
38. Puttlitz, K. J., Sr., "Flip Chip Solder Bump (FCSB) Technology: An Example," pp. 450–476, in *Multichip Module Technologies and Alternatives*, D. A. Doane and P. D. Franzon, eds., Van Nostrand Reinhold, New York, 1993.
39. Lau, J. H., "Reliability Testing and Data Analysis," Workshop at the NEPCON West, Anaheim, Calif., February 1992.

Chapter

3

Failure Analysis

Failure of an electronic component or system is commonly defined as either the loss of its ability to perform the intended function or as a certain deviation from its performance specifications.[1-59] In nature, the failure of electronics is time-dependent, and its rate is traditionally described by a *bathtub curve,* which is formed by three different stages of failure processes, i.e., infant mortality, useful life, and wear-out (see Fig. 2.5). Various statistical models, such as Weibull, exponential, gaussian, lognormal, binomial, or Poisson,[45] have been adopted to assess the failure rate or reliability of electronics systems and components, e.g., MIL-HDBK-217.[34] Some of the models have been delineated in Chap. 2. Although it has been demonstrated that statistical methods, incorporated with lab scale or previous field data, were useful to some extent in predicting the failure rate of electronics, their lack of accounting for the physical insight of failure modes and mechanisms makes it increasingly difficult for them to choose the acceleration model for determining the acceleration factor and to cope with the rapid development of electronic packaging technologies during the past two decades.[15,43] In this chapter the failure of solder interconnects will be discussed from a physics-of-failure aspect. First, an overall introduction to the failure mode and mechanism of electronic packaging will be given, followed by three typical types of SMT solder joints—flip chip bump, leadless, and leaded. Discussions of more advanced solder interconnects, e.g., ball grid array (BGA) or flip chip-on-board (FCOB), can be found in Chap. 5 and Lau[29] and Chap. 6 and Lau,[28] respectively.

3.1   Failure Mode and Mechanism Analysis in Electronic Packaging

The failure mode and the associated failure mechanism of any electronic component may be different by changing the material or configuration of the component, interconnection methods, or any parameters in the manufacturing or assembly processes. The prediction of its reliability requires the understanding of the actual failure mechanism and the interactions of the stress due to the loading and the strength of the entire assembly bonded by dissimilar materials. A failure mode of an electronic component represents a change of its functional status, such as open, short, change of resistance, capacitance, or other electrical parameters. An accurate prediction of the reliability associated with such a failure mode is generally impossible without the prior knowledge of specific failure mechanisms. It is a process where any defect, e.g., crack, void, or any form of damage, nucleates and grows as a function of stresses related to thermal, mechanical, electromagnetic, or chemical loadings. It is also a function of materials, interfaces, and configurations of the component assembly. In general, there may be more than one failure mechanism associated with one failure mode. A design change may eliminate one particular failure mechanism, but could introduce another as well. The identification of all potential mechanisms of one failure mode is generally difficult, but it is even more difficult to develop physical models that account for these mechanisms. However, despite the difficulties and the fact that for most failure mechanisms the understanding of their root causes is still in its infancy, the development of reliability models needs to be based on physics of failure in order to accurately predict the product life.

A brief introduction of common failure mechanisms associated with electronic packages is given here. More details can be found in sources such as Vardaman,[55] Pecht et al.,[43] Hu,[15] Lewis,[31] Sabnis,[46] Minges,[36] Amerasekra and Campbell,[1] Hnatek,[14] and Stojadinovic.[52] Based on the nature of stresses that cause failure, the failure mechanisms of electronics can be classified as mechanical, thermal, chemical, electrical, and radiation failures.[15] Figure 3.1 summarizes some common failure mechanisms of electronic components due to two types of loading mechanisms: overstress and wear-out. The overstress or wear-out stresses can result from various manufacturing, storage, shipping; and operational conditions, such as mechanical/thermal shocks and vibration, thermal fatigue, temperature gradients, electric current/voltage, electromagnetic fields, humidity, dust, or chemical contamination. An example is shown in Fig. 3.2 for typical failures of a plastic IC component due to the previously mentioned loadings. Table 3.1 also lists a number of failure mechanisms due to various types of stresses for typical components of an electronic assembly, e.g., die, interconnects, substrate, lid, and encapsulant.

Recently, numerical modeling such as finite element modeling has been widely used to study the root cause of various failure mechanisms in electronic packaging. While it is a powerful tool for use in understanding the deformation and stresses of structures due to various types of loadings, care needs to be taken when applying it to electronic packages. Some considerations particularly unique to electronic packaging are given as follows:

1. Electronic packages are multiple-material systems, where bi- and tri-material bonded edges and interfaces exist and often introduce stress concentrations from which interfacial damage or cracks are initiated. One major numerical difficulty is associated with the capture of detailed stress/strain distribution near these concentrations with reasonable meshes. Another meshing problem is related to the aspect ratio, where it is recognized that many materials used in electronics packages are of thin layer shapes, such as thin and thick films, or adhesive or solder interlayers. A tremendous number of elements are usually required to accurately model the behavior of those

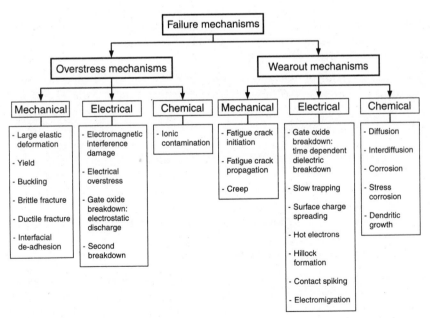

Figure 3.1 Classification of failure mechanisms.[15]

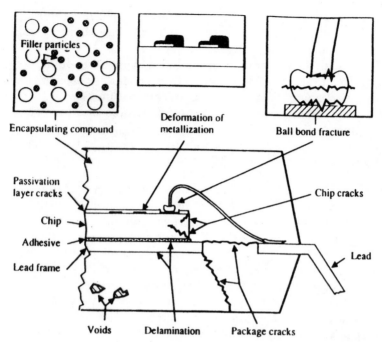

Encapsulating compound

Deformation of metallization

Ball bond fracture

Passivation layer cracks

Chip cracks

Chip

Adhesive

Lead frame

Lead

Voids   Delamination   Package cracks

Figure 3.2 Potential failure mechanisms and sites in a plastic-encapsulated IC component.[15]

thin layers. In addition, the material properties of most interfaces (e.g., intermetallic compounds) and their fracture behavior still remain unknown.

2. Most properties of electronic packaging materials are temperature- and time-dependent. The mechanical behavior of a solder joint, for example, not only depends on temperature but is highly sensitive to the strain rate and its microstructure, which is also a function of temperature, cooling rate, and loading history. Mechanical properties measured using relatively large-size and bulk specimens generally are not the same as those for actual solder interconnects. Experimental techniques for testing actual joint-size specimens are needed for accurate material characterization and development of constitutive equations for numerical modeling. Some required mechanical properties of electronic components for modeling different failure mechanisms are listed in Table 3.2.[15]

3. The variation of dimension of components in electronic packages can be of several orders of magnitude, e.g., 0.08 × 0.1 mm solder bump versus 150 × 150 mm printed wiring board. It is, in general, impossible to incorporate components with dimensions varying several orders of magnitude into one numerical model. Thus, the analysis

TABLE 3.1 Failure Sites, Operational Stresses and Failure Mechanisms[15]

Failure sites		Stresses					
	Contaminant	Range of relative humidity	Range of temperature cycle	Relative humidity	Temperature	Vibration/shock	Voltage/current
Die		J	M, G	J	O	A, M, G	
Die attach			F, G, H, M			M, G, C	
Interconnects							
Flip chip	J	J	M, G	J	B, C, J, N	M, G	N
Tape of TAB			C, E			E	
Bond of TAB	J, K	J	M, G	J, K	A, B, J, K, N	M, G	N, K
Wire	J	J	G, M	J		G, M, E	
Wire bond	J, F	J	G, M, F	J, F		G, M, E, F	
Substrate			M, G			A, B, G, M	
Substrate attach			F, G, M			C, G, M	
Leads	J, K	J	M	J, K	J	M	M, K
Lead seal	F		F, A		F, M, G	F, A	
Lid			M				
Lid seals			F		A, H, F	H	
Encapsulation	J	J		J	J		
Outer lead			B, G, M			G, M	
Board	J	J	A, B, G, M		O	G, M	
Outer solder joint			B, F, G, H, M			F, G, M	

A—Brittle fracture; B—Ductile fracture; C—Yield; D—Buckling; E—Large elastic deformation; F—Interfacial de-adhesion; G—Fatigue crack initiation; H—Creep; I—Wear; J—Corrosion; K—Dendritic growth; L—Interdiffusion; M—Fatigue crack propagation; N—Diffusion; O—Excessive temperature.

TABLE 3.2 Required Material Properties for Reliability Qualification Analysis[15]

	Modulus of Elasticity	Poisson's ratio	Stress-strain relation	CTE	Yield strength	Ultimate strength	Fracture toughness	Thermal conductivity	Fatigue constants	Creep constants	Corrosion constant
Die	1, 2, 7	1, 2, 7		1, 2, 7		1	1	3			6
Die attach	1, 2	1, 2	2	1, 2				3	2	2	
Substrate	1, 2, 4, 7, 9	1, 2, 4, 7, 9		1, 2, 4, 7, 9		9	9	3			
Wire	4, 8	4, 8		4, 8		8					
Bonding pad	4	4		4	8						
Lead	10	10		10			5	3	4		
Encapsulation	1, 7	1, 7		1, 7			5	3			6
Case	9	9		9			5	3			
Solder	10	10	10	10					10	10	
Board	10	10	10	10							

1—Die cracking; 2—Fatigue crack (void) propagation in die attach; 3—Excessive junction temperature; 4—Bond-pad (wire bond) shear fatigue; 5—Encapsulate cracking; 6—Metallization corrosion; 7—Encapsulate delamination (interface); 8—Wire bond rupture; 9—Substrate cracking; 10—Outer-lead solder joint fatigue.

must be broken down into global and local analyses. This increases the time for modeling. Another issue lies in the difference between actual and simplified geometries of the component or interconnect. For example, the actual geometry of a solder joint depends on a number of parameters, such as solder-paste volume, reflow temperature, and flux, which can result in a wide range of different shapes. Oversimplification of geometry may eliminate geometric discontinuities or defects, which, for example, may be intimately associated with a significant part of fatigue life—damage initiation.

4. While the test conditions of electronic packages in the laboratory can be accurately characterized and numerically modeled, those in the field may not be available. The work required to identify the customer-usage profile can become formidable for some industries, e.g., automotive electronics, due to a wide spectrum of customers, geographical locations, ambient temperatures, and humidity. Moreover, the successful conversion of field conditions into quantifiable loadings that can be applied to numerical models poses another difficulty in failure-mechanism modeling of electronic packaging.

3.2 Flip Chip Solder Joints

Wire bonding, tape-automated bonding (TAB), and flip chip solder bump technology are the three main first-level interconnection (or zero-level package) technologies. Wire bonding consists of discrete bonds made at each metal wire end connecting chip and substrate (Fig. 3.3). The flip chip bump technology was first introduced in 1964 and applied to the solid logic technology (SLT) hybrid modules of IBM's System/360.[7] Later in the IC era, the controlled-collapse chip connection (C4) was devised where a pure solder bump, made of 95Pb-5Sn and with a size of 100–125 μm, was restrained from collapsing or wetting-out on the electrode land by using simple, thick-film glass dams, which limited solder flow to the tip of the substrate metallization.[35]

Other applications of the flip chip bump technology were also developed (e.g., thin-film resistor and capacitor network chips[19]). Optimization of solder bumps, such as barrel versus hourglass shapes, has been studied.[38,48] Efforts have also been made to improve the reliability of flip chip bumps by optimizing the materials and processes.[30,32,48,53,56] Thermal fatigue reliability of flip chip bumps has been studied experimentally and numerically.[8,26,37,47,48] Puttlitz[44] has developed flip chip solder bumps with Pb-In and Pb-Sn solders. A review of this technology and its application to VLSI has been given by Goldmann and Totta.[10] Some recent development of this technology can be found in Burkhart,[4] for example.

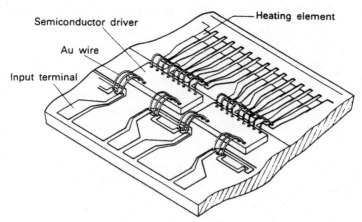

Figure 3.3 Schematic representation of the wire-bonding method for thermal printheads.[16]

3.2.1 Material and structure

Figure 3.4 shows the material system and configuration of a commonly used flip chip bump. Different solder materials have been used for specific applications. Figure 3.5 shows a typical bumped IC used in automotive electronics, and Fig. 3.6 shows two typical flip chip bumps mounted on alumina substrate with 88Pb-10Sn-2Ag (bump A) and 60Sn-40Pb (bump B)[3]. The corresponding microstructures are shown in Fig. 3.7. In addition to the computer industry, the flip chip technol-

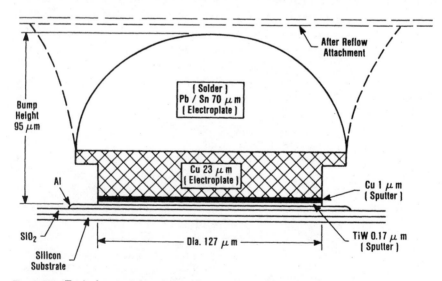

Figure 3.4 Typical material system and structure of flip chip bumps.

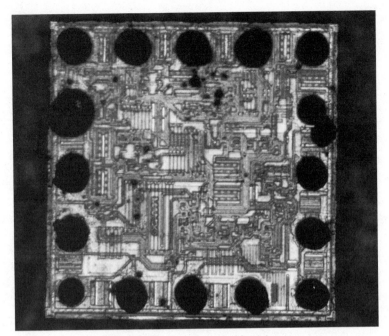

Figure 3.5   A typical IC with flip chip bump solder joints.

ogy has also been applied to other industries. An example is given in Fig. 3.8 for the application to the automotive hybrid ignition module. For electronic modules located in the vicinity of the engine or under-hood where the temperature can reach as high as 140°C, hybrid tech-nologies with flip chip bumps using solder alloys with high melting points are being used to replace the printed circuit board technology in order to ensure the reliability.

3.2.2 Thermal fatigue failure mechanisms

Among various loadings to which the flip chip solder bumps are sub-jected, thermal fatigue is the primary one that causes the bumps to fail. During thermal cycling, the solder bump must accommodate the displacement due to mismatch of the coefficient of thermal expansion (CTE) between the silicon chip (2.4–2.8 ppm/°C) and the substrate (e.g., 6–7 ppm/°C for alumina). The thermally induced displacement results in a complex stress and strain distribution in the bump in that high peel and shear stresses are found near the bonded edges of bump/substrate or bump/chip. Fatigue cracks generally initiate at bonded edges and may propagate along the interface or curve into the solder, leading to failure of the bump.

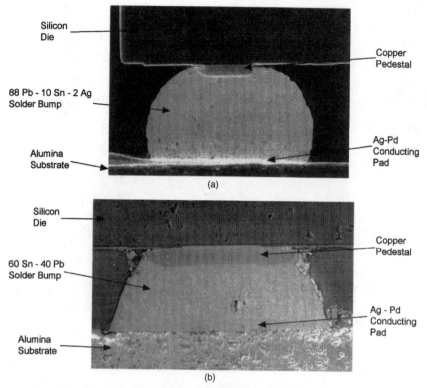

Figure 3.6 Flip chip bumps with different solder alloys.[3]

The thermal fatigue failure of flip chip bumps is also a function of solder material, metallization on substrate and chip, and intermetallic compounds formed at the joined interfaces. Puttlitz[44] has studied the effect of solder material (Pb-Sn and Pb-In), metallization material, and reflow process on the fracture of flip chip bumps. Figure 3.9 shows the fracture patterns of Pb-Sn solder-bumped flip chips (SBFC) attached to gold-capped microsockets on multilayer ceramic (MLC) substrates after thermal cycling between 0 and 100°C with a period of 20 minutes. For lower reflow temperatures or smaller reflow times, the crack initiated at the ball-limiting metallurgy (BLM) periphery and propagated near the interface of bump and chip. However, when the reflow temperature or the number of multiple reflow increased, the crack initiation and propagation shifted to the interface between bump and substrate, termed top surface metallurgy (TSM), with a gold- and nickel-coated metallization. This shift of crack path is due to the growth of gold-tin intermetallics at the higher reflow temperature or

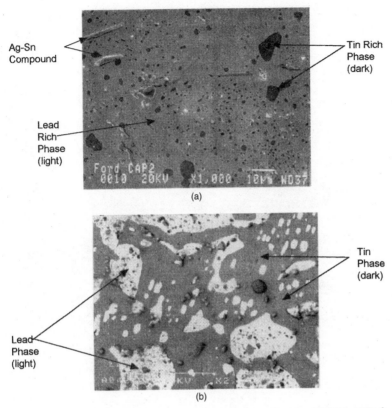

Ag-Sn
Compound

Tin Rich
Phase
(dark)

Lead
Rich
Phase
(light)

(a)

Tin
Phase
(dark)

Lead
Phase
(light)

(b)

Figure 3.7 Microstructure of solder bumps: (a) 88Pb-10Sn-2Ag; (b) 60Sn-40Pb.[3]

the larger number of reflow times. It was also found that the codeposited phosphorus and the thickness of the gold layer plated on top of the nickel could lead to a loss of solder wettability, which would significantly affect the solder strength.

Although Pb-Sn and Pb-In bumps have comparable metallurgical structure, the lead-indium bumps differ in one important aspect, i.e., the Cu_7In_4 intermetallics were found at both BLM and TSM interfaces, but not in between. The transfer mechanism of Cu_7In_4 intermetallics from BLM to TSM is associated with the dissociation into the melt with fast copper diffusion in the joint followed by nucleation and growth at the TSM interface, forming another intermetallics layer on top of the $AuIn_2$ intermetallics. And this has led to more observed failures along the TSM/bump boundary, which is different from that of Pb-Sn solder bumps. However, the fracture could occur in the bump if TSM pads are distorted as a result of multiple reflow (Fig. 3.10).

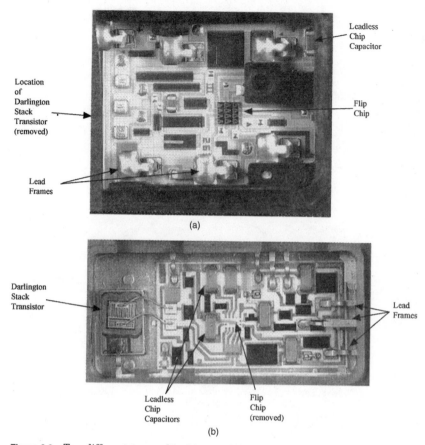

Figure 3.8 Two different types of ignition modules using flip chip solder bump technology: (a) Lexus TFI module; (b) Ford TFI module.[3]

The effect of the shape of 95Pb-5Sn flip chip bumps on failure mechanisms has been studied by Satoh and associates.[48] The fatigue life of the hourglass-shaped bumps was found to be approximately three times that of typical barrel-shaped ones for temperature cycling between −50 to 150°C within a one-hour period. The failure mechanism was identified to be fatigue-dominant crack propagation. Fatigue striations as small as 0.15 μm were observed using scanning electron microscopy (SEM) with a low-acceleration voltage (1–5 kV). The relation between the formation and propagation of striations and slip planes of Pb is shown in Fig. 3.11. The finite element results of the barrel-shaped and hourglass-shaped bumps are shown in Figs. 3.12 and 3.13, respectively. The effective strain contours show that the maximum strain occurs near the silicon chip for the barrel-shaped bumps, whereas the hourglass-shaped bump has the maximum effec-

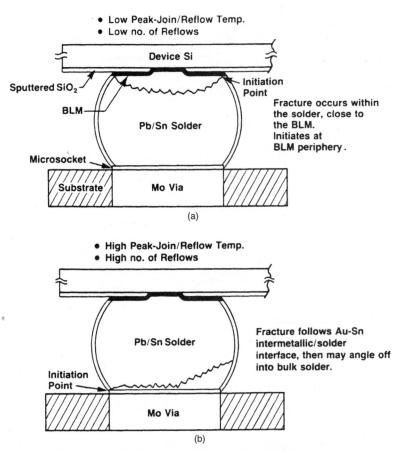

Figure 3.9   Side-view sketch depicting the effect of peak-join/reflow temperature and multiple reflows on the fatigue failure mode of Pb-Sn SBFC connections attached to Au-capped microsockets on MLC substrates: (a) low reflow temperature, few reflows; (b) high reflow temperature, multiple reflows. (Sketch not to scale.[44])

tive strain near its throat. The numerical results were in agreement with the experimental observation of where the crack initiated and propagated. The thermal fatigue life of various barrel- and hourglass-shaped bumps was correlated with the corresponding maximum effective strain range and is shown in Fig. 3.14. It should be pointed out that in order to apply the fatigue life equation in Fig. 3.14, it is necessary to adopt the same constitutive model for the solder and a similar mesh as used by Satoh and associates.[48] Otherwise, the results of prediction could be misleading since the maximum effective strain in the solder bump is highly dependent upon the constitutive model used and the size of the element. A fracture mechanics approach to

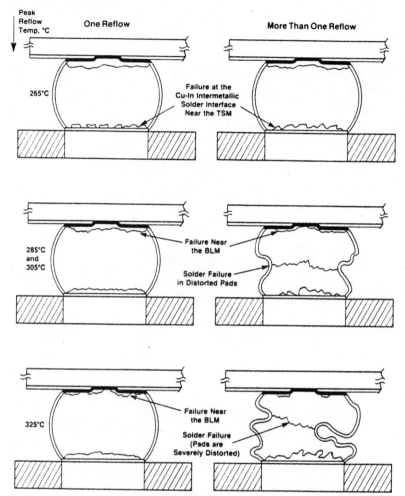

Figure 3.10 Side-view sketch depicting the effect of peak-join/reflow temperature and multiple reflows on the fatigue failure mode of Pb-In SBFC connections attached to Au-capped microsockets on MLC substrates. (Sketch not to scale.[44])

determine the thermal fatigue life of the hourglass-shaped bump is shown in Chap. 6.

3.3   SMT Leadless Chip Resistor (LCR) Solder Joints

In this section the failure analysis and root cause identification of typical LCR solder joints are discussed. Scanning electron microscopy (SEM) and microprobes were used for failure-mechanism identification, and finite element modeling (FEM) was adopted to investigate

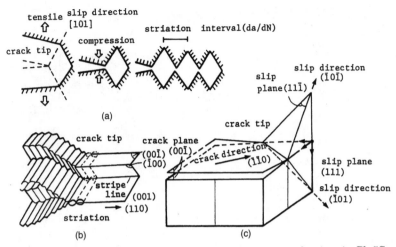

Figure 3.11 Schematic of the thermal fatigue fracture mechanism in Pb-5Sn alloy: (*a*) formation mechanism of Laird-type striation; (*b*) formation of Pb alloy striations; (*c*) estimation of the direction of crack propagation and slip plane.[48]

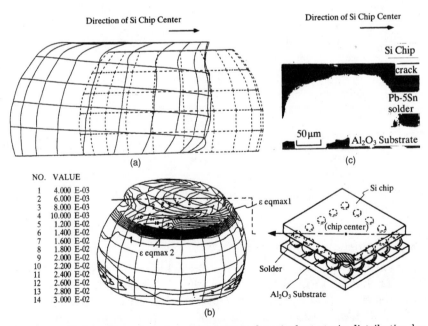

Figure 3.12 Deformation of a microsolder joint and equivalent strain distribution by FEM simulation (at 423 K): (*a*) deformation state; (*b*) equivalent strain contour line; (*c*) thermal fatigue fracture (after 300 cycles; –50 to 150°C, 1 cycle/hr).[48]

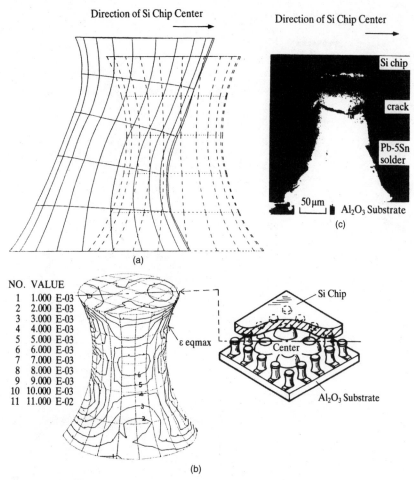

Figure 3.13 Deformation of a microsolder joint by FEM simulation and equivalent strain distribution (at 423 K): (*a*) deformation state; (*b*) equivalent strain contour line; (*c*) thermal fatigue fracture (after 3000 cycles; –50 to 150°C, 1 cycle/hr).[48]

the root cause of failure and identify key design parameters for reliability improvement.

3.3.1 Material and configuration

The material system of a typical LCR solder joint, such as 1206 or 2512, is shown in Fig. 3.15 together with the specific dimensions. The chip resistors are usually attached to the printed wiring boards (PWBs) with eutectic Sn-Pb or 62Sn-36Pb-2Ag solder joints (Fig. 3.16). The shape of the solder joint after reflow depends on the volume of solder, flux, metallization of LCR termination, copper pad size, and reflow temperature profile.[6] As a result of the time-dependent melting and

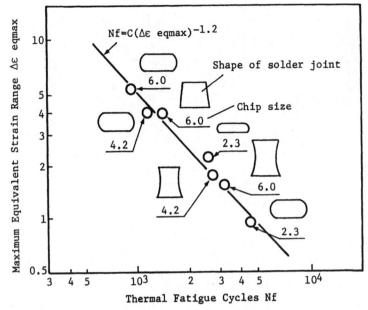

$$Nf = C(\Delta \varepsilon \ eqmax)^{-1.2}$$

(Plot axes)
- Maximum Equivalent Strain Range $\Delta \varepsilon$ eqmax (vertical)
- Thermal Fatigue Cycles Nf (horizontal)

Shape of solder joint

Chip size

6.0

6.0

4.2

2.3

6.0

4.2

2.3

Figure 3.14   Thermal fatigue life of CCB flip chip joints.[48]

solidification process during reflow and the interaction of the flux, solder, and temperature profile (Fig. 3.17), the resulting standoff height of the solder joint is usually between 0.025 and 0.075 mm (1 to 3 mils). Due to different cooling rates, the microstructure of LCR solder joints is different from that of a relatively large bulk specimen. Figure 3.18 shows the comparison of an as-cast bulk, an as-cast joint from a beam specimen, and a joint after thermal cycling for 62Sn-36Pb-2Ag.[13] It is seen that the microstructure of a joint specimen (Fig. 3.18b) is much finer, and thermally induced coarsening is also observed near the crack surface after thermal fatigue (Fig. 3.18c).

3.3.2   Thermal fatigue failure mechanisms

During thermal cycling, the LCR solder joint experiences a complex stress and strain history, and as a result of the low-cycle-fatigue process, cracks initiate and propagate, which leads to the failure of the joint. Different crack patterns may occur depending on where the crack initiates and the strength of the solder and various interfaces (e.g., solder/nickel or solder/copper). In addition, cracks can occur simultaneously at different locations due to high stress concentration and join together later. Figure 3.19 shows a number of stress-concentration sites for different pad/termination configurations. Typical examples of crack patterns in LCR joints include the following:

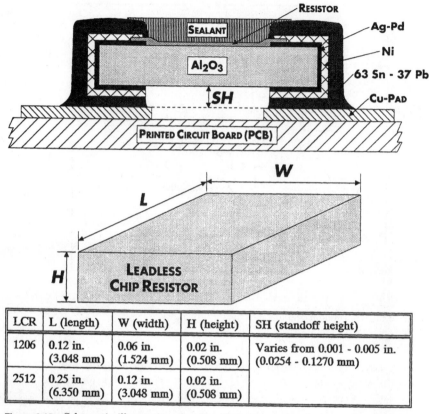

Figure 3.15 Schematic illustration showing the cross-sectional view and approximate dimensions of LCRs.

LCR	L (length)	W (width)	H (height)	SH (standoff height)
1206	0.12 in. (3.048 mm)	0.06 in. (1.524 mm)	0.02 in. (0.508 mm)	Varies from 0.001 - 0.005 in. (0.0254 - 0.1270 mm)
2512	0.25 in. (6.350 mm)	0.12 in. (3.048 mm)	0.02 in. (0.508 mm)	

1. The crack starts from a surface flaw in the fillet (e.g., due to grain boundary sliding) and propagates inside into the solder. This is, in general, rarely observed, but has been seen when the package was subjected to vibrational loadings.

2. The crack starts at the solder/nickel junction at the inner end between the LCR and the board and propagates along the metallization interface of alumina/Pd-Ag or Pd-Ag/Ni. This is usually due to a poor adhesion of these metallization layers to the alumina.

3. The crack starts at the solder/nickel junction at the inner end and curves into the chip resistor, leading to an early brittle fracture. Although not commonly seen, any preexisting surface flaws on the alumina can lead to such an early failure.

4. A main crack starts at the solder/nickel junction at the inner end and propagates, while at the same time local fatigue cracks initiate

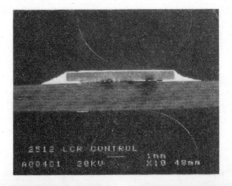

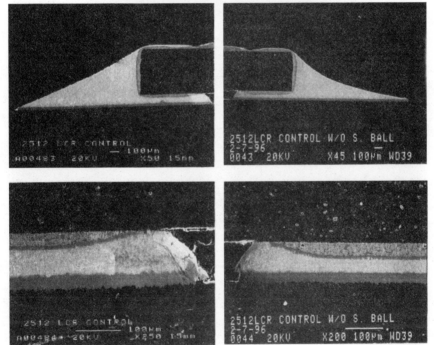

Figure 3.16 Typical 2512 leadless chip resistor attached to PWB with 62Sn-36Pb-2Ag solder joints.[51]

because of large shear strain, particularly for the cases where the standoff height is small. The fracture of the joint between component and board is a result of coalescence of the main crack and those local fatigue cracks (Fig. 3.20).

5. The crack starts at the solder/nickel junction at the inner end and propagates through the solder joint and at approximately a 45° angle into the fillet or vertically near the solder/nickel interface. This is the

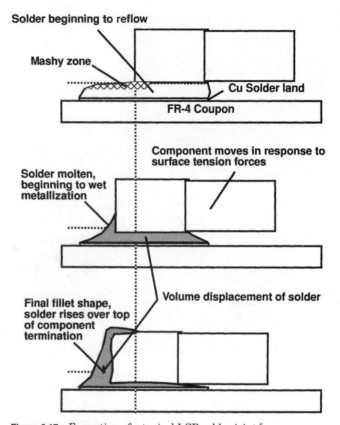

Figure 3.17 Formation of a typical LCR solder joint.[6]

most common failure mechanism seen during lab thermal cycling or under field operating conditions.

Pao and Jung[40] have demonstrated that when using nonlinear finite element modeling, from a fracture-mechanics point of view, both 45° and vertical paths are equally favorable to fatigue crack propagation. However, the shape and size of the fillet and local damage near the corner of LCRs may have a strong effect on the fracture path in the fillet (Figs. 3.21 and 3.22).[11,58] If a small crack initiates in the vicinity of an LCR corner (Fig. 3.21) and is oriented at approximately 45°, and the main crack initiated at the inner end propagates to this corner region, the two cracks would merge and most likely follow the 45° direction into the fillet. Therefore, the small crack at the corner plays an important role in the direction of crack propagation in the fillet, which in turn has an effect on the fatigue life of the joint.

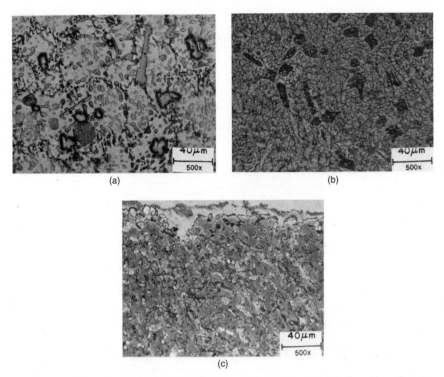

(a)

(b)

(c)

Figure 3.18 Microstructure of 62Sn-36Pb-2Ag solder (etched): (*a*) bulk solder; (*b*) joint before thermal cycling; (*c*) joint after 221 thermal cycles between 40 and 140°C (40 minutes/cycle).[13]

Govila et al.[11] also studied the effect of the radius of curvature (ROC) of LCR corners on the strains in the joint. Figure 3.23 shows such an effect for two typical ROCs, 0.036 and 0.11 mm, on the creep strain distribution near the corner. It is seen that the maximum creep shear strain near the corner with the smaller ROC (sharper corner) is approximately 20 percent higher than that associated with the larger ROC. The results are based upon a nonlinear finite element analysis with a thermal loading simulating a 10-minute ramp-down from 125 to –55°C. This indicates that the ROC of LCR corners may be a design parameter for reliability improvement.

Jih and Pao[18] have performed a nonlinear finite element analysis to study the effect of a number of design parameters, such as solder standoff, solder length, LCR length, LCR height, and fillet shape on the shear strain in the solder between LCR and board. Four groups of joint configurations were also included in the analysis (Fig. 3.24). Figures 3.25 to 3.28 show the total shear strain range, as a result of a two-hour thermal cycling between –55 and 125°C, as a function of solder

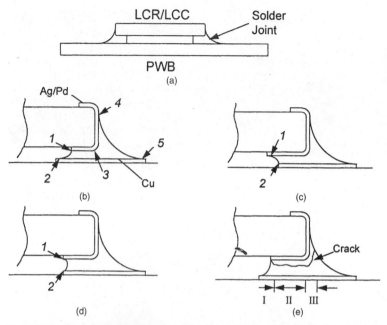

Figure 3.19 (*a*) A typical LCR/LCC solder joint; (*b*) identification of stress concentration sites, in this case site 1 has a higher concentration than site 2; (*c*) site 2 has a higher concentration if the inner end of the Cu-pad is shorter than the termination; (*d*) similar concentration if Cu-pad and termination are lined up; (*e*) the crack path in the solder joint.

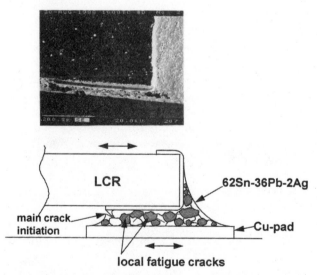

Figure 3.20 Fracture in an LCR solder joint caused by initiation of a main crack and local fatigue cracks, followed by propagation and coalescence of these cracks.

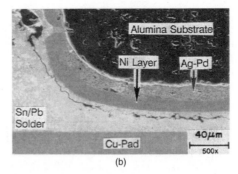

(a)   (b)

Figure 3.21   SEM micrograph showing creep/fatigue-induced damage near the corner of the resistor[11]: (*a*) end view of vendor B, 250 thermal cycles; (*b*) Region A, cracking in solder microstructure.

height, length, LCR length, and height, respectively. It is noted that, in the case of 2512 LCR, the relation between strain range and solder standoff is nonlinear in that the decrease of strain range may not be significant as the standoff is larger than 0.125 mm (5 mils). A similar effect has been revealed on leadless chip capacitor solder joints.[49] A correction factor based upon the effect of the preceding parameters has been obtained for the shear strain range in the joint between LCR and board and is shown in Fig. 3.29.

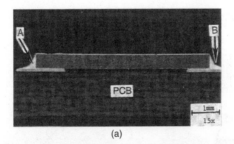

(a)

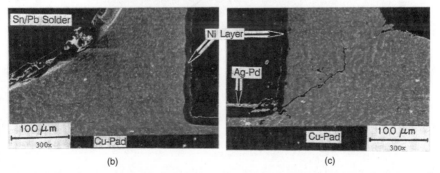

(b)   (c)

Figure 3.22   SEM micrograph showing creep/fatigue-induced damage near the corner of the resistor[11]: (*a*) side view of Vendor A, 250 thermal cycles; (*b*) Region A, solder microstructure; (*c*) Region B, solder microstructure and cracking.

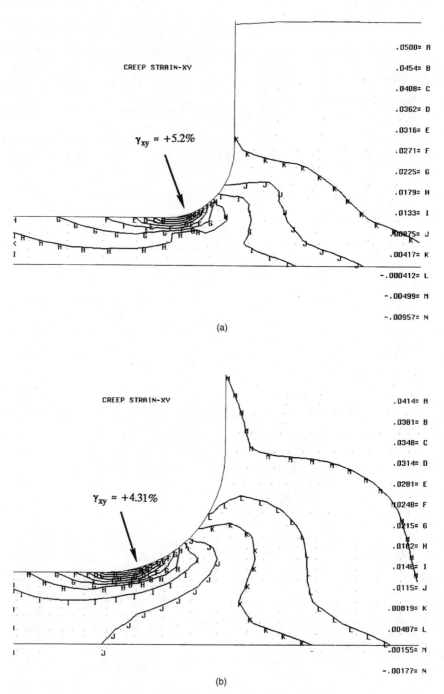

Figure 3.23   Creep shear strain distribution in the vicinity of LCR corner: (a) ROC = 0.036 mm; (b) ROC = 0.11 mm.[11]

The failure of LCR solder joints can be further complicated by the evolution of microstructure and the interaction of creep and fatigue processes during thermal cycling. The strain-assisted microstructural coarsening in 63Sn-37Pb and 62Sn-36Pb-2Ag solder joints (Fig. 3.22c) is well documented and has been demonstrated to have an adverse effect on the fatigue life. The creep-induced ductile fracture and fatigue striations in solder joints are also widely reported in the literature.[9,23,39,48] Figure 3.30 shows a typical creep-fatigue interaction on the fracture surface of a 2512 LCR 62Sn-36Pb-2Ag solder joint.[41] It is clear that both the creep-induced voids and fatigue striations resulted after thermal cycling. And both need to be considered in developing reliability models for LCR solder joints subjected to thermal fatigue.

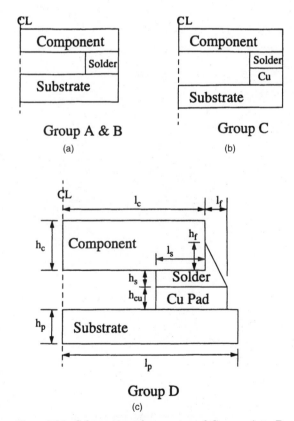

Figure 3.24 Schematics of geometry of Groups A to D LCRs; Group A is based on an analytical model and Groups B to D are for finite element analysis.

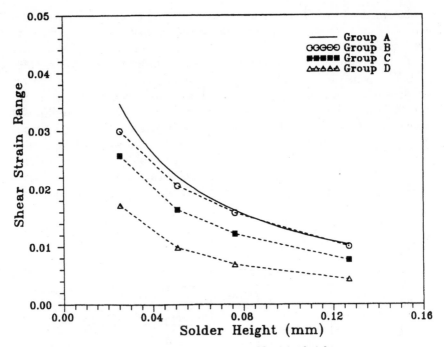

Figure 3.25   The cyclic shear strain range versus solder joint height.

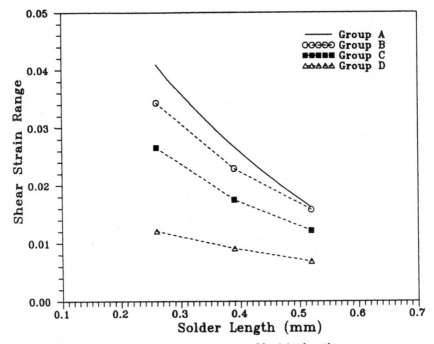

Figure 3.26   The cyclic shear strain range versus solder joint length.

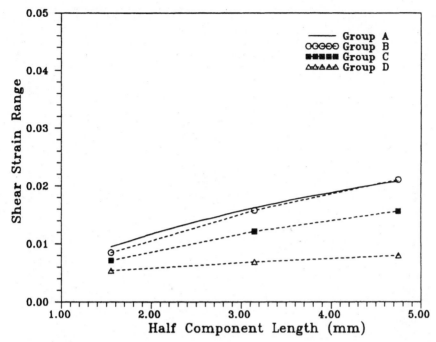

Figure 3.27   The cyclic shear strain range versus component length.

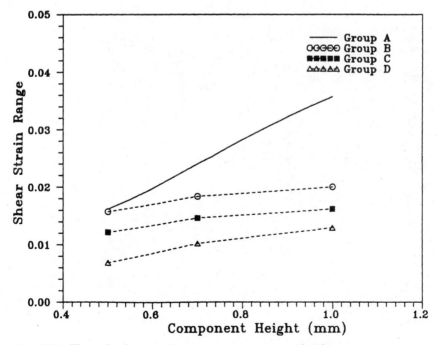

Figure 3.28   The cyclic shear strain range versus component height.

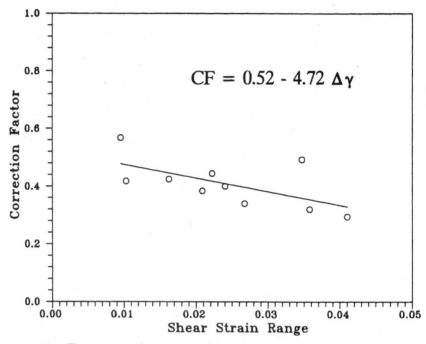

$$CF = 0.52 - 4.72 \; \Delta \gamma$$

Figure 3.29   The correction factor versus shear strain range.

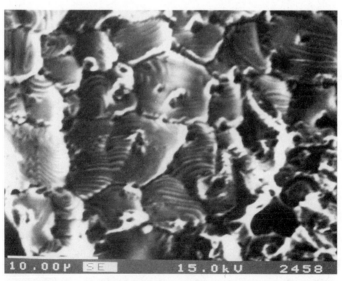

Figure 3.30   An electron microprobe micrograph showing the fatigue striations and creep cavitation of 2512 LCR 63Sn-37Pb solder joint subjected to temperature cycling between –40 and 95°C.

3.4   SMT Leaded Solder Joints

The surface-mounted leaded solder joints are widely used in ceramic and plastic electronics packages. There are three major configurations of the surface-mounted leaded joints: butt, J-type, and gull-wing. A tremendous amount of information regarding these joints exists in the literature.[2,12,17,20,22–25,27,33,42,48,50,54,55] In this section, the failure mechanisms of typical 256-pin 0.4-mm-pitch PQFP (plastic quad flat pack) solder joints will be discussed since fine pitch, gull-wing-leaded packages are among those most commonly seen in today's SMT packages.

3.4.1   Material and geometry

Figure 3.31 shows the geometry and dimension of a 0.4-mm-pitch, 256-pin PQFP with gull-wing leads. The PQFPs are mounted on test boards (Fig. 3.32) with eutectic Sn-Pb solder joints and subjected to temperature cycling between −40 and 125°C.[12,24,25,27] Two types of solder paste flux (water-soluble and no-clean) were also evaluated in terms of thermal fatigue reliability.

The solder joints were mass-reflowed by a forced-convection in-line conveyer infrared soldering system with a peak temperature of 226°C. Immediately after the solder was reflowed, for the water-clean boards an in-line conveyer-type aqueous cleaning system was used with the water temperature at 60–65°C and the pressure set at 420 kPa. The test board layout and SMT assembly process of this 256-pin PQFP are shown in Chap. 8.

3.4.2   Failure mechanisms

The thermal fatigue of leaded solder joints could depend on a number of parameters that are related to material, configuration, and manufacturing. To cite a few, selection of solder paste, surface preparation of copper pad and lead, and control of the temperature profile have a strong effect on the wettability of solder and the resulting joint shape. The stresses and strains in the solder joint result from the global mismatch of the coefficient of thermal expansion (CTE) between package and substrate and the local CTE mismatch between solder and lead or solder and copper pad. Moreover, in contrast to leadless SMT solder joints, the lead stiffness plays an important role in determining the stresses and strains in the joint, which is time-dependent due to the viscoplastic nature of the solder material. With a compliant lead, the time-dependent inelastic response of the solder joint (e.g., creep and stress relaxation) can be significantly reduced so that the life of the solder joint will be increased. However, a certain level of stiffness

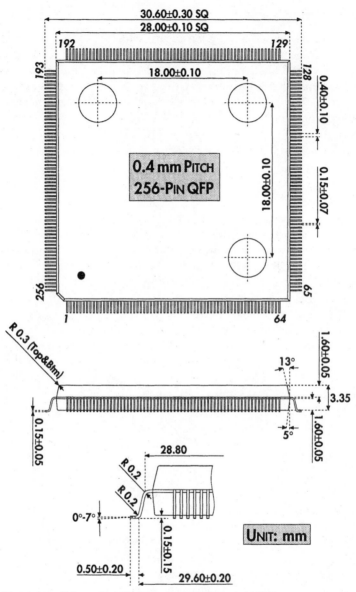

Figure 3.31   Schematics of a 0.4-mm-pitch, 256-pin PQFP.

is required in lead frame forming to achieve good component coplanarity. Therefore, proper design of lead stiffness has a direct impact on
the reliability of the joint.

The typical thermal-cyclic failure of the fine pitch gull-wing solder
joints is shown in Fig. 3.33. The side view of the joint (Fig. 3.33*a*)

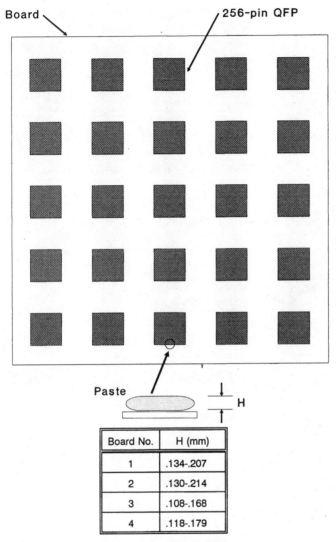

Board   256-pin QFP

Paste   H

Board No.	H (mm)
1	.134-.207
2	.130-.214
3	.108-.168
4	.118-.179

Figure 3.32 Test board layout for 256-pin QPFs.

clearly shows cracking at three different locations near heel and toe. For most cases, cracks generally initiate near the edge of lead/solder or solder/copper interfaces where the stresses are high and the strength may be weakened by the Sn-Cu intermetallic compounds, but toe cracks, due to local CTE mismatch, are sometimes seen. However, it seems that only the heel crack would tend to propagate during thermal cycling, while the toe crack, once initiated, either remains stationary or grows relatively slowly.[12,48] This is because the global CTE mismatch

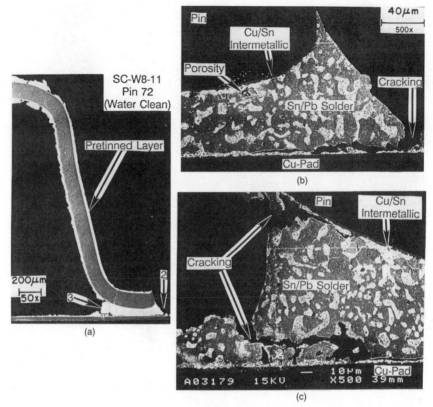

Figure 3.33 Thermal fatigue failure (1605 cycles to failure) of water-clean solder joints: (a) side view; (b) solder microstructure of region 2 at the toe; (c) solder microstructure of region 3 at the heel.[27]

provides an overall fatigue driving force for the heel crack to propagate whereas such a driving force is lacking for the toe crack. The propagation of the crack mainly follows the interface where the crack initiates and rarely curves into the solder.

Also seen in Fig. 3.33 is the coarsened microstructure of 63Sn-37Pb solder joint, particularly in the vicinity of those cracks. More evidence is shown in Fig. 3.34 for a number of cracks at the toe. This indicates that microstructural coarsening may be stress/strain-assisted. Satoh et al.[48] has observed fatigue striations on the fracture surface of PQFP- and SOP (Swiss outline package or small outline package)-leaded solder joints and measured the heel crack growth rate as a function of crack length (Fig. 3.35) and the number of thermal cycles (Fig. 3.36).

The effect of water-clean and no-clean fluxes on the reliability of solder joints is shown in Fig. 3.37. It is seen that the no-clean solder joints have a higher failure rate than the water-clean solder joints for a

THIS IS NOT USED

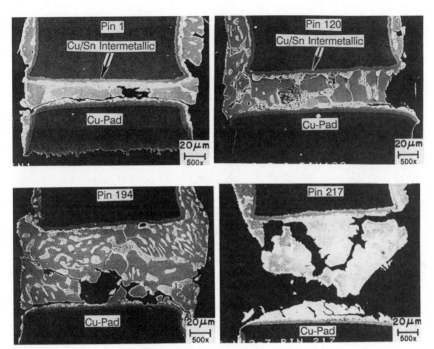

Figure 3.34 Solder cracking near the toe region (end view).[27]

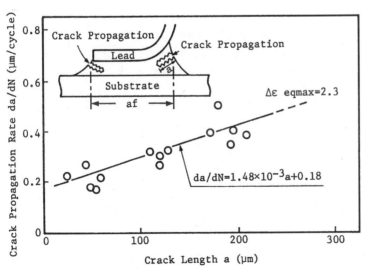

Figure 3.35 Thermal fatigue crack propagation rate versus crack length in PQFP joints.[48]

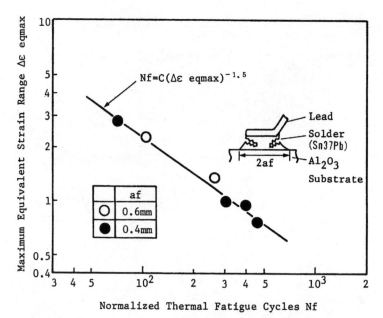

Figure 3.36   Thermal fatigue life of PQFP and SOP microsolder joints.[48]

larger number of cycles. This may be due to the thermochemical reaction of the uncleaned flux residues and contamination on the no-clean solder joints.[27] For more information on the life distribution and failure rate of the 256-pin PQFP solder joints, please read Chap. 8.

3.4.3   Root cause identification and reliability prediction

An analytical model for the corner solder joint subjected to thermal cycling is formulated based on the assumption that the total shear strain rate is the sum of elastic and steady-state creep strain rates. The lead is considered as an elastic spring with its constant determined as (see Kotlowitz[59])

$$K_d = 1.414 \ \frac{K_{11}K_{33}}{\sqrt{K_{11}^2 + K_{33}^2}} \tag{3.1}$$

where K_{11} and K_{33} are the stiffness components in direction 1 (in-plane) and 3 (out-of-plane) defined in Lau et al.[27] The stiffness matrix K_{ij} of the current lead has been determined using the finite element method (Table 3.3).

Considering the thermally induced displacement imposed on the lead and solder joint, we can express the shear stress τ in the solder joint in the following nonlinear differential equation.[5]

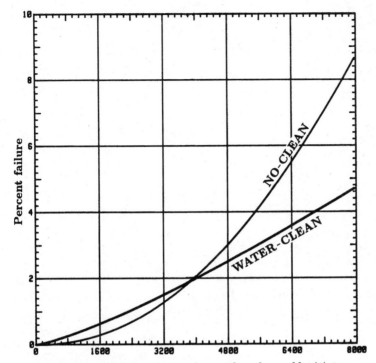

Figure 3.37 Percent failure of water-clean and no-clean solder joints.

$$\frac{d\tau}{dt} = K_G \left[\left(L\Delta\alpha \, \frac{dT}{dt} \right) - h_s C \, |\tau|^2 \, \text{sgn} \, (\tau) + \left(\frac{h_s}{G^2} \, \frac{dG}{dT} \, \frac{dT}{dt} \right) \tau \right] \quad (3.2)$$

where $\Delta\alpha$ is the mismatch of the coefficient of thermal expansion (CTE), L is the half-length of the component, h_s is the average solder standoff, G is solder's shear modulus, T is temperature, C and n are the creep parameters in $d\gamma_{\text{creep}}/dt = C\tau^n$, and K_G is defined as

$$K_G = \frac{1}{\left(\dfrac{h_s}{G} + \dfrac{A_s}{K_d} \right)} \quad (3.3)$$

where A_s is the effective solder area. Equation (3.2) can be solved numerically (e.g., by the Runge-Kutta method). Once the stress is solved, the shear strain can be determined by integrating the creep equation. The constitutive equation for 63Sn-37Pb solder used here follows that given by Knecht and Fox,[21] and the temperature-dependent shear modulus of the solder is taken from Wong et al.[57]

The temperature range has been selected to be from −40 to 125°C to represent the most severe field conditions. A typical one-hour thermal

TABLE 3.3   Stiffness Matrix K_{ij} for the Gull-Wing Lead and Solder Joint in 256-Pin PQFP

(For forces in N/mm and for moments in N/mm/rad)

18.8	−38.1	0	0	0	−3.4	−18.8	38.1	0	0	0	9.6
−38.1	124.7	0	0	0	−12.3	38.1	−124.7	0	0	0	−44.9
0	0	8.3	8.9	3.1	0	0	0	−8.3	4.2	4.8	0
0	0	8.9	12.4	3.6	0	0	0	−8.9	1.8	4.8	0
0	0	3.1	3.6	3.1	0	0	0	−3.1	1.2	−0.3	0
−3.4	−12.3	0	0	0	10.1	3.4	12.3	0	0	0	6.8
−18.8	38.1	0	0	0	3.4	18.8	−38.1	0	0	0	−9.6
38.1	−124.7	0	0	0	12.3	−38.1	124.7	0	0	0	44.9
0	0	−8.3	−8.9	−3.0	0	0	0	8.3	−4.2	−4.8	0
0	0	4.2	1.8	1.2	0	0	0	−4.2	4.9	2.7	0
0	0	4.8	4.8	−0.3	0	0	0	−4.8	2.7	4.8	0
9.6	−44.9	0	0	0	6.8	−9.6	44.9	0	0	0	20.4

cycling between −40 and 125°C with a 20-minute hold and a 10-minute ramp time was simulated; the predicted average shear stress τ, defined as the shear force divided by the pad area A_s, versus time for 20 cycles is shown in Fig. 3.38. Note that the stress level is low (i.e., at 125°C hold time τ ≈ −1.5 MPa and ≈3.2 MPa at −40°C hold) after the cyclic response stabilizes. Because of the low stress level, the creep strain

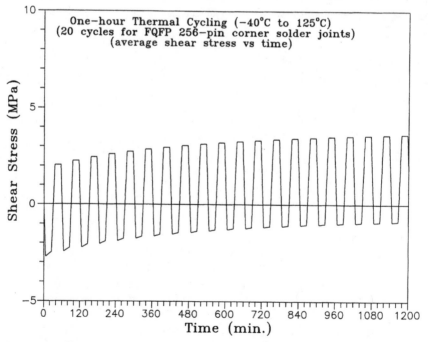

Figure 3.38   Shear stress in the solder joint versus time for 20 thermal cycles.

accumulated at the hold time is insignificant, thus the creep-induced damage is unlikely to occur. Therefore the failure should be dominated by fatigue. This is consistent with the results of Satoh et al.[48] The fact that average stress is low is due to the compliant lead and the entire pad area being considered. However, the stress is not uniformly distributed over the entire pad area but concentrates at the heel and the toe. A smaller pad size was then modeled to simulate the stress concentration. The results are shown in Fig. 3.39 for stress versus time and Fig. 3.40 for strain components versus time. It is seen that stress ranges from –9 MPa at 125°C to 50 MPa at –40°C and stabilizes much faster than the previous case. The stress hardly relaxes at –40°C but relaxes from ≈9.4 to ≈2 MPa in 20 minutes at 125°C. In Fig. 3.40 it is seen that creep strain continues to accumulate over the entire cycle except for the –40°C hold time where the creep shear strain rate apparently vanishes. This implies that the low-temperature hold time can be shortened because little creep damage would accumulate. On the other hand, as the temperature rises, the creep strain continues to increase. This suggests a longer ramp-up time or high-temperature hold time to facilitate the creep damage accumulation.

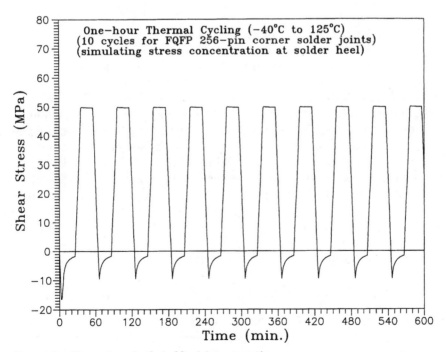

Figure 3.39   Shear stress in the solder joint versus time.

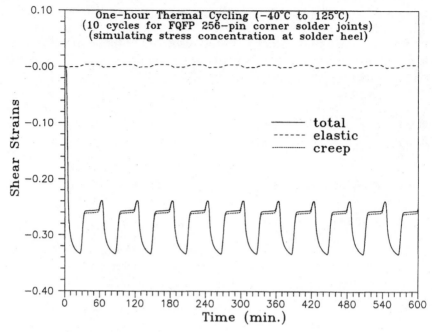

Figure 3.40   Shear strain in the solder joint versus time.

The life prediction approach used here primarily follows the one developed by Kitano and associates.[20] The gull-wing lead was treated as three independent simple beams. The lead foot was modeled as a beam on an elastic-plastic foundation subjected to transverse normal and shear forces. The eutectic solder joint was assumed to be a temperature-dependent elastic-plastic material. Two changes have been made in using this approach. One was to make the boundary condition of the vertical beam and the horizontal one (the lead foot) consistent (i.e., v_{x1} ($x_1 = l_1$) = u_{2x} ($x_2 = 0$) in Eq. (18) of Kitano et al.[20]). This largely improves the accuracy of the results. The other change was, instead of adopting the finite difference methods used by the authors, to implement a finite element algorithm to solve the systems of nonlinear governing equations. Good agreement was obtained between this approach and finite element results.

The failure criterion for the solder joint is based on the von Mises equivalent plastic strain, defined as $\Delta\varepsilon_{eq} = (\Delta\varepsilon_p^2 + \Delta\gamma_p^2/3)^{1/2}$, where $\Delta\varepsilon_p$ is the maximum normal (peel) strain in the solder joint and consists of two parts: one from the in-plane thermal-induced displacement (mismatch) and the other from the out-of-plane displacement between the component and the printed wiring board. The latter is particularly important for leads near the corner. The plastic shear strain range $\Delta\gamma_p$

results from the in-plane shearing of the solder joint. Kitano et al.[20] also performed isothermal and thermal fatigue tests on three different leaded solder joints (gull-wing, J-type, and butt leads). The results are regrouped and plotted in Fig. 3.41 and indicate that, despite the scatter of experimental data, the maximum equivalent plastic strain seems to be an appropriate failure index since it is approximately configuration-independent. The test data are approximated by a power law equation:

$$N_f = 2.37 \times 10^{-3} \, (\Delta\varepsilon_{eq})^{-1/0.39} \tag{3.4}$$

These test data are used to predict the fatigue life of the solder joints in this study.

It was observed that both water-clean and no-clean processes resulted in large variations in a joint's toe and heel heights, defined as a and b in Fig. 3.42, respectively. The variations observed in the heel

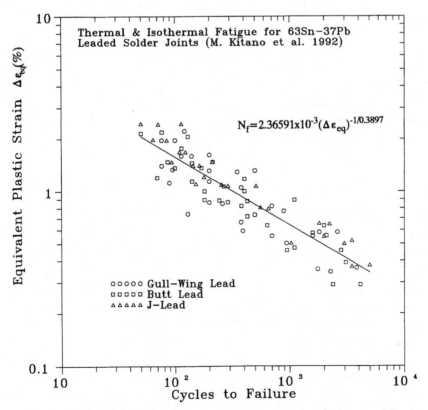

Figure 3.41 Equivalent plastic strain versus number of cycles to failure for different types of leaded solder joints.

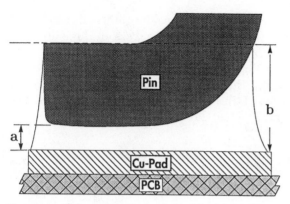

Figure 3.42 Schematics of a leaded solder joint with toe
height a and heel height b.

can range from 0.16 to 0.26 mm, and from 0.003 to 0.1 mm for the toe
height. The effect of these variations on the number of cycles to failure
N_f is shown in Fig. 3.43. It is seen that for toes ranging from 0.005 to
0.045 mm the fatigue life is nearly independent of the heel height in
the range of 0.1 to 0.3 mm. The fatigue life is highly dependent upon
the toe height since the maximum equivalent plastic strain occurs for

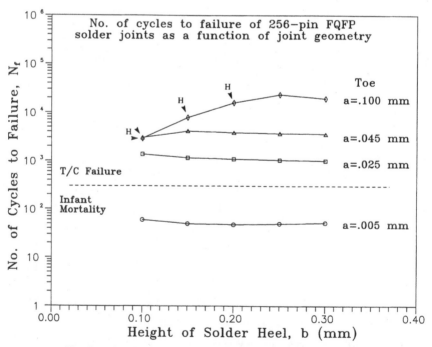

Figure 3.43 Number of cycles to failure versus solder standoff and heel height.

most cases at the toe, except those marked by H in the figure, where the maximum equivalent plastic strain is located at the heel. The maximum equivalent plastic strain at the toe can be attributed to a short foot length of 0.5 mm in the present case, where the deformations (both bending and twisting) are accommodated by the entire joint. As the foot length increases, the effect of toe height decreases, since most of the deformation will be concentrated near the heel.

For the range of toe and heel considered, the predicted fatigue life can vary from ~60 to ~20,000 cycles with corresponding $\Delta\varepsilon_{eq}$ of ~0.02 to ~0.002, respectively. When the toe height is near 0.005 mm, the life drops below 100 cycles, which represents the region of infant mortality, as shown in Fig. 3.43. This provides a manufacturing guideline for controlling the solder configuration against the design life. The range of predicted life, ~60 to ~20,000 cycles, correlates well with the experimental results and corresponds to 0 percent up to 40 percent failure rate of solder joints.

References

1. Amerasekra, E. A., and D. S. Campbell, *Failure Mechanisms in Semiconductor Devices,* John Wiley & Sons, New York, 1987.
2. Barker, D. B., I. Sharif, A. Dasgupta, and M. G. Pecht, "Effect of SMC Lead Dimensional Variabilities on Lead Compliance and Solder Joint Fatigue Life," *ASME Winter Annual Meeting,* 91-WA-EEP-25, Atlanta, GA, Dec. 1–6, 1991.
3. Boswell, B., E. Jih, and Y.-H. Pao, unpublished data, 1996.
4. Burkhart, A., "Recent Developments in Flip Chip Technology," *Surface Mount Technology,* July 1991, pp. 40–44.
5. Clech, J-P., and J. A. Augis, "Engineering Analysis of Thermal Cycling Accelerated Tests for Surface-Mount Attachment Reliability Evaluation," *Proc. 7th IEPS,* Boston, MA, Nov. 1987, pp. 385–410.
6. Conway, P. P., M. R. Kalantary, and D. J. Williams, "Experimental Investigation of the Formation of Surface Mount Solder Joints," *ASME International Mechanical Engineering Congress,* San Francisco, CA, Nov. 12–17, 1995. Also to appear in *ASME Trans., Journal of Electronic Packaging,* 1996.
7. Davis, E. M., W. E. Harding, R. S. Schwartz, and J. J. Corning, "Solid Logic Technology: Versatile, High-Performance Microelectronics," *IBM Journal of Research and Development,* 8:102, 1964.
8. Dudderar, T. D., N. Nir, A. R. Storm, and C. C. Wong, "Isothermal Low-Cycle Fatigue Testing of Microscale Solder Interconnections," *Experimental Mechanics,* March 1992, pp. 11–20.
9. Frear, D. R., W. B. Jones, and K. R. Kinsman, *Solder Mechanics—A State of the Art Assessment,* TMS EMPMD Monograph Series, 1991.
10. Goldmann, L. S., and P. A. Totta, "Area Array Solder Interconnections for VLSI," *Solid State Technology,* June 1983, pp. 91–97.
11. Govila, R. K., E. Jih, Y.-H. Pao, and C. Larner, "Thermal Fatigue Damage in the Solder Joints of Leadless Chip Resistors," *ASME Trans., Journal of Electronic Packaging,* 116:83–88, June, 1994.
12. Govila, R., Y.-H. Pao, C. Larner, J. Lau, S. Twerefour, S. Erasmus, and S. Dolot, "Failure Analysis of No-Clean and Water-Clean Solder Joints of 256-Pin Plastic Quad Flat Pack," *ASME Trans., Journal of Electronic Packaging,* 116:184–190, September, 1994.
13. Govila, R., and Y.-H. Pao, unpublished data, 1996.

14. Hnatek, E. R., *Integrated Circuit Quality and Reliability,* Marcel Dekker, Inc., New York, 1987.
15. Hu, J. M., "Knowledge-Based Qualification Tests for Electronic Components in Harsh Environments," *Quality and Reliability Engineering International,* **10**:377–390, 1994.
16. Ichikawa, K., T. Kubota, and M. Suzuki, "Flip-Chip Joining Technique for 400DPI Thermal Printheads," *OKI Technical Reviews,* **55**(3):18–24, 1988.
17. Jahsman, W. E., "Comparison of Predicted and Measured Lead Stiffness of Surface Mounted Packages," *Proceedings of the 40th ECTC,* Las Vegas, May 20–23, 1990, pp. 926–932.
18. Jih, E., and Y.-H. Pao, "Evaluation of Design Parameters for Leadless Chip Resistors Solder Joints," *ASME Trans., Journal of Electronic Packaging,* **117**:94–99, June, 1995.
19. Kamei, T., M. Nakamura, H. Ariyoshi, and M. Doken, "Hybrid IC Structure Using Solder Reflow Technology," *IEEE Trans. on CHMT,* **2**(2):208–217, 1979.
20. Kitano, M., T. Kumazawa, and S. Kawai, "A New Evaluation Method for Thermal Fatigue Strength of Solder Joint," *The 1st Joint ASME/JSME Conference on Electronic Packaging, Advances in Electronic Packaging,* Milpitas, CA, April 9–12, 1992, pp. 301–308.
21. Knecht, S., and L. R. Fox, "Integrated Matrix Creep: Application to Accelerated Testing and Lifetime Prediction," in *Solder Joint Reliability: Theory and Applications,* J. H. Lau, ed., Van Nostrand Reinhold, New York, 1991.
22. Lau, J. H., "Thermal Stress Analysis of SMT PQFP Packages and Interconnections," *ASME Trans., Journal of Electronic Packaging,* **111**:2–8, March, 1989.
23. Lau, J. H., *Solder Joint Reliability—Theory and Applications,* Van Nostrand Reinhold, New York, 1991.
24. Lau, J., R. Govila, C. Larner, Y.-H. Pao, S. Erasmus, S. Dolot, and V. Solberg, "No-Clean and Water-Clean Mass Reflow Processes of 0.4 Pitch, 256-pin Fine Pitch Quad Flat Packs (QFP)," *IEEE/CHMT International Manufacturing Technology Symposium,* Baltimore, MD, Sept. 28–30, 1992, pp. 305–315.
25. Lau, J., R. Govila, C. Larner, Y.-H. Pao, S. Erasmus, S. Dolot, M. Jalilian, and M. Lancaster, "No-Clean and Solvent-Clean Mass Reflow Processes of 0.4 Pitch, 256-pin Fine Pitch Quad Flat Packs (QFP)," *Circuit World,* **19**(1):19–26, 1992.
26. Lau, J., T. Krulevitch, W. Schar, M. Heydinger, S. Erasmus, J. Gleason, "Experimental and Analytical Studies of Encapsulated Flip Chip Solder Bumps on Surface Laminar Circuit Boards," *Circuit World,* **19**(3):18–24, 1993.
27. Lau, J., Y.-H. Pao, C. Larner, R. Govila, S. Twerefour, D. Gilbert, S. Erasmus, and S. Dolot, "Reliability of 0.4 mm Pitch, 256-Pin Plastic Quad Flat Pack No-Clean and Water-Clean Solder Joints," *Soldering and Surface Mount Technology,* no. 16:42–50, February, 1994.
28. Lau, J. H., *Chip on Board Technologies for Multichip Modules,* Van Nostrand Reinhold, New York, 1994.
29. Lau, J. H., *Ball Grid Array Technology,* McGraw-Hill, New York, 1995.
30. Leibfried, W., "Materials and Processes for Reliable Soldering of Chip Components," *Hybrid Circuits,* no. 4:19–25, 1984.
31. Lewis, R. D., "Understanding Part Failure Mechanisms," *Tutorial Notes of Annual Reliability and Maintainability Symposium,* 1992.
32. Lodge, K. J., and D. J. Pedder, "The Impact of Packaging on the Reliability of Flip Chip Solder Bonded Devices," *Proceedings of the 40th ECTC,* Las Vegas, Nevada, May 20–23, 1990, pp. 470–476.
33. McShane, M., P. Lin, G. Dody, and J. Bigler, "Lead Configuration and Performance for Fine Pitch SMT Reliability," *Proceedings of the National Electronic Packaging and Production Conference,* NEPCON WEST'90, Anaheim, CA, **1**:235–257, February 26–March 1, 1990.
34. MIL-HDBK-217F, *Reliability Prediction of Electronic Equipment,* Revision F, Department of Defense, 1992.
35. Miller, L. F., "Controlled Collapse Reflow Chip Joining," *IBM Journal of Research and Development,* **13**:239, 1969.
36. Minges, M. L., *Packaging, Electronic Materials Handbook,* vol. 1, ASM International, Materials Park, Ohio, 1989.

37. Nir, N., T. D. Dudderar, C. C. Wong, and A. R. Storm, "Fatigue Properties of Micro-electronics Solder Joints," *ASME Winter Annual Meeting*, 90-WA/EEP-28, Dallas, Texas, November 25–30, 1990.
38. Oshima, M., R. Satoh, K. Hirota, and I. Ishi, "New Micro-Soldering Technology and Its Application to VLSI," *Proceedings of the 1st IEEE CHMT Symposium*, Tokyo, Japan, 1984, pp. 165–170.
39. Pao, Y.-H., "A Fracture Mechanics Approach to Thermal Fatigue Life Prediction of Solder Joints," *IEEE Trans. on CHMT*, 15(4):559–570, 1992.
40. Pao, Y.-H., and W.-J. Jung, "Modeling Thermally Induced Cracking of LCC Solder Joints in Automotive Electronic Packages," *Proceedings of Surface Mount International Conference*, San Jose, CA, Aug. 30–Sept. 1, 1994, pp. 73–91.
41. Pao, Y.-H., V. Reddy, E. Jih, D. R. Liu, and W. L. Winterbottom, "An Integrated Reliability Modeling/Testing Approach for Automotive Electronic Packaging," *Proceedings of NEPCON WEST'96*, Anaheim, CA, Feb. 27–29, 1996, pp. 1681–1690.
42. Pecht, M., *Handbook of Electronic Package Design*, Marcel Dekker, Inc., New York, 1991.
43. Pecht, M., A. Dasgupta, J. W. Evans, and J. Y. Evans, *Quality Conformance and Qualification of Microelectronic Packages and Interconnects*, John Wiley & Sons, New York, 1994.
44. Puttlitz, K. J., "Preparation, Structure, and Fracture Modes of Pb-Sn and Pb-In Terminated Flip Chips Attached to Gold Capped Microsockets," *IEEE Trans. on CHMT*, 13(4):647–655, 1990.
45. Rao, S. S., *Reliability-Based Design*, McGraw-Hill, New York, 1992.
46. Sabnis, A. G., *VLSI Reliability*, Academic Press, Inc., San Diego, 1990.
47. Sarihan, V., "Temperature Dependent Viscoplastic Simulation of Controlled Collapse Solder Joint under Thermal Cycling," *ASME Winter Annual Meeting*, 92-WA/EEP-38, Anaheim, CA, Nov. 8–13, 1992.
48. Satoh, R., K. Arakawa, M. Harada, and K. Matsui, "Thermal Fatigue Life of Pb-Sn Alloy Interconnections," *IEEE Trans. on CHMT*, 14(1):224–232, March, 1991.
49. Shah, M. K., "Analysis of Parameters Influencing Stresses in the Solder Joints of Leadless Chip Capacitors," *ASME Trans., Journal of Electronic Packaging*, 112:147–153, June, 1990.
50. Shiratori, M., Q. Yu, and S.-B. Wang, "A Computational and Experimental Hybrid Approach to Creep-Fatigue Behavior of Surface-Mounted Solder Joints," *ASME INTERpack'95, Advances in Electronic Packaging*, EEP-Vol. 10-1:451–457, 1995.
51. Song, X., E. Jih, and Y.-H. Pao, unpublished data, 1996.
52. Stojadinovic, N. D., "Failure Physics of Integrated Circuits," *Microelectronics Reliability*, 23(4):609–707, 1983.
53. Suryanarayana, D., R. Hsiao, T. P. Gall, and J. M. McCreary, "Flip-Chip Solder Bump Fatigue Life Enhanced by Polymer Encapsulation," *Proceedings of the 40th ECTC*, May 20–23, 1990, Las Vegas, Nevada, pp. 338–344.
54. Tummala, R. R., and E. J. Rymaszewski, *Microelectronics Packaging Handbook*, Van Nostrand Reinhold, New York, 1989.
55. Vardaman, J., *Surface Mount Technology—Recent Japanese Development*, The IEEE, Inc., New York, 1993.
56. Warrior, M., "Reliability Improvements in Solder Bump Proceeding for Flip Chips," *Proceedings of the 40th ECTC*, Las Vegas, Nevada, May 20–23, 1990, pp. 460–469.
57. Wong, B., D. E. Helling, and R. W. Clark, "A Creep-Rupture Model for Two-Phase Eutectic Solders," *IEEE Transactions on Components, Hybrids, and Manufacturing Technology*, September, 11(3):284–290, 1988.
58. Yu, Q., and M. Shiratori, "A Study of Mechanical and Thermal Stress Behavior due to Global and Local Thermal Mismatch of Dissimilar Materials in Electronic Packaging," *ASME INTERpack'95, Advances in Electronic Packaging*, EEP-Vol. 10-1:389–394, 1995.
59. Kotlowitz, R. W., "Comparative Compliance of Representative Lead Designs for Surface Mounted Components," *Proceedings of IEEE Electronic Components Conference*, 1989, pp. 791–831.

Design for Reliability

As mentioned in Chap. 2, reliability is generally defined as the probability that a system, equipment, or a component will perform its intended function without failure for a given period of time (e.g., product life cycle), in the environment for which it was designed. Statistical approaches have been prevailing over the past few decades as a major tool for modeling reliability of electronic packaging, but physics-based approaches need to be considered as well in order to account for detailed failure modes and associated mechanisms. The trend nowadays in the electronics, computer, telecommunications, and automotive industries is moving toward the development of a design-for-reliability strategy that combines both statistics and physics approaches in early design stages to enhance reliability and reduce product development cycle time[1–99] (Fig. 1.10).

The design-for-reliability process focuses on evaluating the functionality, cost, manufacturability, and reliability of possible design alternatives in the early design and product development stages prior to the construction of prototype hardware. To implement such a process (in electronic packaging design, for instance) requires a significant amount of numerical modeling and the understanding of electrical, thermal, and mechanical behavior of relevant materials under various test and environmental conditions. Numerical modeling using a number of tools, such as finite element method (FEM), boundary element method (BEM), and computational fluid dynamics (CFD), has been demonstrated to be useful and effective in analyzing the electrical and thermomechanical behavior of electronic assemblies.

However, in addition to accurately specifying the physical dimensions of the packages and loadings, a key factor that determines the accuracy of the model lies in the material properties that are able to account for the intended behavior of the materials in the packages. It also needs to incorporate the failure mechanism of the package and the associated failure criteria for reliability prediction. Examples of the classification of failure modes in electronic packaging and the material properties required for reliability modeling can be found in Tummala and Rymaszewski,[82] Pecht,[66,67] Lau,[42] and Hu.[34]

This chapter will delineate the key elements in a general physics-based reliability approach with the focus on the solder interconnect failure caused by thermomechanical loadings. First, the thermomechanical behavior of solder joints is discussed, followed by a collection of required thermal and mechanical properties of commonly used electronic packaging materials. Then the finite element modeling of solder joints subjected to thermal cycling is discussed. The chapter ends with an introduction of the development of a computer-aided tool that integrates various physics-based approaches for the prediction of electronic packaging reliability. The design for reliability of BGA, CSP, flip chip, and TSOP and PQFP solder joints will be discussed in Chaps. 5, 6, 7, and 8, respectively.

4.1 Thermomechanical Behavior of Solder Joint

Thermal fatigue failure in electronic packages frequently results from cracking in solder joints. For most solder alloys, such as Pb-Sn-based solders, the operating temperature can be as high as 0.5 to 0.8 T_M (melting point temperature) of the alloys. In such high-temperature regimes, a significant amount of creep and creep-induced fracture can occur. When coupled with the fatigue crack growth, it becomes a dominant failure mechanism in the solder joints. Thus, understanding the stress/strain hysteresis behavior is critical in characterizing their fracture behavior.

In this section the thermal-cyclic stress/strain hysteresis response and thermal fatigue behavior of seven solder alloys, 63Sn-37Pb, 96.5Sn-3.5Ag, 90Pb-10Sn, 97Sn-2Cu-0.8Sb-0.2Ag, 97Sn-3Cu, 95.5Sn-4Cu-0.5Ag, and 95Bi-5Sb are discussed. The study of Pb-free solders reflects the increasing demand in the electronics industry for lead solder substitutes.[3,4] The thermomechanical behavior was determined using a bimaterial specimen subjected to temperature cycling.[57] The detailed characteristics of stress relaxation and creep of these solder joints were determined as a function of temperature. The associated secondary creep properties were measured, and the corresponding mechanisms are discussed. From the fatigue experiment, the evolution of the stress/strain hysteresis response as a function of thermal cycles was determined and used, together with the fracture surface examination with scanning electron microscopy (SEM), to identify the failure mechanism.

4.1.1 Solder alloys

Seven solder alloys were selected: 63Sn-37Pb, 96.5Sn-3.5Ag, 90Pb-10Sn, 97Sn-2Cu-0.8Sb-0.2Ag, 97Sn-3Cu, 95.5Sn-4Cu-0.5Ag, and 95Bi-5Sb. Among these solder alloys, 90Pb-10Sn and 63Sn-37Pb are two commonly used Pb-containing solders, while the others are commercially available Pb-free solders. Their melting temperatures are listed in Table 4.1. It is seen that except for the eutectic 63Sn-37Pb and 96.5Sn-3.5Ag, the other five have melting temperatures above 250°C and are usually regarded as high-temperature solders. Studies of high-temperature Pb-containing or Pb-free solders are lacking in the literature, as compared to the availability of extensive research data on solders with low melting temperatures of 130–200°C.

4.1.2 Specimen and test apparatus

The design of the specimen and the test apparatus has been discussed in detail by Pao et al.[57] Only a brief introduction is given here. The specimen consists of two beams, Al_2O_3 and Al 2024-T4, jointed at the ends with solder (Fig. 4.1). It is obvious that under uniform temperature variations both beams are subjected to uniform bending and axial forces. This results in a constant bending curvature in each beam. Assuming uniform shearing in the joint, which has been verified by the finite element analysis treating solder as an elastic/creep material,[61] the average shear stress and strain can be determined by measuring the elastic response of the beams with strain gages.

The specimen is 42 mm long and 5 mm deep. The thicknesses of the Al 2024-T4 and the Al_2O_3 beams are 3 and 1.2 mm, respectively. The solder joint at each end of the specimen is 2 mm wide and 5 mm deep, and the thickness is 0.381 mm (15 mils). Prior to making the joint, the areas to be soldered were sputtered with a thin layer of chromium (approximately 200 nm) to increase the adhesion to the beam and a 3-μm-thick copper film for solderability. These areas were then fluxed and pretinned with the same solder used in the test.

Four strain gages (Micro-Measurements WK-13-062AP-350) were mounted on the aluminum beam, two on the inside surface and two on the outside surface, as shown in Fig. 4.1. The two gages on each surface were connected into a double-quarter Wheatstone bridge. The double-

TABLE 4.1 Melting Temperature of Tested Solder Alloys, °C

State	90Pb-10Sn	97Sn-2Cu-0.8Sb-0.2Ag	97Sn-3Cu	95.5Sn-4Cu-0.5Ag	95Bi-5Sb	63Sn-37Pb	96.5Sn-3.5Ag
Solidus	268	225	227	227	274	183	221
Liquidus	301	257	310	349	296	183	221

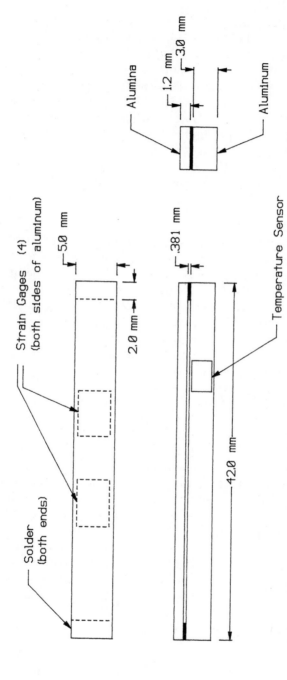

Figure 4.1 Schematic of the beam specimen, associated dimensions, and locations of strain gages and temperature sensor.

Alumina

1.2 mm

3.0 mm

Aluminum

Strain Gages (4)
(both sides of aluminum)

5.0 mm

2.0 mm

.381 mm

Temperature Sensor

42.0 mm

Solder
(both ends)

94

quarter bridge was designed such that the two diagonal active gages measure strains with the same magnitude and the same sign. With this arrangement the output voltage of the bridge is doubled without increasing the noise level. The increase in output signal voltage allows the bridge to resolve smaller strain changes. In addition, a temperature sensor (Micro-Measurements ETG-50A) was bonded on one side of the aluminum to provide feedback to the controller to produce a desired temperature cycle. Two temperature chambers were used to provide two different thermal cycling profiles: one is a 40-minute cycle between 40 and 140°C with a 10-minute ramp rate, and the other has a period of 135 minutes between –40 and 120°C.

4.1.3   Stress/Strain hysteresis response and steady-state creep properties

The measured shear stress/strain hysteresis responses of 63Sn-37Pb, 96.5Sn-3.5Ag, and 90Pb-10Sn to both thermal cycling profiles are shown in Figs. 4.2 to 4.4, respectively, for the first few cycles. And those for 97Sn-2Cu-0.8Sb-0.2Ag, 97Sn-3Cu, 95.5Sn-4Cu-0.5Ag, and 95Bi-5Sb to

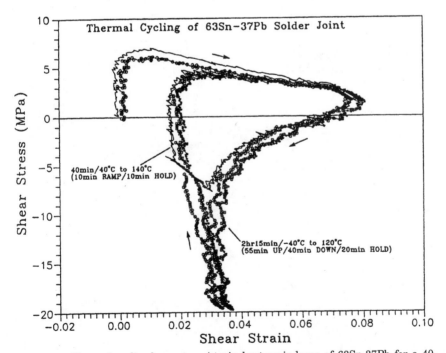

Figure 4.2 Thermal-cyclic shear stress/strain hysteresis loops of 63Sn-37Pb for a 40-minute cycle between 40 and 140°C (10-minute ramp time) and a 135-minute cycle between –40 and 120°C (55-minute ramp-up and 40-minute ramp-down).

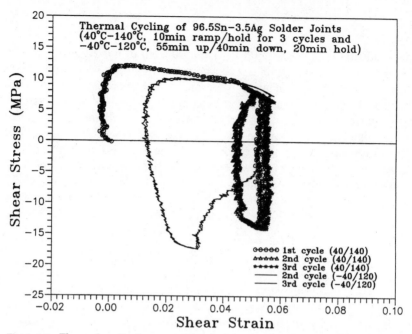

Figure 4.3 Thermal-cyclic shear stress/strain hysteresis loops of 96.5Sn-3.5Ag for a 40-minute cycle between 40 and 140°C (10-minute ramp time) and a 135-minute cycle between –40 and 120°C (55-minute ramp-up and 40-minute ramp-down).

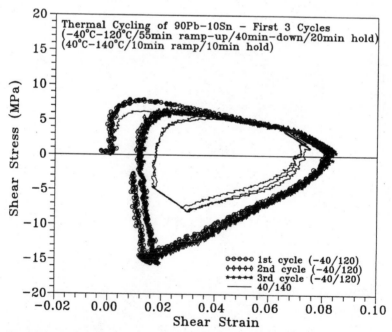

Figure 4.4 Thermal-cyclic shear stress/strain hysteresis loops of 90Pb-10Sn for a 40-minute cycle between 40 and 140°C (10-minute ramp time) and a 135-minute cycle between –40 and 120°C (55-minute ramp-up and 40-minute ramp-down).

the 40-minute thermal cycling are shown in Figs. 4.5 to 4.8, respectively. As can be seen in the figures, the shear stress tends to stabilize very rapidly within three cycles except for the 95Bi-5Sb solder. The fact that 95Bi-5Sb stabilizes at a slower rate is because of a smaller strain range resulting from the test.

From Figs. 4.2 through 4.8 it can be seen that both stress relaxation and creep occur simultaneously at the hold time since the system is neither displacement- nor load-controlled. For thermal cycling between 40 and 140°C both Pb-containing solders have the lowest stress range and the highest strain range, whereas 95Bi-5Sb shows the opposite. As far as the Sn-Cu system is concerned, 97Sn-3Cu behaves quite differently from the other two which are similar in stress/strain range and the shape of the hysteresis loop. In general, the shape of the hysteresis loop reflects how time-dependent plastic deformation flows during the loading and unloading as a result of temperature variation. It is noticed that except for the two Sn-Cu-Ag solders, the others bear no resemblance in hysteresis loops. This indicates that these solders have distinct viscoplastic properties (e.g., steady-state creep during hold time or primary creep during loading and unloading).

Pao and associates[57,58] have developed a method, based on the stress relaxation data at hold time, to determine the secondary creep properties of the solder associated with Norton's equation,

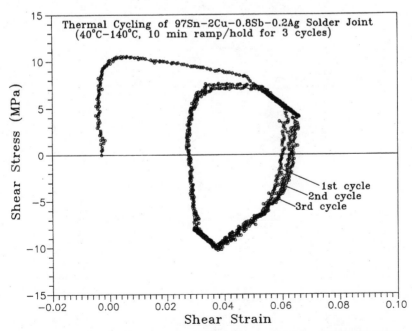

Figure 4.5 Thermal-cyclic shear stress/strain hysteresis loops of 97Sn-2Cu-0.8Sb-0.2Ag for a 40-minute cycle between 40 and 140°C (10-minute ramp time).

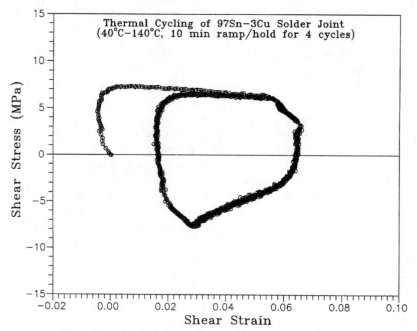

Figure 4.6 Thermal-cyclic shear stress/strain hysteresis loops of 97Sn-3Cu for a 40-minute cycle between 40 and 140°C (10-minute ramp time).

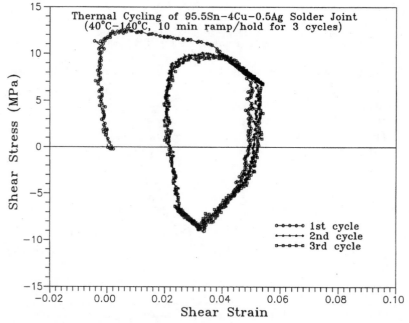

Figure 4.7 Thermal-cyclic shear stress/strain hysteresis loops of 95.5Sn-4Cu-0.5Ag for a 40-minute cycle between 40 and 140°C (10-minute ramp time).

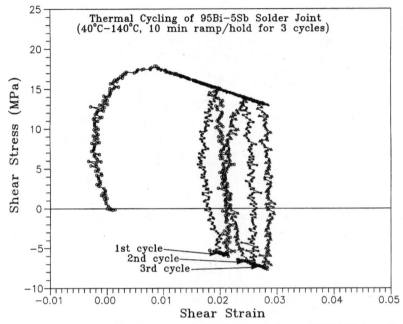

Figure 4.8 Thermal-cyclic shear stress/strain hysteresis loops of 95Bi-5Sb for a 40-minute cycle between 40 and 140°C (10-minute ramp time).

$$\frac{d\gamma_{crp}}{dt} = B^* \exp\left[\frac{-\Delta H}{kT}\right]\tau^n \tag{4.1}$$

where $d\gamma_{crp}/dt$ is the shear creep strain rate, n is the stress exponent, ΔH is the activation energy, k is the Boltzmann's constant, T is temperature in K, and B^* is a material constant. The method is based on the assumption that the deformation at the hold time is controlled by steady-state creep. The creep properties of these solders were determined and are listed in Table 4.2. The creep properties of 95Bi-5Sb could not be accu-

TABLE 4.2   Creep Properties and the Associated Mechanism for 90Pb-10Sn, 97Sn-2Cu-0.8Sb-0.2Ag, 97Sn-3Cu, 95.5Sn-4Cu-0.5Ag, 63Sn-37Pb, and 96.5Sn-3.5Ag Solder Alloys

Property	90Pb-10Sn	97Sn-2Cu-0.8Sb-0.2Ag	97Sn-3Cu	95.5Sn-4Cu-0.5Ag	63Sn-37Pb	96.5Sn-3.5Ag
B^*, 1/MPa^{n}s	100.6	3.031	9.10×10^{-4}	4.229×10^{-12}	0.205	4707.7
ΔH, eV	0.64	0.85	0.51	0.062	0.49	1.34
n, 40°C	4.25	8.91	7.99	8.36	5.25	12.0
n, 140°C	3.03	7.37	6.30	8.36	5.25	8.84
Deformation mechanism	Dislocation core-diffusion	Dislocation glide/climb	Dislocation glide/climb	Athermal, short range Cu clustering	Dislocation glide/climb	Dislocation glide/climb

rately determined because of the lack of sufficient relaxation data at the hold time. A higher stress level or a longer dwell is required.

The determined magnitude of ΔH for 90Pb-10Sn falls in the range of 0.61 to 0.65 eV measured by Lavery[43] on the same solder in the isothermal creep test and is close to that of the dislocation core-diffusion of pure Pb, 0.68 eV.[24] The ΔH of 63Sn-37Pb, 0.49 eV, is also consistent with that reported by Knecht and Fox.[38] The results of 96.5Sn-3.5Ag are in conformity with the previously published data by Darveaux and Banerji.[17] The stress relaxation of 96.5Sn-3.5Ag versus stress is shown in Fig. 4.9, and the dependence of the associated stress exponent on temperature is shown in Fig. 4.10, along with Darveaux and Banerji's data. The comparison of steady-state creep rates at three different temperatures is also shown in Fig. 4.11.

It is noted that the activation energy of the Sn-Cu solders decreases with the weight percentage of Cu and falls down to 0.06 eV when Cu

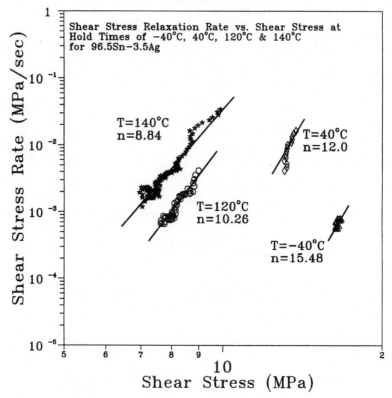

Figure 4.9   Shear stress relaxation rate versus shear stress for 96.5Sn-3.5Ag at four different temperatures.

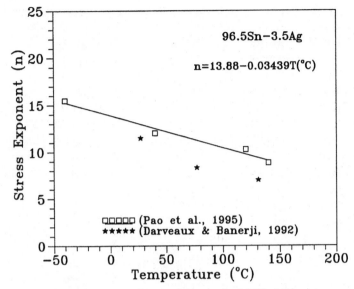

Figure 4.10   Comparison of stress exponent for 96.5Sn-3.5Ag.

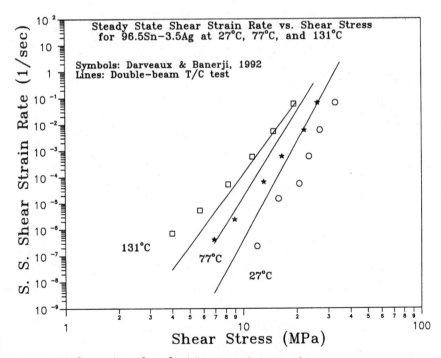

Figure 4.11   Comparison of steady-state creep rate versus stress.

increases to 4 percent. This implies that the creep process could be athermal, which may be due to short-range Cu clustering.[29] The other solder alloys listed in Table 4.2 indicate that the range of ΔH and n correspond to a dislocation-controlled steady-state creep (Region III) (see Hacke et al.[31]).

The steady-state creep rate of Eq. (4.1) is plotted versus stress in Figs. 4.12 and 4.13 for 40 and 140°C, respectively, for 63Sn-37Pb, 90Pb-10Sn, 97Sn-2Cu-0.8Sb-0.2Ag, 97Sn-3Cu, and 95.5Sn-4Cu-0.5Ag. It is seen that 63Sn-37Pb and 90Pb-10Sn show a higher creep rate and are temperature-sensitive, while 95.5Sn-4Cu-0.5Ag has a much lower rate and is almost temperature-insensitive, but has a higher stress exponent than the other two Sn-Cu solders.

4.1.4   Thermal fatigue

During temperature cycling, all solders demonstrate an evolution of shape and area change of hysteresis loops as a function of number of cycles. Figures 4.14 and 4.15 show the evolution of hysteresis loops of 63Sn-37Pb for two different cycling profiles. Similar trends are

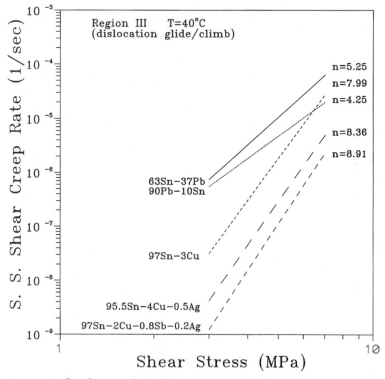

Figure 4.12   Steady-state shear creep rate versus stress at 40°C for 63Sn-37Pb, 90Pb-10Sn, 97Sn-3Cu, 95.5Sn-4Cu-0.5Ag, and 97Sn-2Cu-0.8Sb-0.2Ag.

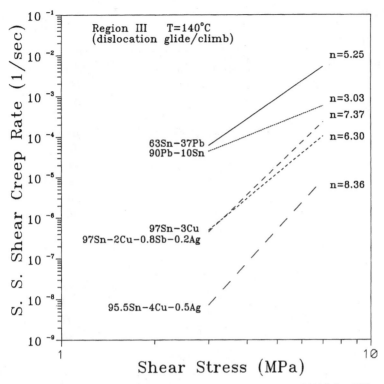

Figure 4.13 Steady-state shear creep rate versus stress at 140°C for 63Sn-37Pb, 90Pb-10Sn, 97Sn-3Cu, 95.5Sn-4Cu-0.5Ag, and 97Sn-2Cu-0.8Sb-0.2Ag.

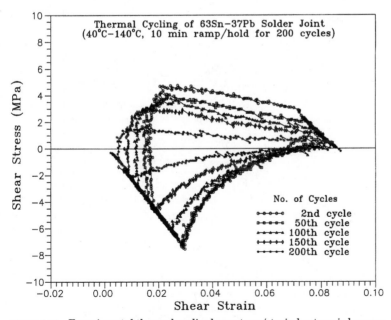

Figure 4.14 Experimental thermal-cyclic shear stress/strain hysteresis loops as a function of number of cycles for 63Sn-37Pb between 40 and 140°C with 10-minute hold and ramp.

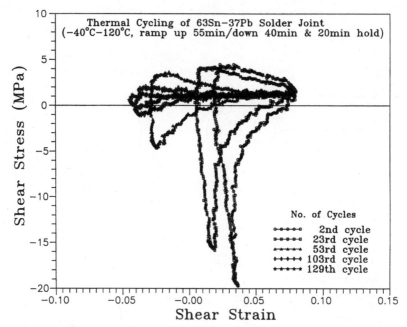

Figure 4.15 Experimental thermal-cyclic shear stress/strain hysteresis loops as a function of number of cycles for 63Sn-37Pb between –40 and 120°C with 20-minute hold, 55-minute ramp-up, and 40-minute ramp-down.

observed for other solder alloys (Figs. 4.16 through 4.23). Such a shape and size change indicates that the solder joint apparently undergoes a softening or weakening process as thermal cycling proceeds. Examination of the fracture surface (e.g., of 90Pb-10Sn, as shown in Fig. 4.24) reveals that this weakening may be attributed primarily to a large amount of secondary cracks and voids generated during the thermal cycling. These defects cause the material to degrade and are reflected in the hysteresis loops.

Figure 4.25 shows the fracture surface of 97Sn-2Cu-0.8Sb-0.2Ag. The crack initiated at the Al_2O_3/solder interface and propagated near and parallel to the interface (Fig. 4.25b). A significant amount of secondary cracks is seen in Fig. 4.25a, c, and d. The large amount of voids may be due to the outgassing of the flux during joint fabrication. However, viewing through a void caused by outgassing shows evidence of creep-induced voids at the grain boundary (Fig. 4.25e). In addition, fatigue striations are also seen in the figure.

The fracture surface of 63Sn-37Pb after 201 cycles (40-minute cycling between 40 and 140°C) is shown in Fig. 4.26. The joint is only partly fractured, indicated by the dashed line in Fig. 4.26a. The crack initiated at the Al_2O_3/solder interface and extended into the solder. The fracture mor-

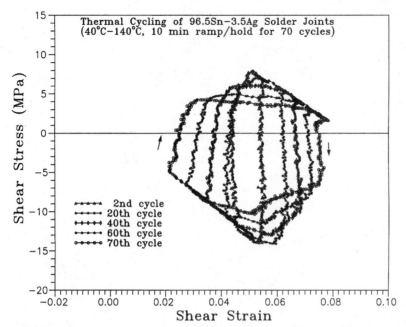

Figure 4.16 Experimental thermal-cyclic shear stress/strain hysteresis loops as a function of number of cycles for 96.5Sn-3.5Ag between 40 and 140°C with 10-minute hold and ramp.

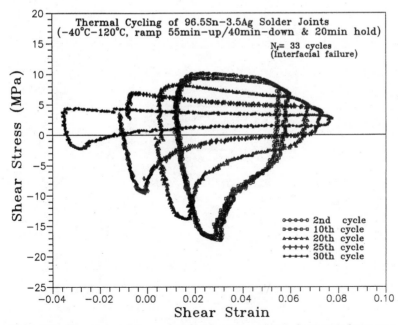

Figure 4.17 Experimental thermal-cyclic shear stress/strain hysteresis loops as a function of number of cycles for 96.5Sn-3.5Ag between −40 and 120°C with 20-minute hold, 55-minute ramp-up, and 40-minute ramp-down.

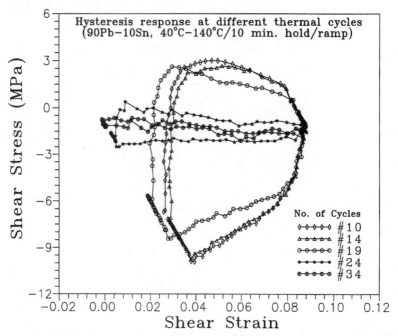

Figure 4.18 Experimental thermal-cyclic shear stress/strain hysteresis loops as a function of number of cycles for 90Pb-10Sn between 40 and 140°C with 10-minute hold and ramp.

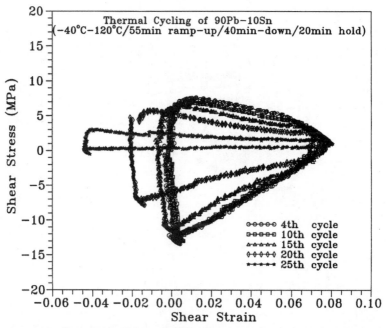

Figure 4.19 Experimental thermal-cyclic shear stress/strain hysteresis loops as a function of number of cycles for 90Pb-10Sn between −40 and 120°C with 20-minute hold, 55-minute ramp-up, and 40-minute ramp-down.

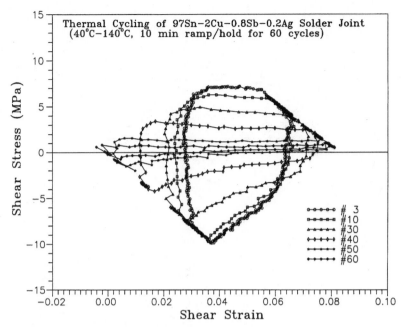

Figure 4.20 Experimental thermal-cyclic shear stress/strain hysteresis loops as a function of number of cycles for 97Sn-2Cu-0.8Sb-0.2Ag between 40 and 140°C with 10-minute hold and ramp.

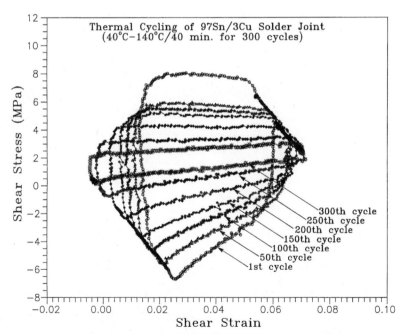

Figure 4.21 Experimental thermal-cyclic shear stress/strain hysteresis loops as a function of number of cycles for 97Sn-3Cu between 40 and 140°C with 10-minute hold and ramp.

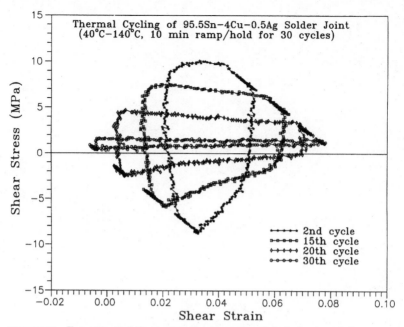

Figure 4.22 Experimental thermal-cyclic shear stress/strain hysteresis loops as a function of number of cycles for 95.5Sn-4Cu-0.5Ag between 40 and 140°C with 10-minute hold and ramp.

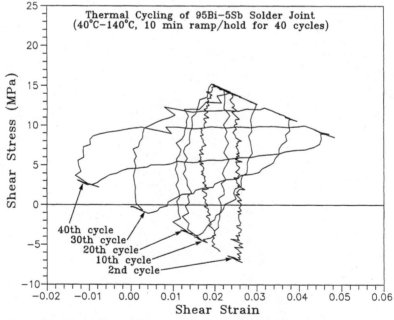

Figure 4.23 Experimental thermal-cyclic shear stress/strain hysteresis loops as a function of number of cycles for 95Bi-5Sb between 40 and 140°C with 10-minute hold and ramp.

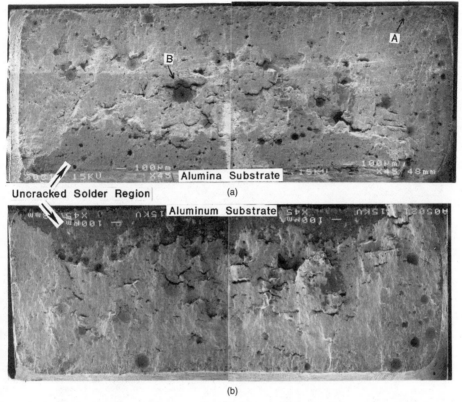

Figure 4.24 Fracture surface of 90Pb-10Sn solder joint subjected to thermal cycling between 40 and 140°C with 10-minute hold and ramp after 44 cycles: (*a*) on alumina beam; (*b*) on aluminum beam.[61]

phology is quite different from those of 90Pb-10Sn, Sn-Ag, and Sn-Cu solders in that there are more creep-induced dimples and less secondary cracks and fatigue striations. A magnified view of the thermal fatigue cracked region (region 1 in Fig. 4.26*a*) and pulled fracture after the test (region 2) is shown in Fig. 4.27, where the microstructure coarsening is seen in region 1, as compared to region 2. This indicates that the coarsening only occurs at the high-strain field near the crack tip.

The number of cycles to failure and the associated failure mechanism as a result of SEM examination are listed in Table 4.3. 97Sn-3Cu apparently has a much longer life than 90Pb-10Sn, 96.5Sn-3.5Ag, the other two Sn-Cu solders, and 95Bi-5Sb. It is interesting to note that 63Sn-37Pb, though having a lower melting point of 183°C, shows a much higher creep/fatigue resistance than the other solders, except for 97Sn-3Cu. This implies that the higher steady-state creep rate of 63Sn-37Pb, as shown in Figs. 4.12 and 4.13, does not necessarily sug-

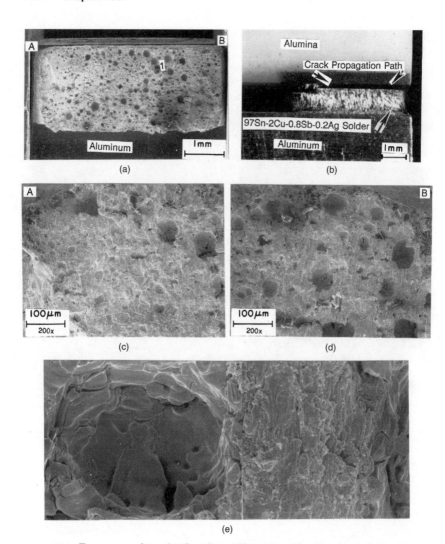

Figure 4.25 Fracture surface of 97Sn-2Cu-0.8Sb-0.2Ag solder joint subjected to thermal cycling between 40 and 140°C with 10-minute hold and ramp: (*a*) fracture surface after 45 thermal cycles on aluminum beam; (*b*) front view showing crack path near the alumina beam; (*c*) magnification of region A in (*a*); (*d*) magnification of region B in (*a*); (*e*) magnification of region 1 in (*a*).

gest that the material would fail at a faster rate. The corresponding mechanism associated with the crack propagation is essentially creep-fatigue-coupled.

The technique developed here is particularly useful for studying the thermomechanical response over a wide range of solder alloys and for

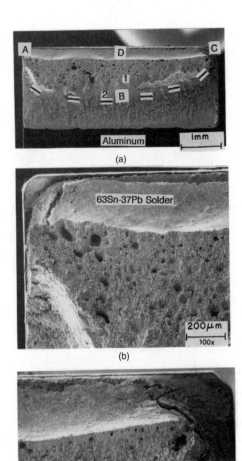

(a)

(b)

(c)

Figure 4.26 Fracture surface of 63Sn-37Pb solder joint subjected to thermal cycling between 40 and 140°C with 10-minute hold and ramp: (a) fracture surface after 201 thermal cycles on aluminum beam; (b) magnification of region A in (a); (c) magnification of region C in (a).

evaluating their thermal fatigue strength and associated failure mechanisms. The experimentally determined hysteresis loops and their evolution as a function of temperature cycle are also important for developing viscoplastic constitutive equations and damage models. The current test results of testing seven different solder alloys show that the lower-melting-point 63Sn-37Pb ($T_M = 183$°C) has a thermal fatigue life approximately three to ten times longer than 96.5Sn-3.5Ag, 90Pb-10Sn, 95Bi-5Sb, 97Sn-2Cu-0.8Sb-0.2Ag, and 95.5Sn-4Cu-0.5Ag,

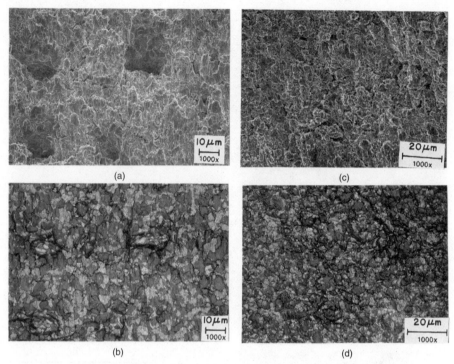

Figure 4.27   Fracture surfaces of 63Sn-37Pb: (a) region 1 in Fig. 4.26 (a) corresponding to the fatigue cracked area; (b) back scatter electronic image (BSEI) showing crack-induced coarsened microstructure; (c) region 2 in Fig. 4.26 (a) corresponding to the uncracked ligament; (d) the associated BSEI of region 2 showing noncoarsened microstructure.

which all have a higher T_M. In addition, it has been demonstrated that 97Sn-3Cu exhibits greater creep/fatigue resistance than 90Pb-10Sn and thus may be considered as a high-temperature lead solder substitute. However, more studies are needed to assess the fatigue strength at different strain ranges and temperature cycling profiles.

4.2   Material Properties for Thermomechanical Analysis

Thermomechanical properties of interconnects play an important role in determining the reliability of electronic packaging, especially for surface-mounted assemblies, because most of the interconnects serve not only as electrical but as mechanical connections. This section reviews a number of thermal and mechanical properties for commonly used materials in electronic packaging that are related to reliability modeling.

TABLE 4.3   Number of Cycles to Failure and the Associated Failure Mechanisms for the 40-Minute Cycle

Solder	Initial shear strain range (%)	Initial shear stress range (MPa)	Final shear strain range (%)	Final shear stress range (MPa)	Number of cycles to failure (N_f)	Primary failure mechanisms
90Pb-10Sn	5.6	13.2	8.8	2.5	20 (interface*) 45 (cohesive†)	Major crack (one end) Trans/intergranular Secondary cracking Creep/fatigue
97Sn-2Cu-0.8Sb-0.2Ag	3.8	17.4	8.2	1.3	60 (interface)	Similar to 90Pb-10Sn One end
97Sn-3Cu	5.2	14.8	7.6	1.5	100 (interface) 300 (cohesive)	Similar to 90Pb-10Sn Two ends
95.5Sn-4Cu-0.5Ag	3.3	18.8	8.1	0.6	30 (interface)	Similar to 90Pb-10Sn One end
95Bi-5Sb	0.9	22.0	5.6	6.8	50 (interface and cohesive)	Major crack (one end) Transgranular Secondary cracking Fatigue only
63Sn-37Pb	6.4	12.3	8.1	3.6	>200 (interface and cohesive)	Major cracks (two ends) Intergranular Secondary cracking Coarsening Creep/fatigue
96.5Sn-3.5Ag	1.2	22.0	>5.6	<15	>70 (interface and cohesive)	One end major crack

* Interface crack is defined as a crack propagating either along or near the interface between the beam and the solder.
† Cohesive crack is defined as a crack propagating in the solder joint and relatively away from the interface.

In contrast to most review articles, which only discuss material properties at room temperature, an effort has been made to identify the temperature dependence of the materials. More than 50 properties are included here as a function of temperature, and the coefficients of the polynomials used to fit the temperature variations are given in the tables. The thermal and mechanical properties compiled here include Young's modulus, yield stress, tensile strength, thermal conductivity, coefficient of thermal expansion (CTE), power-law creep for solder joints, stress-to-rupture time for solder joints, and fracture toughness.

General electronic material properties can be found in numerous works published during the past 20 years (e.g., Wild,[92] ASM,[6] Tummala and Rymaszewski,[88] Dally,[16] Frear et al.,[21] Lau,[42] Pecht,[66] Allenby,[3,4] and Glazer[25]). Properties of packaging materials have been continuously updated and are increasingly important to the modeling of thermomechanical behavior and reliability. Most of the data listed here are based on recent publications. Note, however, that these properties are, in general, functions of test conditions, such as test method, instrumentation accuracy, specimen preparation, microstructure, and human factors. The temperature dependence of these properties can be approximated with polynomials as

$$Y(T) = a_0 + a_1 T + a_2 T^2 + a_3 T^3 + \cdots \qquad (4.2)$$

where Y is the property; a_i, $i = 0, 1, 2, 3, \ldots$ are constants, and T is temperature in °C.

Due to the complexity of material composition, cooling rate, and heat treatment, solder joints have the most uncertain thermal and mechanical properties compared to most other interconnection materials. Moreover, the deformation of solder joints subjected to thermal or isothermal loadings are not only temperature-dependent, through elastic and thermal properties, but time-dependent, i.e., viscoplastic, as well. One important property associated with the viscoplastic behavior is the creep behavior. Most solders even exhibit creep behavior at relatively low temperatures, such as –40°C. Therefore, a focus has been placed on properties of various solder alloys.

Another important property related to the reliability of electronic packages is fracture toughness. Fracture toughness represents the capability of materials to resist fracture and is a useful property in design-for-reliability. Over the past few years, the application of fracture mechanics to packaging reliability prediction has increased significantly in many areas (e.g., pop-corning of plastic packages, solder joint fatigue, and coating/film delamination). Thus, a table of fracture toughness for a number of material systems is included to address the need.

4.2.1   Young's modulus, yield stress, and tensile strength

More than 30 packaging materials, such as die, interconnect, substrate, and molding compound, were selected, and the Young's modulus, yield stress, and tensile strength for each are listed in Tables 4.4–4.6, respectively, in terms of the coefficients of the polynomials used in Eq. (4.2). For data originally measured as shear yield stress or strength, the relation $\sigma = \sqrt{3}\tau$, based on the von Mises criterion, has been employed to convert them into the tensile components. Some temperature-dependent Young's moduli and yield stresses for solders are also shown in Figs. 4.28 and 4.29, respectively.

The reference column in these tables gives the original source of the data. Some unpublished data from the material database of the CAIR (Computer Aided Interconnect Reliability) system developed at Ford Research Laboratory[64] has been used and cited.

TABLE 4.4   Temperature Constants a_i of Young's Modulus (MPa) of Some Electronics Packaging Materials

Materials	a_0	a_1	a_2	a_3	Reference
Si	163,280	−7.5487			Frost & Ashby
GaAs	84,950				Pecht*
AlN	327,000				Pecht
BeO	345,000				Pecht
SiC	412,000				Pecht
Al$_2$O$_3$ (96%)	304,598	−45.57			Pao et al.[57]
Diamond	785,000				CAIR
Mo	355,000	−51.348			Frost & Ashby
W	345,000				Pecht
Ti	110,000				McClintock & Argon
Au	82,700				Pecht
Silicone rubber	5				CAIR
Epoxy-Novolac-Cast (adhesive)	3,100				Pecht
Epoxy-Novolac-Mold (molding)	352				CAIR
FR-4	17,200				Pecht
Polyimide	4,200				Pecht
Silicone	0.2 (>T_g)				Adhihetty &
	63 (<T_g)				Padmanabhan
Silicone	2,210				Pecht
Invar	141,000				ASM[6]
Kovar (29Ni17Co54Fe)	138,000				Pecht
Stainless steel 304	193,000				Pecht
Steel (1018-1020)	207,000				Pecht
Cu	114,041	−44.93			Pao et al.[51]
Brass (65/34Zn)	103,420				Machine Design
Al-6061 T4	69,000				Pecht
Al-2024 T4	74,141	−39.16			Pao et al.[57]
Ni	205,000				Pecht
Alloy-42 (42Ni58Fe)	145,000				Pecht
63Sn-37Pb	32,000	−88			Wong et al.
90Pb-10Sn	10,819	−47.23			Cole & Caulfield
96.5Sn-3.5Ag	52,708	−67.14	−0.0587		CAIR
42Sn-58Bi	41,336	−103	1.457	−0.01	CAIR
80Au-20Sn	59,000				Morgan
77.2Sn-20Sn-2.8Ag	38,610				Lee et al.

* All properties from Pecht here are referred to his book.[66] For example, the Young's modulus of 42Sn-58Bi is $41,336 - 103T + 1.457T^2 - 0.01T^3$ (MPa) and T is temperature in °C.

4.2.2   Thermal conductivity and coefficient of thermal expansion

Thermal conductivity and coefficient of thermal expansion (CTE), also called thermal coefficient of expansion (TCE), are two other important properties for determining the thermomechanical response of electronic devices and packages subjected to mechanical or thermal loadings. The former is related to the transient and steady-state heat transfer in the packages and, in turn, the temperature distribution. The latter gives rise to deformation of the package as a result of temperature change for bonded materials with different CTE values. Tables 4.7 and 4.8 list thermal conductivity and CTE, respectively, for the selected materials.

TABLE 4.5   Temperature Constants a_i of Coefficient of Yield Stress σ_y (MPa) of Some Electronic Packaging Materials

Materials	a_0	a_1	a_2	a_3	Reference
Si	34.45				CAIR
GaAs	42.03				CAIR
AlN	392				CAIR
BeO	166.7				CAIR
SiC	234.4				CAIR
Al_2O_3 (96%)	196				CAIR
Mo	552				Pecht
W	809				Machine Design
Ti	138				Machine Design
Au	207				Pecht
Epoxy-Novolac-Mold (molding)	60				Bar-Cohen & Mix
Polyimide	105				CAIR
Silicone	36				Bar-Cohen & Mix
Invar	275.6				McClintock & Argon
Kovar (29Ni17Co54Fe)	343				Pecht
Stainless steel 304	289				Pecht
Steel (1018-1020)	207				Pecht
Cu	52				CAIR
Brass (65/34Zn)	131–290				Machine Design
Al-6061 T4	145				Pecht
Al-2024 T4	324				Pecht
Ni	103–206				Machine Design
Alloy-42 (42Ni58Fe)	276				Pecht
62Sn-36Pb-2Ag	48.40	−0.246			Tomlinson
63Sn-37Pb	34.43	−0.306			Cole & Caulfield
90Pb-10Sn	14.55	−0.049			Cole & Caulfield
96.5Sn-3.5Ag	48.88	−0.2583			Harada & Satoh
42Sn-58Bi	35.11	−0.119			Tomlinson
100Sn	55.69	−0.3162			Tomlinson
99Sn-1Cu	42.51	−0.16			Tomlinson
95Sn-5Sb	45.95	−0.21			Tomlinson

* For example, the yield stress of the 42Sn-58Bi is $35.11–0.119T$ (MPa) and T is temperature in °C.

It is seen that more temperature-dependent data are available in the literature for these two properties. The temperature-dependent thermal conductivity is also shown in Fig. 4.30 for a range of −50 to 150°C, and Figs. 4.31 and 4.32 show the CTE as a function of temperature for the same range.

4.2.3   Power-law creep

The creep and stress relaxation of solder joints are associated with thermal fatigue failure of electronics packages. While there are a variety of creep mechanisms that exist in the solder joint when subjected to thermomechanical loadings, power law (or steady state or secondary

TABLE 4.6   Temperature Constants a_i of Coefficient of Tensile Strength σ_{ult} (MPa) of Some Electronic Packaging Materials

Materials	a_0	a_1	a_2	a_3	Reference
AlN	192.9				CAIR
BeO	152				CAIR
SiC	17.24				Pecht
Al_2O_3 (96%)	323				CAIR
Mo	655				CAIR
W	940				CAIR
Ti	477				Tummala & Rymaszewski
Au	221				Pecht
Silicone rubber	4.5				Pecht
Epoxy-Novolac-Cast (adhesive)	69				Pecht
Epoxy-Novolac-Mold (molding)	36				Pecht
FR-4	276				Pecht
Polyimide	65				Pecht
Silicone	10.3				Pecht
Invar	489				McClintock & Argon
Kovar (29Ni17Co54Fe)	552				Pecht
Stainless steel 304	579				Pecht
Steel (1018-1020)	379				Pecht
Cu	220				CAIR
Brass (65/34Zn)	345–379				Machine Design
Al-6061 T4	241.3				Pecht
Al-2024 T4	469				Pecht
Ni	379–552				Machine Design
Alloy-42 (42Ni58Fe)	672				CAIR
62Sn-2Ag-36Pb	53.263	−0.0996			Tomlinson
63Sn-37Pb	46.367	−0.4			Cole & Caulfield
90Pb-10Sn	24.033	−0.166			Cole & Caulfield
96.5Sn-3.5Ag	34.605	−0.1536			CAIR
42Sn-58Bi	34.73	−0.039			Tomlinson
100Sn	59.84	−0.2468			Tomlinson
77.2Sn-20In-2.8Ag	46.9				Lee et al.
91Sn-9Zn	103				McCormack & Jin
86Sn-9Zn-5In	62.1				McCormack & Jin
81Sn-9Zn-10In	55.2				McCormack & Jin
99Sn-1Cu	45.51	−0.0845			Tomlinson
95Sn-5Sb	45.00	−0.200			Mahidhara et al.
95Sn-5Sb	53.38	−0.106			Tomlinson

* For example, the tensile strength of the 42Sn-58Bi is 34.73–0.039T (MPa) and T is temperature in °C.

creep) has been widely used to characterize solder's inelastic, time-dependent behavior under thermal cycling. Also as a simple viscoplastic constitutive equation, the power law equation can be expressed as [rewriting Eq. (4.1)]

$$\frac{d\gamma_{creep}}{dt} = C^* \exp\left(\frac{-\Delta H}{k(T + 273)}\right)\tau^n \qquad (4.3)$$

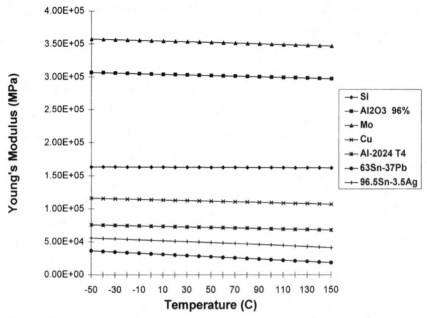

Figure 4.28 Young's modulus versus temperature for several electronics packaging materials.

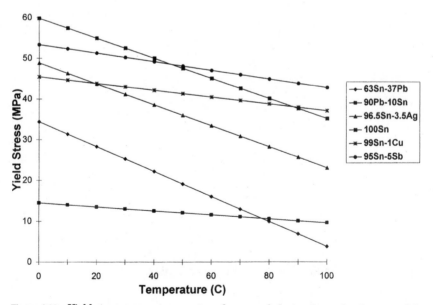

Figure 4.29 Yield stress versus temperature for several electronics packaging materials.

where $d\gamma_{creep}/dt$ = shear creep strain rate in 1/sec

C^* = frequency constant

k = Boltzmann's constant (8.63×10^{-5} eV/K)

T = temperature (°C)

τ = shear stress in MPa

ΔH = activation energy (eV)

n = shear stress exponent; for $n = n(T)$, a linear relation is assumed, i.e., $n = a_0 + a_1 T$ (°C).

The creep properties of eight different solder alloys are listed in Table 4.9. It is noted that the stress exponent is temperature-dependent for some solders. The power law equation is in general true only for a range of applied stresses. For a wider stress range, several power law equations with different sets of parameters, usually termed regions I to IV, are needed to account for the behaviors. Other approximations,

TABLE 4.7 Temperature Constants a_i of Thermal Conductivity (W/m − K) of Some Electronic Packaging Materials

Materials	a_0	a_1	a_2	a_3	Reference
Si	165.43	−0.6415	1.008×10^{-3}		Touloukian et al.
GaAs	33.67	−0.01			Touloukian et al.
AlN	140				Kuramoto et al.
BeO	298.77	−0.9621	1.201×10^{-3}		Touloukian et al.
SiC	270				Charles
Al$_2$O$_3$ (96%)	22.7				Pecht
Diamond	2627.27	−12.3843	3.079×10^{-2}		Touloukian et al.
Mo	139.05	−0.0399			Touloukian et al.
W	182.37	−0.1756	1.303×10^{-4}		Touloukian et al.
Ti	22.375	−0.01934	3.163×10^{-5}		Touloukian et al.
Au	318				Pecht
Silicone rubber	0.38				Pecht
Epoxy-Novolac-Cast (adhesive)	0.87				Pecht
Epoxy-Novolac-Mold (molding)	0.52				CAIR
FR-4	0.18				Pecht
Polyimide	8				Pecht
Silicone	7				Pecht
Invar	16.4				Tummala & Rymaszewski
Kovar (29Ni17Co54Fe)	16.3				Pecht
Stainless steel 304	6				Tummala & Rymaszewski
Steel (1018-1020)	50				Pecht
Cu	400.24	−0.0733	6.7×10^{-5}		Touloukian et al.
Brass (65/34Zn)	116				Machine Design
Al-6061 T4	154				ASM (1983)
Al-2024 T4	121				Dally
Ni	90.7				Pecht
Alloy-42 (42Ni58Fe)	15.9				Tummala & Rymaszewski
63Sn-37Pb	50.6				CAIR
90Pb-10Sn	36				ITRI
77.2Sn-20In-2.8Ag	53.5				Lee et al.
80Au-20Sn	251				Pecht

TABLE 4.8   Temperature Constants a_i of Coefficient of Thermal Expansion (ppm/°C) of Some Electronic Packaging Materials

Materials	a_0	a_1	a_2	a_3	Reference
Si	2.478	0.006639	-8.95×10^{-6}		Touloukian et al.
GaAs	5.595	0.005528	-6.6×10^{-6}		Touloukian et al.
AlN	4.5				CAIR
BeO	6.167	0.005832			Touloukian et al.
SiC	3.137	0.009531	-2.137×10^{-5}		Touloukian et al.
Al_2O_3 (96%)	4.880	0.01665	-6.579×10^{-5}	1.05×10^{-7}	Pao et al. (1992)
Diamond	0.8759	0.007995	-6.32×10^{-6}		Touloukian et al.
Mo	4.779	0.000734	2.678×10^{-6}		Touloukian et al.
W	4.498	-0.000188	2.512×10^{-6}		Touloukian et al.
Ti	8.455	0.007944	-6.22×10^{-6}		Touloukian et al.
Au	14.10	0.005584			Touloukian et al.
Silicone rubber	40				Pecht
Epoxy-Novolac-Cast (adhesive)	122				Pecht
Epoxy-Novolac-Mold (molding)	11				CAIR
FR-4	16				Pecht
Polyimide	47				Pecht
Silicone	262				Pecht
Invar	1.393	0.002416	7.495×10^{-5}		Chanchani & Hall
Kovar (29Ni17Co54Fe)	5.87				CAIR
Stainless steel 304	15.46	0.0183	-4.66×10^{-5}		Raeder
Steel (1018-1020)	11.57	0.01494	-1.739×10^{-5}		Raeder
Cu	16.28	0.01144	-1.064×10^{-5}		Touloukian et al.
Brass (70/30Zn)	17.45	0.01002	-3.228×10^{-5}		Raeder
Al-6061 T4	22.51	0.00958			Raeder
Al-2024 T4	18.88	0.02369			Pao et al.
Ni	12.29	0.01612	-2.995×10^{-5}		Raeder
Alloy-42 (42Ni58Fe)	4.45				Pecht
63Sn-37Pb	24.7				CAIR
90Pb-10Sn	28.59	0.015			CRC
96.5Sn-3.5Ag	21.85	0.02039			Harada & Satoh
77.2Sn-20In-2.8Ag	28				Lee et al.
42Sn-58Bi	13.78				ITRI
42Sn-58Bi	14.9				Seyyedi
43Sn-43Pb-14Bi	24.0				Seyyedi
48Sn-52In	28.0				Seyyedi
40Sn-20Pb-40In	28.0				Seyyedi
80Au-20Sn	16				Pecht

such as hyperbolic sine functions, have also been applied to describe solder's creep behavior, see for examples, Darveaux and Banerji[17] and Lau.[97,98,99]

4.2.4   Stress-to-rupture time for solder joints

A property of failure directly resulting from creep is *creep strength,* defined as the time required to sustain an applied constant stress to the point where final failure occurs. Such data have been reported by ITRI,[35] and equations based on the logarithmic function, as expressed

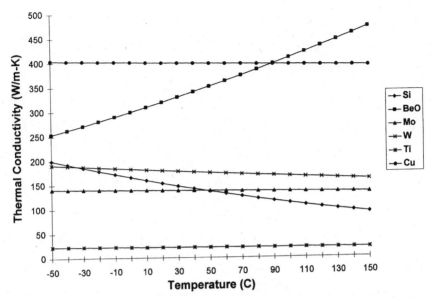

Figure 4.30   Thermal conductivity versus temperature for several electronics packaging materials.

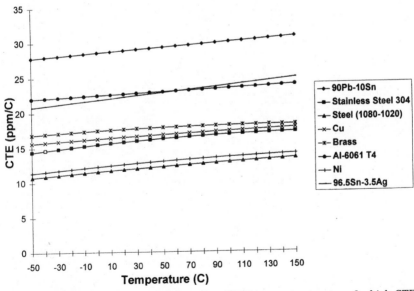

Figure 4.31   Coefficient of thermal expansion (CTE) versus temperature for high-CTE materials.

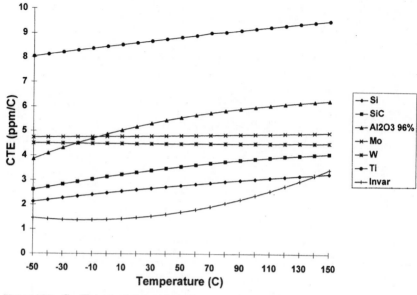

Figure 4.32   Coefficient of thermal expansion (CTE) versus temperature for low-CTE materials.

in Eqs. (4.4) and (4.5) for 20 and 100°C, respectively, are used to fit the test data. The fitted constants are listed in Table 4.10.

$$\text{Shear stress} = C + D \log_{10} (\text{tr}20)$$

$$\text{Shear stress} = E + F \log_{10} (\text{tr}100)$$

$$\text{tr}20 = 10^{[(\text{shear stress} - C)/D]} \tag{4.4}$$

$$\text{tr}100 = 10^{[(\text{shear stress} - E)/F]} \tag{4.5}$$

TABLE 4.9   Coefficients of Power-Law Creep for Some Solder Alloys

Solder alloys	C^* (1/sec-MPa^n)	ΔH (eV)	a_0	a_1	Reference
63Sn-37Pb	0.2046	0.494	5.248		Pao et al.[63]
90Pb-10Sn	100.6	0.643	4.739	−0.0122	Murty & Turlik
95Pb-5Sn	1.936	1.03	11.03		Darveaux & Banerji
96.5Sn-3.5Ag	0.0001927	0.697	8.67		Darveaux & Banerji
48Sn-52In	1.094×10^5	0.73	3.3		Freer Goldstein & Morris
48Sn-52In	19,492	0.991	3.22		Mei & Morris[51]
42Sn-58Bi	5.54×10^{-7}	0.731	4.05		Mei & Morris[52]
80Au-20Sn	0.0000196	0.82	2.55		Morgan
97Sn-3Cu	0.0009101	0.514	8.66	−0.01685	Pao et al.[61]

where C, D, E, F = constants in Table 4.10
$\quad\quad$ τ = shear stress in MPa
$\quad\quad$ tr20 = stress-to-rupture hours at 20°C
$\quad\quad$ tr100 = stress-to-rupture hours at 100°C.

4.2.5 Fracture toughness

In the reliability analysis of electronics packages, the fracture toughness is used to assess the ability of an existing crack in the structure to propagate under the prescribed loading. Although it may not be directly related to the fatigue propagation of cracks, it represents a threshold value beyond which the crack will propagate in an unstable manner, which eventually leads to catastrophic failure. It is thus always useful to set this property as the upper bound in the fatigue analysis for the associated driving force K, J, or G to ensure the failure criterion $K, J, G \geq K_c, J_c, G_c$ is not violated.

The fracture toughness values of a number of materials are listed in Table 4.11. Half of them are solder joints since most of the failures in electronics packaging are associated with cracking in the solder joint. Different fracture parameters are found in the literature, e.g., K_c, J_c, or G_c, depending on the specimen used, the deformation associated with the test, and the way authors choose to interpret the experimental results. Therefore, the type of specimen used is also listed as a reference. Most of the data for solder joints are for room temperature; more data at different temperatures, ranging from −40 to 160°C, are needed since this property is generally temperature-dependent.

4.3 Constitutive Equations and Finite Element Modeling of Solder Joints

During the past two decades, research on solder joint reliability has been focused on: (1) studying the effects of temperature, strain range,

TABLE 4.10 Coefficients of Creep Strength for Some Solder Alloys

Solder alloys	C	D	E	F
63Sn-37Pb	13.2	−3.47	1.92	−0.475
90Pb-10Sn	12.85	−3.22	3.99	−0.97
96.5Sn-3.5Ag	32.05	−1.99	17.2	−1.83
42Sn-58Bi	17.5	−4.72	3.09	−0.76
97Sn-3Cu	15.6	−2.53	5.47	−1.01
5Sn-93.5Pb-1.5Ag	23.8	−2.86	9.62	−0.82
95Pb-5Ag	32.13	−3.35	—	—
10Sn-88Pb-2Ag	12.85	−3.22	3.99	−0.97

SOURCE: ITRI.[35]

TABLE 4.11 Fracture Toughness K_c (MPa – m$^{1/2}$) or J_c (J/m^2) and Fracture Energy G_c (J/m^2) of Solder Joints, Intermetallics, Molding Compound, and PWB

Material	K_c (MPa – m$^{1/2}$)	J_c (J/m^2)	G_c (J/m^2)	Specimen	Reference
40Sn-40In-20Pb	4.20–4.50			CT ($h^* = 0.254$ mm)	Frear & Viaco
50Sn-50In	3.80–4.00			CT ($h = 0.254$ mm)	Frear & Viaco
63Sn-37Pb		39–87 (RT)		CT ($h = 0.076$ mm)	Pao
63Sn-37Pb		1167–1865 (RT)		CT ($h = 0.254$ mm)	Pao
63Sn-37Pb		800–1200 (RT)		DCB* ($h = 0.254$ mm)	Yamada
63Sn-37Pb		108 (RT)		mini. single lap-shear ($h = 0.5$ mm)	Clough et al.
63Sn-37Pb	8.36 (RT)			CT (bulk)	Logsdon et al.
63Sn-37Pb	6.00–8.00 (RT)			CT ($h = 0.254$ mm)	Pratt et al.
60Sn-40Pb	5.50–6.70 (RT)			CT ($h = 0.254$ mm)	Frear & Viaco
50Pb-50In	6.70–9.60 (RT)			CT ($h = 0.254$ mm)	Frear & Viaco
95Sn-5Sb	4.70–11.0 (RT, 60°C)			CT ($h = 0.254$ mm)	Frear & Viaco
96.5Sn-3.5Ag		1.65×10^5 (RT)		Four-point bend ($h = 0.16$–0.37 mm)	Stromswold et al.
97Sn-3Ag	0.75–10.5 (RT)			CT ($h = 0.254$ mm)	Frear & Viaco
Cu$_6$Sn$_5$	1.4			Indentation	Fields et al.
Cu$_3$Sn	1.7			Indentation	Fields et al.
Epoxy (glass-filled)	1.37–2.20 (RT)			CT	Wu
Epoxy mold (Nitto-Denko MP7150M)	2.8 (50°C) 0.2 (150°C)			Three-point bend (bulk)	Pfeil et al.
Epoxy mold (Sumitomo Bakelite EME7320)	3.0 (50°C) 0.6 (150°C)			Three-point bend (bulk)	Pfeil et al.
Epoxy mold	3.16 (50°C) 2.53 (150°C) 0.316 (220°C)			Three-point bend (bulk)	Swada et al.
Epoxy mold	1.79–2.47 (–196°C) 2.30–3.73 (20°C) 1.98–4.05 (85°C) 0.12–0.68 (155°C)			Single-edge notch bend (bulk)	Sauber et al.
FR-4			1000–2800	MMB*	Goodelle et al.
Polyimide (Kapton H)			75–180 (RT)	T-peel	Stoffel et al.
Polyimide (Kapton H, modified)			1100–1200 (RT)	T-peel	Stoffel et al.
PWB/Cu-Mo-Cu			175–1052 (RT)	DCB, MMF*, ENF*	Liechti

* h: solder joint thickness; DCB: double cantilever beam; MMF: mixed mode flexure; ENF: end-notch flexure; MMB: mixed mode bending; RT: room temperature; CT: compact tension.

strain rate, and frequency on the solder fatigue life through isothermal and thermal fatigue lab tests,[77,90] (2) developing constitutive equations to model various deformation modes of solder joints,[9,36,38,50,80] (3) developing empirical life prediction models,[18] and (4) developing accelerated thermal cycling tests to simulate the service conditions.[19,45]

In each of these areas, a detailed understanding is generally required of the deformation behavior of solder joints under prescribed thermal history. For complicated joint configurations, numerical techniques, such as finite element methods (FEM) or boundary element methods (BEM), provide a powerful tool to serve this purpose. Finite element modeling considering solder joints as elastic or elastic-plastic has been applied to model the deformation under thermal conditions.[42] In addition to plasticity, creep effect for 37Pb/63Sn solder has been included in modeling its behavior under thermal cycling.[17,38,56,97]

4.3.1 Constitutive equations

The mechanical behavior of Pb/Sn solder joints under thermal cycling can be very involved. In high-homologous-temperature regimes ($\geq 0.5 T_M$) primary and secondary (steady-state) creep are the dominant deformation modes and are accompanied by stress relaxation, provided the joint, such as a leadless surface-mounted device, is primarily under strain-control conditions. At lower temperatures, alternative plastic deformation takes place since creep would be suppressed. Knecht and Fox[38] have developed constitutive equations for eutectic solder that account for both time-independent plasticity and steady-state creep. Similar formulation was employed here for both 63Sn-37Pb and 90Pb-10Sn solders. It is assumed that the time-independent part of the total strain rate consists of an elastic strain rate and a plastic strain rate. In terms of pure shear, the time-independent strain γ_{indep} can be expressed as

$$\gamma_{indep} = \gamma_e + \gamma_{pl} = \frac{\tau}{G} + A \left[\frac{\tau}{\tau_p} \right]^m \qquad (4.6)$$

where τ is the shear stress, G is the shear modulus, A and m are material constants, and τ_p depends on temperature and is assumed to be linearly dependent upon temperature, i.e., $\tau_p = C_1 T(\mathrm{K}) + C_2$. The rate of change of γ_{indep} is obtained by direct differentiation of Eq. (4.6) with respect to time. The time-dependent part of the shear strain rate $\dot{\gamma}_{cr}$ is assumed to be steady state and follows a modified Norton law, i.e., bilinear in the logarithmic stress-strain rate plot. Therefore, the constitutive equation for pure shear can be written as

$$\frac{d\gamma_{\text{tot}}}{dt} = \frac{d\gamma_e}{dt} + \frac{d\gamma_{\text{pl}}}{dt} + \frac{d\gamma_{\text{cr}}}{dt}$$

$$= \left[\frac{1}{G} + \frac{Am}{\tau_p^m}\,\tau^{m-1}\right]\frac{d\tau}{dt} - \frac{\tau}{G^2}\frac{dG}{dt} + C_o\left[\left(\frac{\tau}{\tau_o}\right)^{n_1} + \left(\frac{\tau}{\tau_o}\right)^{n_2}\right] \quad (4.7)$$

where $\tau_o = [\tau^* \exp (U/T)]^{1/(n_2 - n_1)}$ and $C_o = k\tau_o^{n_1} \exp (-Q_c/T)$. The material constants A, m, C_1, C_2, τ^*, U, k, Q_c, n_1, and n_2 for both solders are listed in Table 4.12, where those for eutectic solder were taken from Knecht and Fox,[38] and those for 90Pb-10Sn were determined from the experimental data of Lavery[43] and Medla.[53] The generalized stress-strain relation that reduces to Eq. (4.7) is

$$\frac{d\epsilon_{ij}}{dt} = \frac{1+\nu}{E}\frac{d\sigma_{ij}}{dt} - \frac{\nu}{E}\frac{d\sigma_{kk}}{dt}\,\delta_{ij} + \left(\frac{\nu}{E^2}\,\sigma_{kk}\delta_{ij} - \frac{1+\nu}{E^2}\,\sigma_{ij}\right)\frac{dE}{dt}$$

$$+ \frac{\alpha Am}{2\tau_p}\left(\frac{\sigma_e}{\sqrt{3}\tau_p}\right)^{m-1}\frac{dS_{ij}}{dt} \quad (4.8)$$

$$+ \frac{C_o}{2\tau_o}\left[\left(\frac{\sigma_e}{\sqrt{3}\tau_o}\right)^{n_1-1} + \left(\frac{\sigma_e}{\sqrt{3}\tau_o}\right)^{n_2-1}\right]S_{ij}$$

where $d\epsilon_{ij}/dt$ is the strain rate tensor, E is the Young's modulus, ν is the Poisson's ratio, σ_{ij} is the stress tensor, S_{ij} is the deviatoric stress tensor defined as $\sigma_{ij} - \sigma_{kk}\delta_{ij}/3$, δ_{ij} is the Kronecker delta, and σ_e is the effective stress and $=(3J_2)^{1/2}$, where $J_2 = S_{ij}S_{ij}/2$.

For eutectic solder the dependence of E on the temperature is assumed to follow the relation derived by Wong and associates[94] as $E = -88T$ (°C) + 32,000 MPa, and for 90Pb-10Sn, that was assumed to be the same as pure Pb.[24] The Poisson's ratios for both solders were taken to be 0.4. The J_2 criterion is used where $J_2 > 0$ is for loading ($\alpha = 1$) and $J_2 < 0$ is for unloading ($\alpha = 0$). In the case of pure shear, J_2 is replaced by $\tau(d\tau/dt)$.

4.3.2 Finite element analysis

The preceding constitutive equations were implemented in a finite element program ABAQUS with a user subroutine incorporating part of the creep strain rate since such specific form of the constitutive equation was not included in the material library of the program. Linear

TABLE 4.12   Material Parameters for 37Pb/63Sn and 90Pb/10Sn

Solder	n_1	n_2	k	Q_c	τ^*	U	A	m	C_1	C_2
37Pb/63Sn	2	7.1	8.31	5790	7.42	3940	1	2	−2.067	913.04
90Pb/10Sn	0	4.0	1.00	0	1.08	7300	1	2	−0.0827	66.28

four-node plane strain elements were used. To examine the behavior of the implemented constitutive equations, a single element subjected to pure shear was tested, and the results were compared with the analytical predictions based on the Runge-Kutta method. The comparison for the eutectic solder is shown in Fig. 4.33. The imposed thermal cycle had a temperature range between –25 and 125°C with a ramp rate of 30°C/h and a two-hour hold time. Figure 4.33a shows the shear stress versus time for two complete cycles, and Fig. 4.33b shows the variation of shear stress with temperature. It is seen that an excellent agreement has been attained between the FEM results and the analytical predictions, which demonstrates that the implementation of the constitutive equations has been successful.

Next, the experiment by Hall[32] on the deformation of 60Sn-40Pb solder joints under thermal cycling was modeled. In the experiment, leadless ceramic chip carriers (LCCCs) were mounted on printed wire boards with eutectic solder and were subjected to thermal cycling that was identical to that used in the verification test. Only half of the test specimen was modeled because of the symmetry about the neutral axis. The finite element mesh is shown in Fig. 4.34a. The material properties and physical dimensions of the specimen used in the finite element analysis followed those reported by Hall.[32] The FEM results are shown in Fig. 4.34b, c, and d, together with the experimental results. Figure 4.34b shows the bend angle versus temperature, where bend angle is defined as the slope of the bending deflection of the PWB and CCC at the center of the solder joint. Figure 4.34c shows the shear

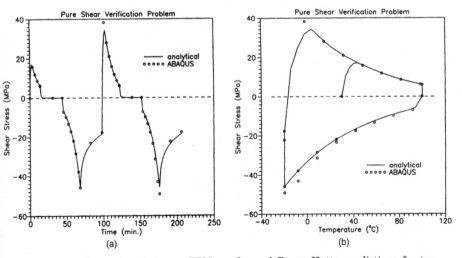

Figure 4.33 Comparison between FEM results and Runge-Kutta predictions for two complete cycles: (a) shear stress versus time; (b) shear stress versus temperature.

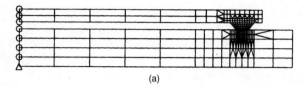

(a)

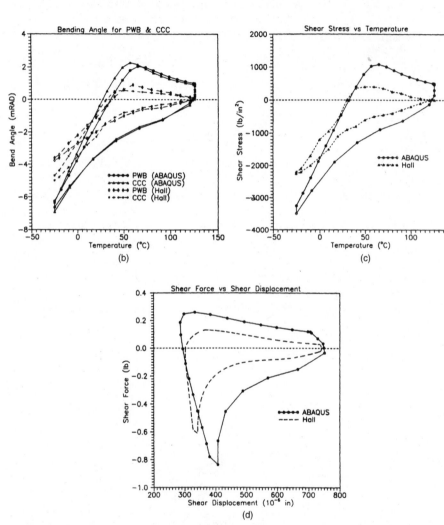

Figure 4.34 Comparison between FEM results and experimental data by Hall[32]: (*a*) FE model; (*b*) bend angles of PWB and CCC versus temperature; (*c*) average shear stress in the solder joint versus temperature; (*d*) shear force versus shear displacement.

stress in the solder joint versus temperature, and the shear force–shear displacement loop is shown in Fig. 4.34*d*. Shear force is determined by multiplying the average shear stress by the area of the solder joint, and shear displacement is determined by multiplying the average shear strain by the joint height. Despite the discrepancies between the FEM and experimental results, the trends shown in Fig. 4.34*b*, *c*, and *d* are similar, which indicates the constitutive equations implemented were able to capture the key deformation modes, i.e., plasticity and creep, as functions of temperature. The discrepancies between the FEM and experimental results may be attributed to the difference between the parameters in the constitutive equation and the material used in the experiment, since those parameters were determined from separate tests. Another possibility may be due to the assumptions employed in deriving the experimental data based on a simplified plate theory.

The implemented constitutive equation for 90Pb-10Sn solder was first applied to model the behavior of the bimaterial beam specimen[57] for verification. The comparison of hysteresis response to thermal cycling between 40 and 140°C between the experimental measurement and finite element prediction is shown in Fig. 4.35. It is seen that good

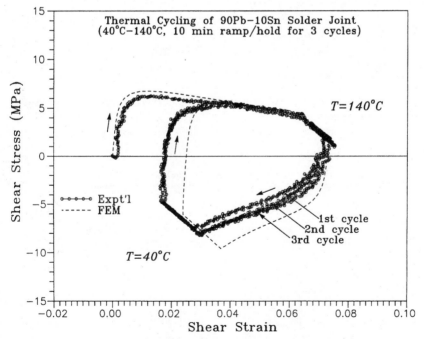

Figure 4.35 Comparison of 90Pb-10Sn hysteresis response between FEM and experimental results.

agreement has been achieved, and the constitutive equation was then applied to model the thermal-cyclic behavior of two solder layers in a typical transistor stack.

The stack consists of a silicon chip, a beryllium oxide (BeO) heat spreader, a copper interlayer, and an aluminum baseplate. 90Pb-10Sn solder was used to connect silicon to BeO and BeO to Cu. The stack was subjected to a temperature cycle between –40 and 140°C, with a 18°C/min ramp rate and a 20-minute hold time. This temperature range corresponds to 0.43 to 0.75 T_M of the 90Pb-10Sn solder. The length of the silicon and the BeO was 10.16 and 22.86 mm, respectively. The elastic properties and physical dimensions of the stack are listed in Table 4.13. The FE mesh is shown in Fig. 4.36a. The boundary conditions were imposed such that the model was constrained at the far end of the aluminum baseplate and had a symmetric condition on the left side.

The results of shear and peel stresses versus distance from the free edge for one complete cycle are shown in Fig. 4.36b and c for the top solder layer and in Fig. 4.36d and e for the second layer. Several observations can be made here: (1) The stress state of the solder joint in the stack is by no means pure shear but highly multiaxial; both shear and peel stress concentrations occur in the vicinity of the free edge; any crack, once initiated in such an area and intending to grow, is subjected to a complicated combination of driving forces. (2) While the peel stress of the top solder layer concentrates more closely to the edge (within 1 mm) than that of the second layer, the magnitude is in the same order as the shear stress; the fact that the peel stress near the edge is tensile at high temperatures and compressive at low temperatures indicates that the peel stress may play an important role in assisting or retarding the crack propagation. (3) Generally, the stresses in the second solder layer are higher than those in the top layer due to shorter distance

TABLE 4.13   Elastic Properties and Thickness of the Transistor Stack

Materials	Thickness (mm), h	Young's modulus (MPa), E	Poisson's ratio ν	Coeff. of thermal expansion (CTE) (ppm/°C), α
Silicon	0.254	180,908	0.42	2.80
90Pb/10Sn	0.051	Eq. (a)	0.40	21.1
BeO	0.635	344,737	0.26	5.06
90Pb/10Sn	0.152	Eq. (a)	0.40	21.1
Copper	0.635	Eq. (b)	0.34	17.0
Aluminum	1.626	Eq. (c)	0.33	23.8

90Pb/10Sn:	$E = 20{,}440[1 - 0.76(\text{T}-300)/601]$	MPa	(T in K)	Eq. (a)
Copper:	$E = 112{,}828[1 - 0.54(\text{T}-300)/1356]$	MPa	(T in K)	Eq. (b)
Aluminum:	$E = 67{,}564[1 - 0.50(\text{T}-300)/933]$	MPa	(T in K)	Eq. (c)

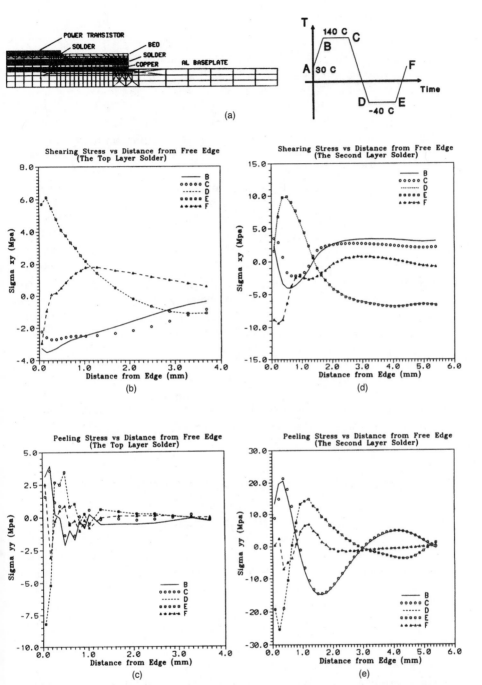

Figure 4.36 FEM results of transistor stack: (*a*) FE model and temperature cycle; (*b*) shear stress versus distance from free edge in the top solder layer; (*c*) peel stress versus distance from free edge in the top solder layer; (*d*) shear stress versus distance from free edge in the second solder layer; (*e*) peel stress versus distance from free edge in the second solder layer.

from the neutral axis and a smaller CTE mismatch. (4) Stress relaxation at high-temperature hold times B and C is more pronounced for shear stress (approximately 20 to 30 percent) than in low-temperature hold times where the change of stress is hardly seen. (5) In contrast to the shear stress, the peel stress in both layers shows little relaxation at both temperature extremes. (6) The stress distribution along the distance from the free edge reverses its sign when temperature changes from high (B and C) to low (D and E).

The distribution and nature of the peel stress in the solder layer may have a significant effect on the propagation of a crack, provided it is initiated at the edge. Moreover, a similar stress-strain loop for each temperature cycle that can apply to every point in the solder does not exist, since not only the magnitude of the stresses changes with temperature but their distributions vary as well. Another interesting feature observed was that when the solder layer becomes relatively thin and highly constrained, the singular behavior of the stresses near the bonded edge of dissimilar materials tends to be suppressed as a result of the sophisticated interactions from the adjacent boundaries and is no longer governed by the asymptotic solutions. Thus, such a decrease of the dominance region of the asymptotic solutions implies that it may fail to control either the initiation or subsequent growth of small scale cracks at the edge. More complicated theories, together with fully accounted stress distributions in such regimes, are needed to describe the behavior of such thermally induced cracks.

4.4 Design for Reliability in Harsh Environments

Any industry that is related to electronics has been developing products with extensive applications of state-of-the-art electronics technology. Take the automotive industry as an example: electronics have been widely used in various areas, such as control of the engine, powertrain, automatic transmission, antilock braking system, traction, suspension, air bag, navigation, and entertainment equipment. Such growth in electronics applications results in an increasing demand in the number and quality of electronic devices, as well as in higher packaging density and cost. For example, the automotive industry in Japan sees a demand for a packaging density ten times higher than today's in next generation's powertrain control modules. In the United States, the current average price per vehicle spent on automotive electronics is $1200, with an expected 100 percent increase by the end of this century. As the amount of electronics increases, the technologies used to produce and package them, namely, surface-mount technologies (SMTs) such as fine-pitch quad flat pack (QFP), hybrid IC, ball grid array (BGA), chip-on-board (COB), CSP, flip chip, and multichip module (MCM), must also evolve to meet the needs of the harsh underhood environment. Therefore it is essential to develop new methodologies to design and manufacture these electronics products so that reliability, cost-effectiveness, and shorter cycle time are ensured.

One major reliability issue facing today's electronic packaging technologies is solder interconnect failure due to thermal and vibrational fatigue. An extensive amount of research has been focused on this issue for the past three decades, and some understanding of solder fatigue failure and capabilities of predicting life was obtained.[21,42,74] However, in order for these methodologies to be applied to the design and manufacturing of actual products, an integrated approach capable of dealing from board layout to solder joint life prediction is needed for design engineers.

This section discusses the development of a computer-aided integrated system by Ford for evaluating thermal fatigue life of solder interconnects. The system is developed based on a UNIX workstation with a user interface toolkit, SUIT (Simple User Interface Toolkit). The system uses a commercial software package as the front end for the PCB layout and thermal analysis. The results give the temperature history of each component used by the system to determine the time-dependent deformation and stress/strain distribution of the selected interconnects. The fatigue life of the interconnects can then be predicted based on built-in physics-based failure criteria.

Three major types of interconnects are considered in the current version of the system: *leaded solder joints,* such as gull-wing, J-bend, and butt; *leadless solder joints,* such as chip capacitor, resistor, carrier or flip chip bump; and *solder interlayers* in multilayered transistor stacks. A fully automated 3D finite element program has also been incorporated to assist the design of component lead frames. In addition, a comprehensive material database of thermal and mechanical properties for commonly used electronics materials has been developed and implemented. This system is intended to assist in realizing the concept of "prevention mode" in the product design process to reduce cycle time and meet future reliability requirements.[93]

4.4.1 System structure

The system was developed on a UNIX-based workstation. It uses PCB/Explorer (from Pacific Numerix Corporation) as the front end for PCB/component layout and board-level thermal/vibration analysis. The system makes use of a variety of software tools and languages, including LISP, C++, FORTRAN, and PATRAN Command Language, depending on the needs of different system requirements. Most of the data structures and user interface routines are modeled using SUIT developed by the University of Virginia. The architecture and key elements of the system are shown in Fig. 4.37. Two other commercial programs, PATRAN (from PDA Engineering) and ABAQUS (from HKS, Inc.), are also incorporated for automated finite element analysis of lead frame stiffness. The board and component information and results of associated thermal analysis from the front-end software (PCB/Explorer) can be directly transported into the system for reliability analysis. Any modifications and changes made in the system can be transferred back for updated thermal analysis. All these functions are supported by a built-in material database. The loose interdependency of the system components shown in Fig. 4.37 allows for a modular assembly of heterogenous programs to be used to solve reliability problems. It also possesses the flexibility for system updates and adding new functions.

4.4.2 Capability and applications

The system currently provides six major functions:

1. PCB layout and component placement
2. Heat transfer analysis
3. Thermal fatigue life prediction of leadless and leaded solder joints
4. Interfacial thermal stress analysis of multilayered transistor stacks
5. Automated 3D finite element modeling of lead frame

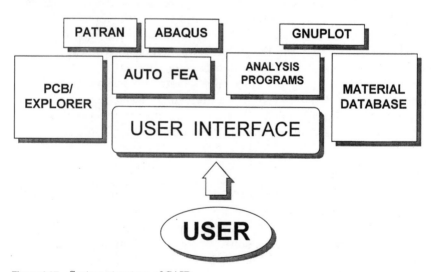

Figure 4.37   System structure of CAIR.

6. A comprehensive material database for mechanical and thermal design

A flow chart is shown in Fig. 4.38 to illustrate how these functions are tied in with the PCB design process. It is seen that a wide spectrum of design elements ranging from component placement and thermal management to solder joint fatigue life prediction can be analyzed in an integrated fashion.

Although the board layout is made with PCB/Explorer, it may also be generated using other CAD systems. A typical PCB layout brought into the system is shown in Fig. 4.39. The thermal analysis determines the temperature field of the board and is used to pinpoint hot spots or critical components for a local, detailed analysis of interconnect reliability.

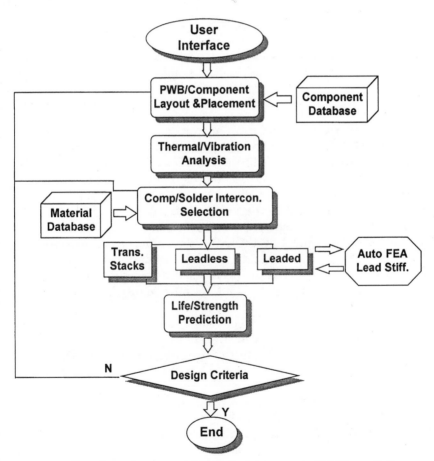

Figure 4.38   Flow chart of system capabilities for interconnect reliability prediction.

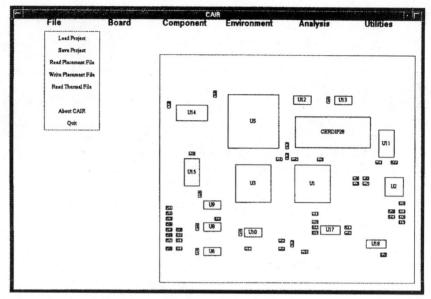

Figure 4.39   A typical PWB input from PCB/Explorer to CAIR.

Both leaded and leadless surface-mount (SMT) solder joints are considered and treated as inelastic materials. Different failure criteria are implemented to estimate the life of solder joints under prescribed temperature loadings. The nonlinear and temperature-dependent properties of commonly used solder alloys have been carefully determined or measured and compiled in the material database (see Sec. 4.2). The failure criteria for the leadless solder joints are based upon the total strain range determined by the given temperature cycle profile and treating solder as an elastic and viscoplastic material. Thermal cycling test data by Sherry and Hall[76] were used to correlate the predicted strain range. As shown in Fig. 4.40, a Coffin-Manson-type of relation exists between the total strain range and thermal fatigue life. It is noted that all the eleven tests had different temperature extremes, hold times, and ramp rates. The effect of these parameters on fatigue life is intrinsically embedded in the calculation of total strain range. Therefore, the Coffin-Manson parameters are constant, as shown in Fig. 4.40.

In the analysis of leaded solder joints, the approach used is based on a modified version of the method developed by Kitano et al.[37] The leads are treated as simple beams, and the solder joint is modeled as a temperature-dependent elastic-plastic material. The time-dependent

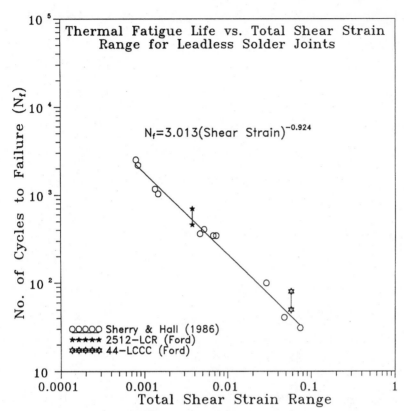

Figure 4.40 Thermal fatigue life of leadless 63Sn-37Pb solder joints versus total strain range predicted by CAIR.

effect is only considered via a frequency-dependent parameter in the constitutive equation. Creep and stress relaxation are not considered for the leaded solder joints primarily because of the relatively low stress and strain levels due to the existence of compliant leads. The equivalent strain, a collective measure of normal and shear strain components, is determined in the joint along the lead foot length, and the maximum value is used to correlate the fatigue life. Figure 4.41 shows the test results from Kitano and associates[37] on three different types of lead, together with the data by Uegai et al.[89] Similarly, a Coffin-Manson equation can be used to correlate the test data with the predicted maximum equivalent strain.

The life prediction equations shown in Figs. 4.40 and 4.41 are implemented in the system for leadless and leaded 63Sn-37Pb solder joints, respectively. In Fig. 4.42 the predicted hysteresis loop and fatigue life of a typical leadless component are shown based on a temperature pro-

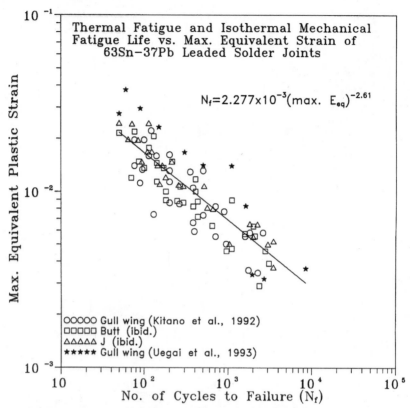

Figure 4.41 Thermal fatigue life of several types of leaded 63Sn-37Pb solder joints versus maximum equivalent plastic strain range predicted by CAIR.

file resulting from the thermal analysis. A similar example of a gull-wing leaded joint is shown in Fig. 4.43, where a plot of the distribution of equivalent strains is shown at the top left corner, and the definition of lead material and dimension is shown at the lower right corner.

In the multilayered transistor stack analysis, the thermal stresses and deformation, particularly the interfacial shear and peel stresses, of the structure subjected to nonuniform temperature change are analyzed with a specific built-in finite element code for both plane stress and plane strain. At the current stage, the system allows for the analysis of two to ten layered structures, i.e., a maximum of nine interfaces, with different lengths and temperature changes for each layer. The length of each layer is stipulated such that it cannot be longer than the one immediately below it. In other words, the current allowable configurations are of a pyramidal shape, which represents most of the commonly used transistor stacks. An example of a five-layered transistor stack is shown in Fig. 4.44a.

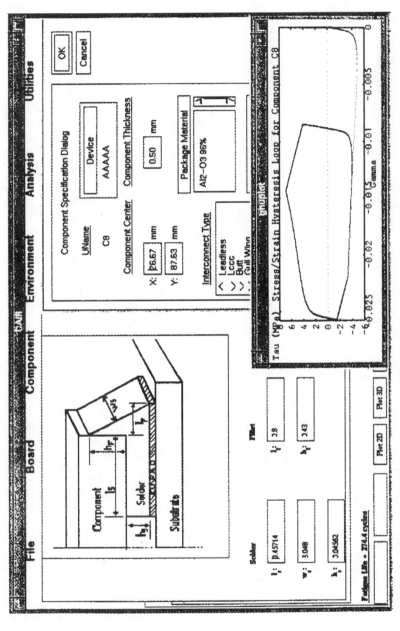

Figure 4.42 Screen display of the predicted shear stress/strain hysteresis loop of a typical leadless solder joint subjected to thermal cycling.

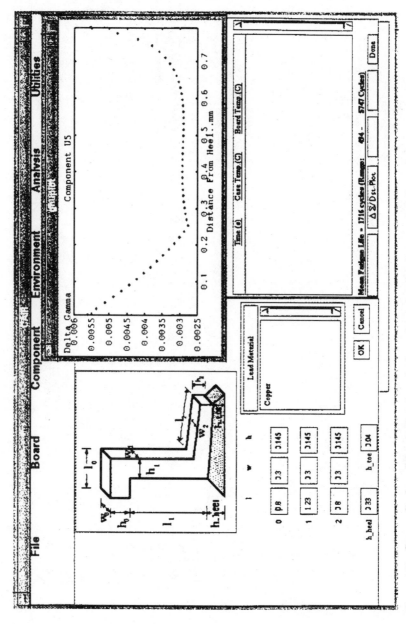

Figure 4.43 Screen display of the predicted equivalent plastic strain of a typical gull-wing leaded solder joint along the foot length.

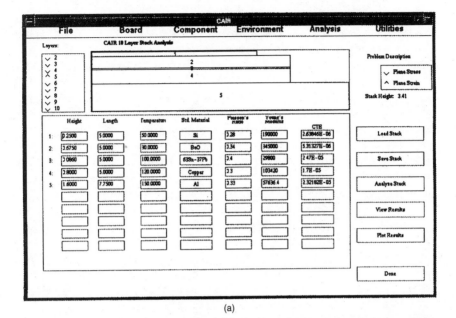

(a)

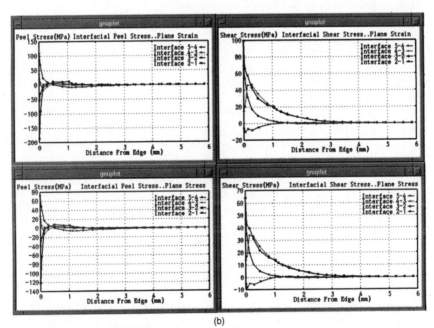

(b)

Figure 4.44   Thermal stress analysis of transistor stacks: (*a*) definition of a five-layered stack; (*b*) interfacial peel and shear stresses for plane stress and plane strain.

The mesh generator automatically divides each layer into five sub-layers with various thicknesses, such as 1/10, 2/10, 4/10, 2/10, and 1/10 of the thickness of that layer. The finer mesh near the interface tends to capture more details of the interfacial peel and shear stresses which are usually more complex than those inside the layer. For each step with its length no longer than 4.2 times the average thickness, defined by the total thickness of the stack divided by the number of layers, five layers of elements with a geometric progression factor of 1.5 are generated starting from the free edge of the step. As such, the maximum size of the elements is limited to approximately 1.6 times the average thickness. For steps longer than 4.2 times the average thickness, additional elements will be added to that step based on a rule of $(L - 4.2t_{avg})/1.2t_{avg}$, where L is the step length and t_{avg} is the average thickness of the stack. This results in additional elements no larger than $1.2t_{avg}$. For the remaining length of the layer, the first five elements have a similar geometric progression factor of 1.5, while the rest have a constant length of 1.2 times the average thickness. Figure 4.44b shows the associated interfacial shear and peel stresses in each layer. This capability provides engineers with a useful tool for rapid assessment of a number of design parameters, such as material, dimension, and temperature.

In order to facilitate the PCB design process, sometimes it is necessary to make transparent to engineers those sophisticated modeling computations that require significant amounts of input, output, and file transfer between a number of software programs. A fully automated finite element analysis for packaging subcomponents has been implemented. Parametric 3D FE models are used to determine the lead stiffness of the widely used gull-wing and J-bend leads in SMT packages. For any given set of lead dimensions and material properties, as shown in Fig. 4.45 for a typical gull-wing lead, the implemented automatic finite element module, based on PATRAN and ABAQUS, determines the 12×12 stiffness matrix (Fig. 4.46), the corresponding 12 fundamental deformed shapes (Fig. 4.47), and the associated von Mises stress distributions. The stiffness of the component lead frame plays an important role in thermal and vibrational fatigue of solder joints. In thermal fatigue, the lead stiffness is one of the key elements that determine the magnitude of stress and strain in the solder joint and in turn the fatigue life of the joint. The fact that creep and stress relaxation are not as pronounced as in leadless solder joints is associated with the lead stiffness—the smaller the stiffness, the slower the stress relaxation and creep. In addition, the lead stiffness also dominates the coplanarity in the lead-forming process. Thus, it is crucial for engineers to design the lead stiffness properly.

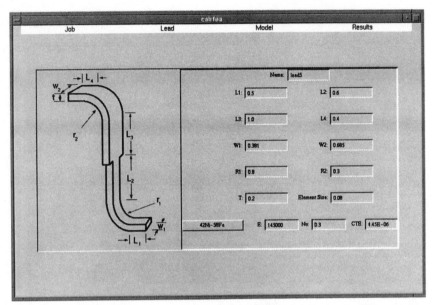

Figure 4.45   Screen display of the input menu for 3D FEA of gull-wing leadframe.

```
TEST OF GULL-WING JOINT

The stiffness matrix K =

  35.44  -81.08    .00    .00    .00   38.30  -35.44   81.08    .00    .00    .00   13.49
 -81.08  285.17    .00    .00    .00 -191.78   81.08 -285.17    .00    .00    .00 -121.06
    .00     .00  26.43  28.45  28.30    .00     .00     .00 -26.43  50.83  23.24    .00
    .00     .00  28.45  60.81  36.29    .00     .00     .00 -28.45  24.55  19.19    .00
    .00     .00  28.30  36.29  50.26    .00     .00     .00 -28.30  48.61   4.92    .00
  38.30 -191.78    .00    .00    .00  162.21  -38.30  191.78    .00    .00    .00   96.86
 -35.44   81.08    .00    .00    .00  -38.30   35.44  -81.08    .00    .00    .00  -13.49
  81.08 -285.17    .00    .00    .00  191.78  -81.08  285.17    .00    .00    .00  121.06
    .00     .00 -26.43 -28.45 -28.30    .00     .00     .00  26.43 -50.83 -23.24    .00
    .00     .00  50.83  24.55  48.61    .00     .00     .00 -50.83 127.95  50.52    .00
    .00     .00  23.24  19.19   4.92    .00     .00     .00 -23.24  50.52  40.39    .00
  13.49 -121.06    .00    .00    .00   96.86  -13.49  121.06    .00    .00    .00   98.72
```

Figure 4.46   The predicted 12×12 stiffness matrix for the gull-wing lead defined in Fig. 4.45.

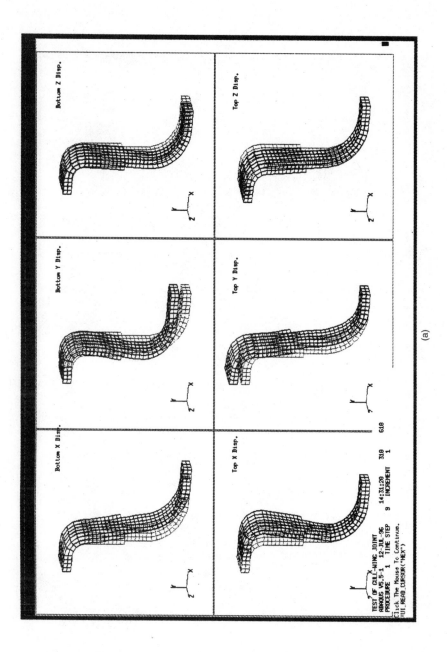

Bottom Z Disp.

Top Z Disp.

Bottom Y Disp.

Top Y Disp.

Bottom X Disp.

Top X Disp.

(a)

TEST OF CLI_HINC JOINT
ABAQUS V5.5-1   12-JUL-96   14:31:20   318   618
PROCEDURE   1   TIME STEP   9   INCREMENT   1
Click The Mouse To Continue.
IUI_READ_DUPSDR("HEY")

144

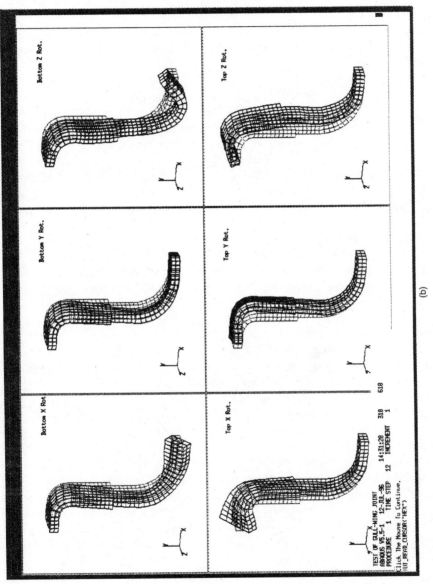

(b)

Figure 4.47 The predicted twelve fundamental deformation modes of the gull-wing lead-frame: (*a*) six translational modes; (*b*) six rotational modes.

145

The material database developed and implemented is intended to support the internal functions for mechanical design. The database is divided into four major categories: *Die, Interconnect, Substrate,* and *Packaging* materials. For each material, over 40 thermal and mechanical properties are listed, such as melting temperature, density, elastic modulus, strength, hardness, fracture toughness, creep and plasticity parameters, CTE, thermal conductivity, and so forth. Figure 4.48*a* shows the material database interface, the left column gives the list of materials, the center column gives the types of properties, and the right column gives the values of the selected properties. A unique feature of this database is that any property can be specified as a function of other variables. This feature is essential for accuracy since many properties are temperature-dependent. A screen plot of Young's modulus versus temperature for three different solder alloys is shown in Fig. 4.48*b*. Moreover, a reference ID is assigned to each property to document the source of that property so that it can be easily tracked and updated. Hundreds of documents, papers, books, and reports have been reviewed to identify appropriate values for those material properties. Over 4500 mechanical and thermal properties of approximately 300 materials commonly used in electronics packaging have been implemented in the current system. The database can be used either in connection with the previously mentioned functions or as a guideline for material comparison and selection.

The future development of the system will be driven by users' needs for specific packaging designs or capabilities, the trend of packaging technology, and reliability requirements. Examples are automated, parametric FE modeling of packaging components, failure-mode analysis via neural network, wire bond design, MCM, BGA, PTH, thin and thick films reliability, vibration/thermal-coupled effect on solder joint fatigue life, reflow simulation, prediction of solder joint shape, design of accelerated thermal cycling tests, and manufacturability assessment. In addition, the current functions will be continuously improved, and the material database and failure data will be updated on a regular basis. Other methodologies, such as DOE (design of experiment), fracture mechanics–based fatigue models, and improved material-modeling capabilities of PCB in vibration analysis will also be incorporated. Although the system is designed for more general purpose applications, its modular structure of knowledge sources is flexible enough to be tailored for analyzing any specific interconnect configurations under prescribed environmental and loading conditions.

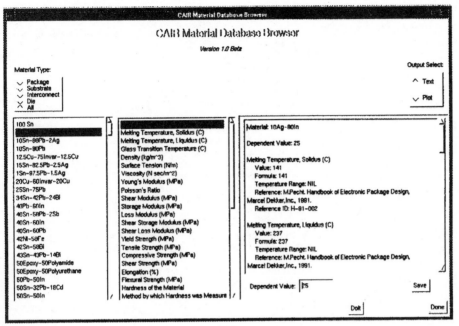

(a)

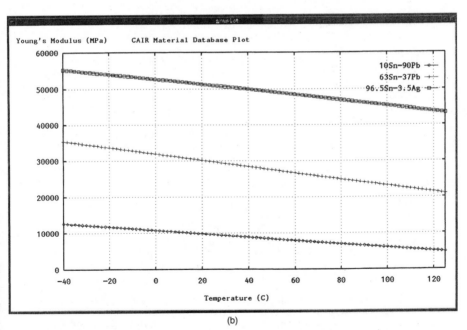

(b)

Figure 4.48 Material database: (*a*) menu display showing the materials (left column), types of properties (center column), and values of the selected properties (right column); (*b*) screen plot of Young's modulus versus temperature for three different solders.

References

1. ABAQUS, Hibbit, Karlsson & Sorensen, Inc., Pawtucket, Rhode Island 02860, USA.
2. Adhihetty, I. S., and R. P. Padmanabhan, "The Role of Thermal Analysis in Package Materials Characterization," *Proceedings of INTERpack'95, ASME Advances in Electronic Packaging,* **2**:695–705, 1995.
3. Allenby, B. R. et al., "An Assessment of the Use of Lead in Electronics Assembly— Part I," *Circuit World,* **19**(2):18–24, 1993.
4. Allenby, B. R. et al., "An Assessment of the Use of Lead in Electronics Assembly— Part II," *Circuit World,* **19**(3):25–31, 1993.
5. *Metals Reference Book,* 2d ed., ASM International, 1983.
6. *Electronic Materials Handbook,* vol. 1. ASM International, 1989.
7. *ASM Metals Handbook,* 10th ed., **2**:1167, 1990.
8. Bar-Cohen, A., and D. E. Mix, "Thermal Modeling of Plastic IC Packages, Mechanical Behavior of Materials and Structures in Microelectronics," *Material Research Society Symp. Proceedings,* **226**:80, 1991.
9. Busso, E. P., M. Kitano, and T. Kumazawa, "A Visco-Plastic Constitutive Model for 60/40 Tin-Lead Solder Used in IC Package Joints," *Journal of Engineering Materials and Technology,* **114**:331–337, 1992.
10. Chanchani, R., and P. M. Hall, "Temperature Dependence of Thermal Expansion of Ceramics and Metals for Electronic Packages," *IEEE Transactions on Components, Hybrids, and Manufacturing Technology,* **13**(4):743–750, 1990.
11. Charles, H. K., "Electronic Materials and Structures for Multichip Modules," *Journal of Electronic Packaging,* Trans. of ASME, **114**:226–231, June 1992.
12. Clough, R. B., A. J. Shapiro, A. J. Bayba, and G. K. Lucey Jr., "Boundary Layer Fracture in Composite Solder Joints," *Journal of Electronic Packaging,* Trans. of ASME, **117**:270–274, December 1995.
13. Cole, M., and T. Caulfield, "Constant Strain Rate Tensile Properties of Various Lead Based Solder Alloys at 0, 50, 100°C," *Scripta Metallurgica et Materialia,* **27**:903–908, 1992.
14. Coors Ceramics Company, Grand Junction, CO 81505.
15. *CRC Handbook of Tables for Applied Engineering Science,* 2d ed., Chemical Rubber Co., 1973, p. 331.
16. Dally, J. W., *Packaging of Electronic Systems,* McGraw-Hill, New York, 1990.
17. Darveaux, R., and K. Banerji, "Constitutive Relations for Tin-Based Solder Joints," *Proc. 42nd Electronic Components Technology Conference,* San Diego, Calif., May 18–20, 1992, pp. 538–551.
18. Engelmaier, W., "Functional Cycles and Surface Mounting Attachment Reliability," *Proceedings,* ISHM, Silver Spring, Md., Oct. 1984, pp. 87–114.
19. Engelmaier, W., and A. I. Attarwala, "Surface-Mount Attachment Reliability of Clip-Leaded Ceramic Chip Carriers on FR-4 Circuit Boards," *IEEE Trans. Comp. Hybrids, Manuf. Technol.,* **12**(2):284–296, 1989.
20. Fields, R. J., S. J. Low III, and G. K. Lucey Jr., *The Metal Science of Joining,* M. J. Cieslak, J. H. Perepezko, S. Kang, and M. E. Glicksman, eds., TMS, Warrendale, Pa., 1992, p. 165.
21. Frear, D. R., W. B. Jones, and K. R. Kinsman, *Solder Mechanics—A State of the Art Assessment,* TMS EMPMD Monograph Series, 1991.
22. Frear, D. R., and P. T. Vianco, "Intermetallic Growth and Mechanical Behavior of Low and High Melting Temperature Solder Alloys," *Metallurgical and Materials Transactions A,* **25A**:1509–1523, July 1994.
23. Freer Goldstein, J. L., and J. W. Morris Jr., "Microstructure Development of Eutectic Bi-Sn and Eutectic In-Sn During High Temperature Deformation," *Journal of Electronics Materials,* **23**(5):477–486, 1994.
24. Frost, H. J., and M. F. Ashby, *Deformation—Mechanism Maps,* Pergamon Press, New York, 1982.
25. Glazer, J., "Microstructure and Mechanical Properties of Pb-Free Solder Alloys for Low-Cost Electronic Assembly: A Review," *123rd TMS Annual Meeting,* San Francisco, February 28–March 4, 1994.

26. Goodelle, J. P., R. A. Pearson, and T. Y. Wu, "Interlaminar Fracture Toughness of a Glass-Filled FR-4 Epoxy Composite as a Function of Mode-Mixity," *Application of Fracture Mechanics in Electronic Packaging and Materials,* ASME, EEP-Vol. 11/MD-Vol. **64:**163–169, 1995.

27. Guo, Z., P. Hacke, A. F. Sprecher, and H. Conrad, "Effect of Composition on the Low-Cycle Fatigue of Pb Alloy Solder Joints," *Proceedings of 40th Electronic Components & Technology Conference,* 1990, p. 496.

28. Guo, Z., A. Sprecher Jr., H. Conrad, and M. Kim, "Monotonic Properties and Low-Cycle Fatigue of Several Soft Solder Alloy Systems," *Materials Developments in Microelectronic Packaging Conference Proceedings,* Montreal, Quebec, Canada, August 19–22, 1991, pp. 155–162.

29. Guo, Z., Y.-H. Pao, and H. Conrad, "Plastic Deformation Kinetics of 95.5Sn4Cu0.5Ag Solder Joints," ASME Transactions, *Journal of Electronics Packaging,* **117:**100–104, 1995.

30. Hacke, P., A. F. Sprecher, and H. Conrad, "Computer Simulation of Thermo-mechanical Fatigue of Solder Joints Including Microstructure Coarsening," presented at *ASME WAM,* Anaheim, Nov. 8–13, 1992.

31. Hacke, P. L., A. F. Sprecher, and H. Conrad, "Thermo-Mechanical Fatigue of 63Sn-37Pb Solder Joints," *Thermal Stress and Strain in Microelectronic Packaging,* J. Lau, ed., Van Nostrand Reinhold, New York, 1993.

32. Hall, P. M., "Force, Moments, and Displacements During Thermal Chamber Cycling of Leadless Ceramic Chip Carriers Surface Mounted to Printed Wiring Boards," *IEEE Trans. Comp. Hybrids, Manuf. Technol.,* **7**(4):314–327, 1984.

33. Harada, M., and R. Satoh, "Mechanical Characteristics of 96.5Sn/3.5Ag Solder in Micro-Bonding," *Proc. 40th Electronic Components Technology Conference,* May 20–23, 1990, pp. 510–517.

34. Hu, J. M., "Knowledge-Based Qualification Tests for Electronic Components in Harsh Environments," *Quality and Reliability Engineering International,* **10:**377–390, 1994.

35. ITRI (International Tin Research Institute), *Solder Alloy Data,* Publication # 656, 1986.

36. Kashyap, B. P., and G. S. Murty, "Experimental Constitutive Relations for the High Temperature Deformation of a Pb-Sn Eutectic Alloy," *Mater. Sci. and Eng.,* **50:** 205–213, 1981.

37. Kitano, M., T. Kumazawa, and S. Kawal, "An Evaluation Method for Thermal Fatigue Strength of Solder Joint," ASME *Advances in Electronic Packaging,* **1:**301–308, 1992.

38. Knecht, S., and L. Fox, "Integrated Matrix Creep: Application to Accelerated Testing and Lifetime Prediction," *Solder Joint Reliability: Theory and Application,* J. H. Lau, ed., Van Nostrand Reinhold, New York, 1991, pp. 508–544.

39. Krishna, K. K., *Mechanical Behavior of Thin Layers of Solder Loaded in Shear,* M.S. Thesis, The Ohio State University, 1993.

40. Kuramoto, N., H. Taniguchi, and I. Aso, "Translucent ALN Ceramic Substrate," *IEEE Transactions on CHMT,* **9**(4):386–390, 1986.

41. Lau, J. H., and D. W. Rice, "Solder Joint Fatigue in Surface Mount Technology: State of the Art," *Solid State Technology,* **28:**91–104, October 1985.

42. Lau, J. H., *Solder Joint Reliability—Theory and Applications,* Van Nostrand Reinhold, New York, 1991.

43. Lavery, P. R., *An Investigation of the Thermal and Mechanical Properties of High Lead-Low Tin Alloy Solders,* M.S. Thesis, Dartmouth College, Hanover, N.H., 1987.

44. Lee, N.-C., M. Xiao, K. J. Lawless, J. A. Slattery, and J. R. Sovinsky, "Solder Paste for Tomorrow's Electronics Manufacturing Technology," *Journal of Electronics Manufacturing,* **4:**181–202, 1994.

45. Liljestrand, L.-G., and L.-O. Andersson, "Thermal Fatigue Cycling of Surface Mounted PWB Assemblies in Telecom Equipment," *Circuit World,* **14**(3):69–73, 1988.

46. Logsdon, W. A., P. K. Liaw, and M. A. Burke, "Fracture Behavior of 63Sn-37Pb Solder," *Engineering Fracture Mechanics,* **36**(2):183–218, 1990.

47. Mahidhara, R. K., S. M. L. Sastry, I. Turlik, and K. L. Murty, "Deformation and Fracture Behavior of Sn-5%Sb Solder," *Scripta Metallurgica et Materialia*, **31**(9): 1145–1150, 1994.
48. *Machine Design,* Materials Reference Issue, Penton/IPC, April 1984.
49. McClintock, F. A., and A. S. Argon, *Mechanical Behavior of Materials,* Addison-Wesley, N.Y., 1966.
50. McDowell, D. L., M. P. Miller, and D. C. Brooks, "A Unified Creep-Plasticity Theory for Solder Alloys," *Fatigue of Electronic Materials,* ASTM 1153, S. A. Schroeder and M. R. Mitchell, eds., pp. 42–59.
51. Mei, Z., and J. W. Morris, Jr., "Superplastic Creep of Low Melting Point Solder Joints," *Journal of Electronic Materials,* **21**(4):401–407, 1992.
52. Mei, Z., and J. W. Morris Jr., "Characterization of Eutectic Sn-Bi Solder Joints," *Journal of Electronic Materials,* **21**(6):599–607, 1992.
53. Medla, A. A., *Low Temperature Mechanical Properties of Lead-Tin Solder Alloys,* M.S. Thesis, Dartmouth College, Hanover, N.H., 1990.
54. Morgan, H. S., "Thermal Stresses in Layered Electrical Assemblies Bonded with Solder," *Journal of Electronic Packaging,* Trans. of ASME, **113**:350–354, June 1991.
55. Murty, K. L., and L. Turlik, "Deformation Mechanisms in PbSn Alloys: Application to Solder Reliability in Electronic Packaging," *Advances in Electronic Packaging,* ASME, EEP-Vol. 1-1, **1**:309–318, 1992.
56. Pao, Y.-H., K.-L. Chen, and A.-Y. Kuo, "A Nonlinear and Time Dependent Finite Element Analysis of Solder Joints in Surface Mounted Components Under Thermal Cycling," *Mechanical Behavior of Materials and Structures in Micro-electronics,* Material Research Society Symp. Proceed., **226**:23–28, 1991.
57. Pao, Y.-H., S. Badgley, R. Govila, L. Baumgartner, R. Allor, and R. Cooper, "Measurements of Mechanical Behavior of High Lead Lead-Tin Solder Joints Subjected to Thermal Cycling," *Journal of Electronic Packaging,* Trans. of ASME, **114**:135–145, 1992.
58. Pao, Y.-H., R. Govila, and S. Badgley, "Thermal Fatigue Fracture of 90Pb/10Sn Solder Joints," Proceedings of the 1992 Joint ASME/JSME Conference on Electronic Packaging, *Advances in Electronic Packaging 1992,* pp. 291–300.
59. Pao, Y.-H., E. Jih, B. E. Artz, and L. W. Cathey, "A Note on the Implementation of Temperature Dependent Coefficient of Thermal Expansion (CTE) in ABAQUS," ASME Trans., *Journal of Electronic Packaging,* **114**(4):470–472, 1992.
60. Pao, Y.-H., "A Fracture Mechanics Approach to Thermal Fatigue Life Prediction of Solder Joints," *IEEE Transaction on Components, Hybrids, & Manufacturing Technology,* **15**(4):559–570, August 1992.
61. Pao, Y.-H., R. Govila, S. Badgley, and E. Jih, "An Experimental and Finite Element Study of Thermal Fatigue Fracture of PbSn Solder Joints," ASME Trans., *Journal of Electronic Packaging,* **115**(1):1–8, 1993.
62. Pao, Y.-H., S. Badgley, E. Jih, R. Govila, and J. Browning, "Constitutive Behavior and Low Cycle Thermal Fatigue of 97Sn-3Cu Solder Joints," ASME Trans., *Journal of Electronic Packaging,* **115**(2):147–152, 1993.
63. Pao, Y.-H., S. Badgley, R. Govila, and E. Jih, "An Experimental and Modeling Study of Thermal Cyclic Behavior of Sn-Cu and Sn-Pb Solder Joints," MRS Symposium Proceedings, *Electronic Packaging Materials Science VII,* P. Borgesen et al., eds., **323**:153–158, 1994.
64. Pao, Y.-H., E. Jih, and V. Reddy, "Thermal Reliability Prediction of Automotive Electronic Packaging," SAE International Congress and Exposition, Detroit, February 27–March 2, 1995.
65. Pao, Y.-H., S. Badgley, R. Govila, and E. Jih, to be published in ASME Transactions, *Journal of Electronic Packaging,* 1996.
66. Pecht, M., *Handbook of Electronic Package Design,* Marcel Dekker, Inc., New York, 1991.
67. Pecht, M., *Integrated Circuit, and Multichip Module Package Design Guidelines—A Focus on Reliability,* John Wiley & Sons, New York, 1994.
68. Pfeil, M. C., V. H. Kenner, and C. H. Popelar, "Fracture Toughness of Filled Epoxy Molding Compounds," *Application of Fracture Mechanics in Electronic Packaging and Materials,* ASME, EEP-Vol. 11/MD-Vol. **64**:263–270, 1995.

69. Pratt, R. E., E. I. Stromswold, and D. J. Quesnel, "Model I Fracture Toughness Testing of Eutectic Sn-Pb Solder Joints," *Journal of Electronic Materials,* **23**(4):375–381, 1993.
70. Raeder, C. H., *Partially Constrained Thermomechanical Fatigue of Eutectic Tin-Bismuth/Copper Solder Joints,* Ph.D. Dissertation, Rensselaer Polytechnic Institute, Troy, New York, 1995.
71. Satoh, R., Ph.D. Dissertation, Hokkaido University (in Japanese), 1987.
72. Satoh, R., K. Arakawa, M. Harada, and K. Matsui, "Thermal Fatigue Life of Pb-Sn Alloy Interconnections," *IEEE Trans. CHMT,* **14**(1):224–232, 1991.
73. Sauber, J., L. Lee, S. Hsu, and Hongsmatip, "Fracture Properties of Molding Compound Materials for IC Plastic Packaging," *Proceedings of 44th ECTC,* Washington, D.C., May 1–4, 1994, pp. 164–170.
74. Schroeder, S. A., and M. R. Mitchell, *Fatigue of Electronic Materials,* ASTM STP 1153, 1994.
75. Seyyedi, J., "Thermal Fatigue Behavior of Low Melting Point Solder Joints," *Soldering & Surface Mount Technology,* **13**:25–32, February 1993.
76. Sherry, W. M., and P. M. Hall, "Materials, Structures, and Mechanics of Solder Joints for Surface Mount Microelectronics," *Proceedings, Third International Conference on Interconnection Technology in Electronics,* Fellbach, Germany, February 18–20, 1986, pp. 47–81.
77. Solomon, H. D., "Creep, Strain Rate Sensitivity and Low Cycle Fatigue of 60/40 Solder," *Brazing and Soldering,* **11**:68–75, 1986.
78. Stoffel, N., C. Zhang, and E. J. Kramer, "Adhesion of Polyimide Laminates," *Application of Fracture Mechanics in Electronic Packaging and Materials,* ASME, EEP-Vol. 11/MD-Vol. 64:79–84, 1995.
79. Stromswold, E. I., R. E. Pratt, and D. J. Quesnel, "The Effect of Substrate Surface Roughness on the Fracture Toughness of Cu/96.5Sn-3.5Ag Solder Joints," *Journal of Electronic Materials,* **23**(10):375–381, 1994.
80. Subrahamanyan, R., J. R. Wilcox, and C.-Y. Li, "A Damage Integral Approach to Thermal Fatigue of Solder Joints," *IEEE Trans.,* CHMT, **12**:480–491, 1989.
81. Swada, K., et al., "Simplified and Practical Estimation of Package Cracking during Reflow Soldering Process," *IEEE/IRPS,* 1994, pp. 114–119.
82. Tummala, R. R., and E. J. Rymaszewski, *Microelectronic Packaging Handbook,* Van Nostrand Reinhold, New York, 1989.
83. Tomlinson, W. J., and A. Fullylove, "Strength of Tin-Based Soldered Joints," *Journal of Materials Science,* **27**:5777–5782, 1992.
84. Touloukian, Y. S., et al., "Thermal Conductivity Metallic Elements and Alloys," *Thermophysical Properties of Matter,* vol. 1, IFI/PLENUM, 1970.
85. Touloukian, Y. S., et al., "Thermal Conductivity Nonmetallic Solids," *Thermophysical Properties of Matter,* vol. 2, IFI/PLENUM, 1970.
86. Touloukian, Y. S., et al., "Thermal Expansion Metallic Elements and Alloys," *Thermophysical Properties of Matter,* vol. 12, IFI/PLENUM, 1975.
87. Touloukian, Y. S., et al., "Thermal Expansion Nonmetallic Solids," *Thermophysical Properties of Matter,* vol. 13, IFI/PLENUM, 1977.
88. Tummala, R. R., and E. J. Rymaszewski, *Microelectronics Packaging Handbook,* Van Nostrand Reinhold, New York, 1989.
89. Uegai, Y., S. Tani, A. Inoue, S. Yoshioka, and K. Tamura, "A Method of Fatigue Life Prediction for Surface-Mount Solder Joints of Electronic Devices by Mechanical Fatigue Test," ASME *Advances in Electronic Packaging,* **1**:493–498, 1993.
90. Vaynman, S., "Effect of Strain Rate on Fatigue of Low-Tin Lead-Base Solder," *IEEE Trans. Comp. Hybrids, Manuf. Technol.,* **12**(4):469–472, 1989.
91. Watson, K. A., and K. M. Liechti, "Adhesion Measurements of Printed Wiring Board Assemblies," *Application of Experimental Mechanics to Electronic Packaging,* ASME, EEP-Vol. 13/MD-Vol. 214:25–29, 1995.
92. Wild, R. N., "Properties of Some Low Melt Fusible Solder Alloys," *Proceedings of Internepcon,* 1971, pp. 81–92.
93. Winterbottom, W. L., "Electronic Packaging Design: A Mechanistic Approach for Harsh Environment," EEP-vol. 10-1, *Advances in Electronic Packaging,* ASME, pp. 341–343, 1995.

94. Wong, B., D. E. Helling, and R. W. Clark, "A Creep-Rupture Model for Two-Phase Eutectic Solders," *IEEE Trans. CHMT,* **11:**284–290, September 1988.
95. Wu, T. Y., and G. H. Thiel, "Fracture Toughness of Flip-Chip Encapsulants," *Application of Fracture Mechanics in Electronic Packaging and Materials,* ASME, EEP-Vol. 11/MD-Vol. 64:205–210, 1995.
96. Yamada, S. E., "A Fracture Mechanics Approach to Soldered Joint Cracking," *IEEE Transactions on Components, Hybrids, and Manufacturing Technology,* **12**(1): 99–104, 1989.
97. Lau, J. H., "Bending and Twisting of 63Sn37P$_b$ Solder Interconnects with Creep," *ASME Transactions of ASME, Journal of Electronic Packaging,* pp. 154–157, June 1994.
98. Lau, J. H., "Creep of Solder Interconnections Under Combined Loads," *IEEE Transactions on Components, Hybrids, and Manufacturing Technology,* pp. 794–798, December 1993.
99. Lau, J. H., "Creep of 96.5Sn 3.5Ag Solder Interconnects," *Soldering & Surface Mount Technology,* **15:**45–49, September 1993.

Solder Joint Reliability of BGA Assemblies

As the trend toward higher input/output, higher performance, and higher board manufacturing yield continues, ball grid array (BGA) becomes the package of choice.[1-141] In this chapter the solder joint reliability of the most common BGA (i.e., CBGA, PBGA, DBGA, and TBGA) assemblies under thermal, mechanical, and vibrational conditions will be discussed. Also, a no-clean mass reflow assembly process of PBGAs is provided.

5.1 Thermal Reliability of CBGA Assemblies

The solder joint reliability of three commonly available ceramic BGAs (CBGA, CCGA, and DBGA) on FR-4 PCB are presented in this section.

5.1.1 Solder joint reliability of IBM's CBGA assemblies

Figure 5.1 schematically shows the cross sections of IBM's ceramic BGAs on a FR-4 PCB. The one on the right is IBM's ceramic ball grid array (CBGA). The two on the left are IBM's ceramic column grid array (CCGA). In all cases, the solder ball (or column) pitch was 1.27 mm. In order to prevent the solder balls (or columns) from collapse, i.e., to maintain the standoff, and to reflow at lower temperatures, i.e., to be SMT compatible, two different solder interconnects were used. They were the 90wt%Pb/10wt%Sn solder balls (or columns) with a melting point higher than 300°C, and the 63wt%Sn/37wt%Pb solder with a melting point of 183°C. The solder ball's diameter was 35 mils or 0.9 mm and the solder column's diameter was 20 mils or 0.5 mm.

The solder joint life distribution of different sized CBGAs on FR-4 PCB has been reported extensively.[1-5] It has been shown by IBM that

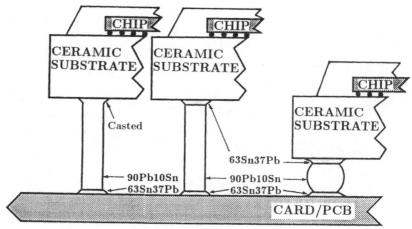

Figure 5.1   IBM's ceramic ball grid array (CBGA).

for a 25 × 25 mm CBGA assembly the solder joints are reliable for use under most operating conditions.

A typical thermal cycling failure mode of the CBGA solder joint is shown in Fig. 5.2. It can be seen that cracks occurred in the 63wt%Sn/37wt%Pb solder and not in the 90wt%Pb/10wt%Sn solder ball. Also, SEM of many failed samples by IBM shows that the fully cracked solder joints are on the PCB side.

Figure 5.2   Thermal cycling failure mode of IBM's CBGA.[9]

A detailed finite element analysis has been performed by Corhin[58] of IBM (Fig. 5.3). It can be seen that the maximum plastic strain occurred at the eutectic solder joints near the FR-4 PCB and the ceramic carrier. For this case, the plastic strain in the solder joint on the PCB side was larger than that on the carrier side. This confirmed the experimental results very well.

5.1.2 Solder joint reliability of IBM's CCGA assemblies

The length of the solder column (Fig. 5.1) varied from 50 mils (1.27 mm) to 87 mils (2.2 mm), or even longer. From a solder joint reliability point of view, the taller the better, because that increases the compli-

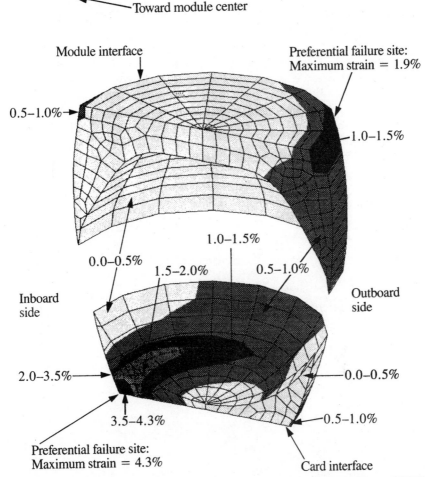

Figure 5.3 Equivalent plastic strain distribution in the eutectic solder portions of IBM's CBGA.[58]

ance of the structure and reduces the stress and strain in the interconnects. However, from an electrical performance point of view, the shorter the better, because that reduces the signal path and consequently lowers the inductance. In this section, we will look at IBM's results for a 32×32 mm CCGA with 2.2-mm-tall solder columns as well as 1.27-mm-tall solder columns.

Figure 5.4 shows some of the daisy-chain designs of some of the columns with similar distance from the neutral point (DNP) of the 25×25 (625-column) cofired, 19-layer ceramic substrate.[1] The substrate dimensions were $32 \times 32 \times 2.9$ mm. An aluminum cap was attached to the package using a nonhermetic seal of Sylgard 577.[1] The interior gap between the substrate and cap was filled with thermal grease. The 87-mil CCGA carrier was assembled on a FR-4 PCB with a thickness of 1.83 mm. The solder volume per pad was 7700 cubic mils. The assemblies were subjected to thermal cycling with a condition: 0 to 100°C, three cycles per hour.

The life (lognormal) distribution of the 32×32 mm CCGA (87-mil column) solder joints is shown in Fig. 5.5. It can be seen that the 50 percent failure point occurred at about 5011 cycles, which is very reliable, for most of the applications. By using the simple shear theory and the Coffin-Manson low-cycle thermal fatigue concept, the life distribution of the 32×32 mm (50-mil column) solder joints can be predicted with a deacceleration factor of 87/50, or 2.9, which is also shown in Fig. 5.5. The 50 percent failure point for the 50-mil solder columns occurred at about 1720 cycles. Again, this is reliable for most of the computer and related applications. A typical failure mode of the CCGA is shown in Fig. 5.6. It can be seen that for CCGA assemblies, the solder failure occurred at the

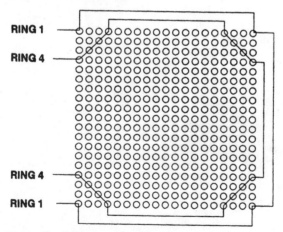

Figure 5.4 Schematic of the daisy-chain design of the columns of similar DNP of the 625-pin CCGA.[1]

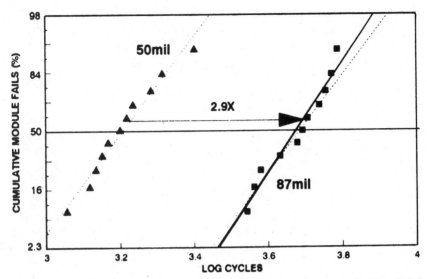

Figure 5.5 Life (lognormal) distribution of the 87-mil solder column tested at 0 to 100°C, three cycles per hour. The life distribution of the 50-mil solder column is obtained with the Coffin-Manson equation.[1]

Figure 5.6 Failure mode of the CCGA assembly. (It fails at the high-lead column.)[12]

high-lead solder. This is because of the thermal expansion mismatch between the ceramics carrier and the FR-4 substrate. This mismatch produces a relative displacement (and consequently a bending moment) of the solder column which is the driving force for solder failure.

5.1.3 Solder joint reliability of Kyocera's DBGA assemblies

Kyocera's DBGA (dimple ball grid array) is quite different from IBM's CBGA. The DBGA uses only one kind of solder[88,89] (Sn/Pb/Bi with a melting point of approximately 183°C) to make reliable solder balls for SMT-compatible PCB assemblies. This solder has a better thermal fatigue life than the 63wt%Sn/37wt%Pb solder. Also, the DBGA adds an extra layer of green sheet (with 0.55-mm diameter, 0.2-mm deep holes in the tape) to the substrate. This cylindrical hole, which is called a dimple, is screen-printed with solder paste, and the solder ball sits inside of the dimple. The DBGA-ball-attach process flow is shown in Fig. 5.7. The cross section of the solder ball is shown in Figs. 5.8a and 5.8b. It can be seen that the interface area is enclosed with an alumina ceramic wall. The solder around the interface area is therefore difficult to deform.

Figures 5.9a and 5.9b show the life (Weibull) distribution of the solder joints of different DBGA PCB assemblies tested at different temperature cycling conditions.[88,89] In Fig. 5.9a, the life distribution of a 35 × 35 mm DBGA solder joint subjected to a temperature cycling condition of 0 to 100°C with a 20-minute dwell time and one cycle per hour,

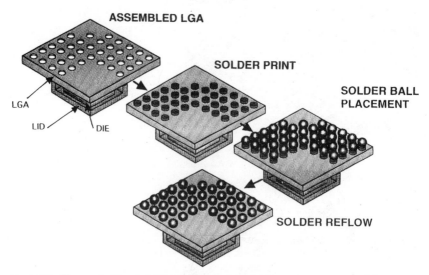

Figure 5.7 Kyocera's Dimpled BGA (DBGA) solder-ball-attach process flow.[88,89]

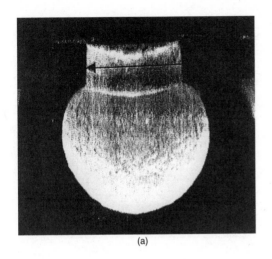

(a)

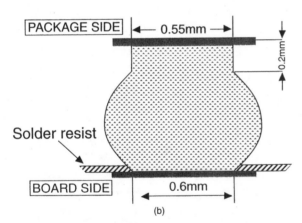

(b)

Figure 5.8 (*a*) Kyocera's Dimpled BGA (DBGA). (*b*) Kyocera's optimized Dimpled BGA (DBGA).[88,89]

is presented. It can be seen that the 50 percent failure point occurred at about 1800 cycles which is reliable for most of the computer and related applications. The failure mode is at the solder joint near the interface of the ceramic carrier (Fig. 5.10). Also Fig. 5.9*a* shows that there is no solder joint failure for the 21 × 21 mm DBGA at 2500 cycles or for the 25 × 25 mm DBGA at 2000 cycles.

In Fig. 5.9*b*, the life distribution of 21 × 21 mm and 35 × 35 mm DBGA solder joints subjected to a temperature cycling condition of −40 to 125°C, one cycle per hour with a 20-minute dwell time, is presented. Also, the results of a conventional CBGA with the same carrier sizes are presented. It can be seen that, for both carrier sizes, the sol-

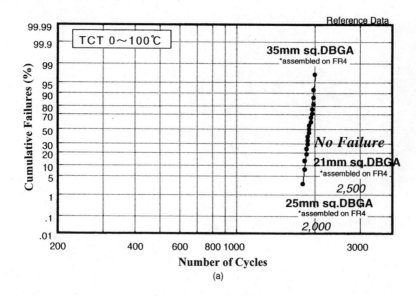

(a)

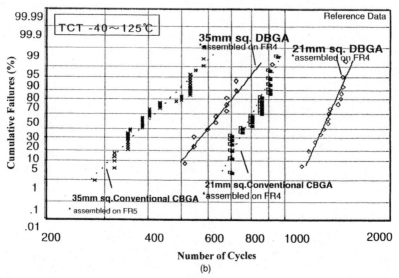

(b)

Figure 5.9 (*a*) Thermal cycling results of DBGA (0 to 100°C, one cycle per hour). (*b*) Thermal cycling results of DBGA and CBGA (−40 to 125°C, one cycle per hour).[88,89]

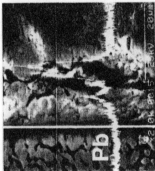

Figure 5.10   Thermal cycling failure model of the
DBGA.[88,89]

der joint reliability of the DBGA is better than that of the conventional
CBGA, even though the CBGA's solder joints are reliable for most of
the computer applications.

5.2   Thermal Reliability of PBGA Assemblies

There is a fundamental difference between the well-known PQFP
(plastic quad flat pack) and the plastic ball grid array (PBGA) packages. The PQFP has a leadframe, which is formed into a gull-wing
shape, and when the PQFP is assembled on the PCB, this gull-wing
lead provides the compliance and reduces the stress and strain in the
solder joint. Thus the solder joint of the PQFP is very reliable for most
applications. The PBGA does not have a leadframe but has an organic
substrate, and the interconnection to the PCB is through solder balls
only. Consequently, the compliance is very small and the solder joint
reliability of PBGAs is under scrutiny.

In this section, the solder joint reliability of different plastic ball grid
array (PBGA) packages on FR-4 PCBs under thermal, mechanical, and

vibrational conditions will be discussed. Both theoretical and experimental results will be presented.

5.2.1 Motorola's solder joint reliability data and analysis

Figure 5.11 shows a conventional full-matrix PBGA and its major components.[65] Three different PBGA configurations and four different die sizes (Table 5.1) were tested at two different temperature cycling conditions: (1) 0 to 100°C, two cycles per hour, with a 5-minute dwell time and 10-minute ramp-up and ramp-down, and (2) −40 to 125°C, one cycle per hour, with a 15-minute dwell time and 15-minute ramp time.

Figure 5.12 shows the solder joint life (Weibull) distributions of the PBGAs on FR-4 PCB with a solder-mask-defined (SMD) copper pad and a non-SMD (NSMD) copper pad (Fig. 5.13) tested at the cycling condition of 0 to 100°C. For the 15 × 15 = 225-pin PBGA, the thermal fatigue reliability of the SMD solder joint is slightly better than that of the NSMD. This is because the standoff height of the SMD solder joint is a little taller than that of the NSMD solder joint. It is noted that the copper pad on the PCB with the SMD solder joint is larger than that with the NSMD. This could introduce PCB routing problems.

The effect of PBGA package size is also shown in Fig. 5.12. It can be seen that the life distribution of the 225-pin (27 × 27 mm body size and 1.5-mm pitch) is not as good as that of the 19 × 19 = 361-pin (25 × 25 mm body size and 1.27-mm pitch). This is because the 361-pin PBGA's body size is smaller than that of the 225-pin PBGA.

Figure 5.14 shows the SMD solder joint life distributions of the 361-pin PBGA with different die sizes and test conditions. It can be seen that the smaller the silicon die (11.4 mm versus 13.5 mm for 0 to 100°C) the better the solder joint reliability. This is due to the local thermal expansion mismatch between the silicon die (~2.5 ppm/°C) and the bismaleimide triazine (BT) substrate (~15 ppm/°C). Nonlinear finite element analysis by Lee and Lau[112] verified these results. The effect of the temperature cycling range (13.5-mm die) is also shown in

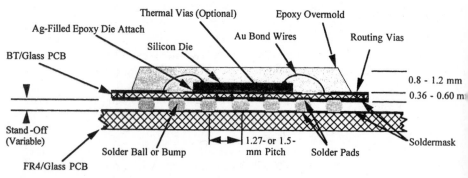

Figure 5.11   Cross section of a Motorola PBGA assembly.[65]

TABLE 5.1 Configuration of the Three PBGA Package Types Used in Motorola's Study (All Dimensions Are in mm)[65]

# of Pins	Pitch	Array	BT board size	Die size	Molded body thickness
119	1.27	7 × 17	14 × 22	7.7 × 11.1	0.90
225	1.50	15 × 15	27 × 27	10.2 × 10.2	1.20
361	1.27	19 × 19	25 × 25	11.4 × 11.4 and 13.5 × 13.5	0.80

Fig. 5.14. It can be seen that the smaller the temperature range (0 to 100°C versus −40 to 125°C) the better the solder joint life, since solder is very sensitive to temperature.

Figure 5.15 shows the effect of the thickness of the PBGA BT substrate on the solder joint reliability. It can be seen that under the temperature condition of 0 to 100°C, the life distribution of the thicker BT substrate is much better than that of the thinner BT substrate. This is because the bending deflection of the PBGA of the thicker substrate is much smaller than that of the thinner substrate. (The bending stiffness of the substrate is proportional to the cube of its thickness.) Nonlinear finite element analysis by Lee and Lau[112] confirmed these results.

Figure 5.16 shows the effect of composite solder ball on the 225-pin PBGA solder joint reliability (tested at 0 to 100°C cycling). It can be seen that the thermal fatigue life of the eutectic solder ball is not as good as that of the composite solder joint. This is because the standoff height of the composite solder is taller (Fig. 5.17).

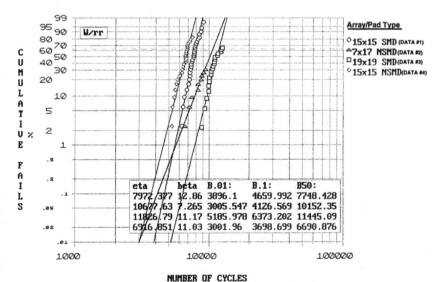

Figure 5.12 Thermal cycling life distribution of a 225-pin solder joint tested at 0 to 100°C, two cycles per hour (effect of SMD, NSMD, and body size). Motorola's results.[65]

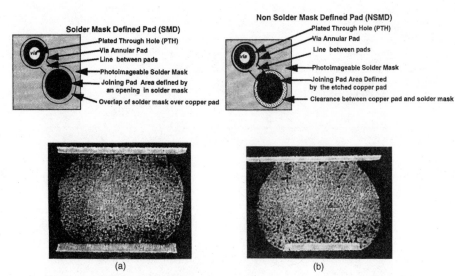

Figure 5.13   (*a*) SMD pads versus (*b*) NSMD pads.

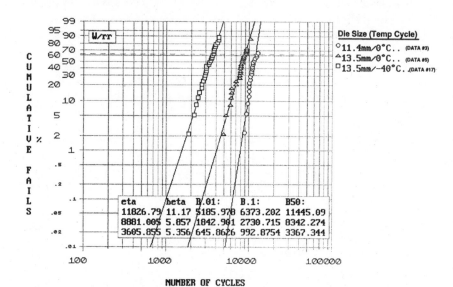

NUMBER OF CYCLES

Figure 5.14   Life distribution of a 361-pin PBGA solder joint (effect of die size and temperature range). Motorola's results.[65]

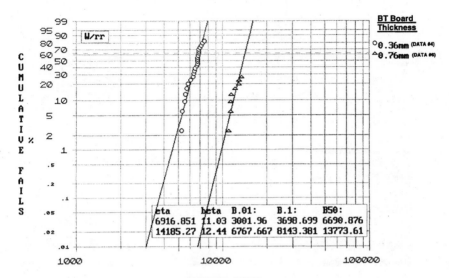

Figure 5.15   Life distribution of a 225-pin PBGA solder joint tested at 0 to 100°C, two cycles per hour (effect of BT thickness). Motorola's results.[65]

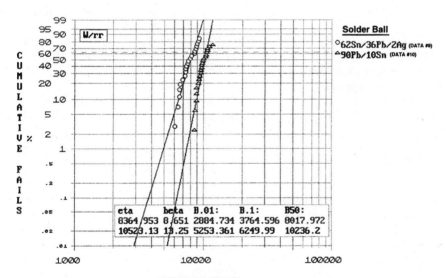

Figure 5.16   Life distribution of a 225-pin PBGA solder joint tested at 0 to 100°C, two cycles per hour (effect of all eutectic bumps versus composite bumps). Motorola's results.[65]

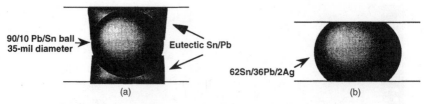

90/10 Pb/Sn ball
35-mil diameter

Eutectic Sn/Pb

62Sn/36Pb/2Ag

(a)

(b)

Figure 5.17   (a) Composite bumps versus (b) all eutectic bumps.

5.2.2   Temperature-dependent elastoplastic analysis of full-matrix PBGA solder joints

Figure 5.18a shows the bottom view of a 1.5-mm pitch 225-pin FM-PBGA (full-matrix PBGA). Figure 5.19 shows the cross section along the diagonal of the FM-PBGA assembly. The package size is 27 × 27 mm and the thickness of the BT substrate is 0.27 mm. The molding compound is 1.3 mm thick. Because of symmetry, only one-half is modeled (Fig. 5.20). There are 2896 plane strain elements and 9224 nodes. The material properties of the FM-PBGA are shown in Table 5.2. It can be seen that all the material properties are assumed to be constant except the eutectic solder, which is assumed to be temperature-dependent. The nonlinear temperature-dependent stress-strain curves are shown in Figs. 5.21a and b. The temperature-dependent Young's modulus is shown in Fig. 5.22.

The temperature condition imposed on the FM-PBGA is shown in Fig. 5.23. For elastoplastic analysis, the temperature input to the finite element program is from −40 to 125°C. There are 10 temperature increments and each increment has about 10 iterations.

The deformation of the FM-PBGA is shown in Fig. 5.24a and b. It can be seen that the maximum deflection occurs at the middle of the package with a value of 0.069 mm. This is due to the thermal expansion mismatch of the molding compound, silicon chip, BT substrate, solder joints, and the FR-4 PCB.

The von Mises stress and equivalent plastic strain range of the solder balls along the diagonal direction is shown in Figs. 5.25 and 5.26. It can

TABLE 5.2   Material Properties for the Full-Matrix and Perimeter-Arrayed PBGA PCB Assemblies[111]

	Young's modulus (GPa)	Poisson's ratio	TCE (ppm/°C)
Chip (Si)	131	0.3	2.8
Molding compound	16	0.25	15
BT substrate	26 (xy)	0.39 (xz, yz)	15 (xy)
	11 (z)	0.11 (xy)	52 (z)
Solder ball	Temperature dependent	0.4	21
FR-4	22 (xy)	0.28 (xz, yz)	18 (xy)
PCB	10 (z)	0.11 (xy)	70 (z)

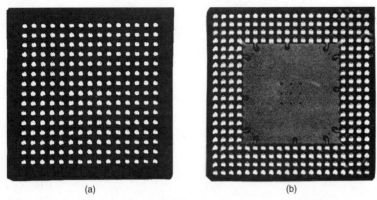

Figure 5.18 (a) 225-pin full-matrix PBGA (1.5-mm pitch), (b) 256-pin perimeter-arrayed PBGA (1.27-mm pitch). Both packages have a 27 × 27 mm body size.[111]

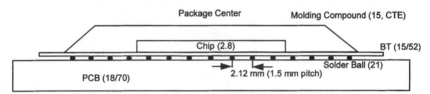

Figure 5.19 Schematic of the full-matrix PBGA along the diagonal direction.[111]

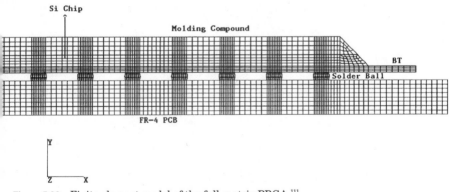

Figure 5.20 Finite element model of the full-matrix PBGA.[111]

be seen that the balls directly beneath the chip edges have the maximum stress (12.92 MPa) and strain range (0.039). This is due to the local thermal expansion mismatch between the chip, molding compound, and BT substrate.[105] The stress and strain at the corner balls are very small because the global thermal mismatch between the BT, molding compound, and the FR-4 PCB is very small and they are far away from the chip. Detailed stress and strain distributions in the maximum solder

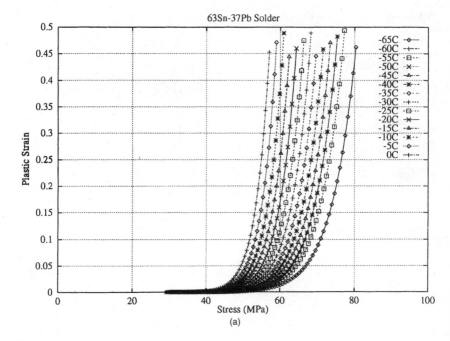

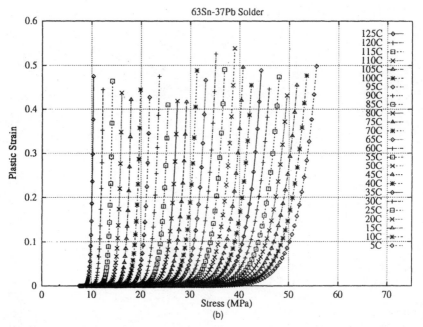

Figure 5.21 (a) Stress-strain curves of the eutectic solder at various temperatures. (b) Stress-strain curves of the eutectic solder at various temperatures.[111]

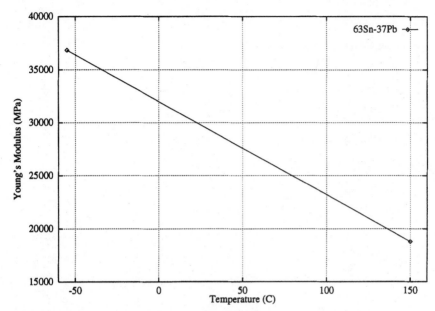

Figure 5.22   Young's modulus of the eutectic solder at various temperatures.[111]

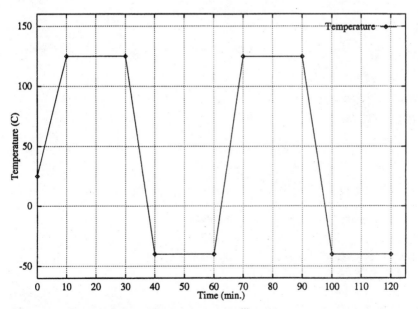

Figure 5.23   Thermal cycling temperature profile.[111]

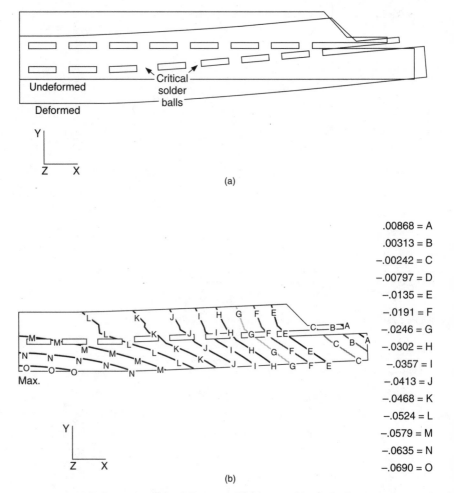

Figure 5.24 (a) Deformation of the full-matrix PBGA assembly. (b) Deflection contours of the full-matrix PBGA assembly.[111]

ball are shown in Figs. 5.27 and 5.28, respectively. The maximum values concentrate at the corners and decrease rapidly in the solder joint.

5.2.3 Temperature-dependent elasto-plastic-creep analysis of full-matrix PBGA solder joints

The Norton's creep relation, Eq. (4.1), described in Sec. 4.1.3 has been used for the elasto-plastic-creep analysis of the 225-pin FM-PBGA solder joints. For eutectic solder, the material constants for the Norton equation are: $B^* = 0.205$ $1/MPa^{5.25}sec$, $H = 0.49$ eV, and $n = 5.25$ (Table 4.2). The temperature-dependent strain-strain curves and Young's

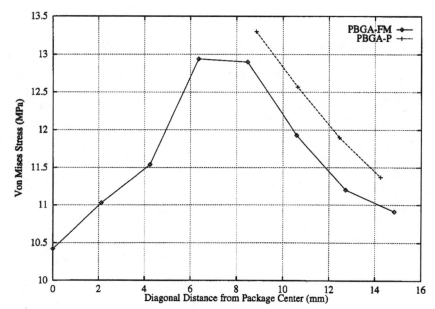

Figure 5.25   von Mises stress distribution along the diagonal solder balls.[111]

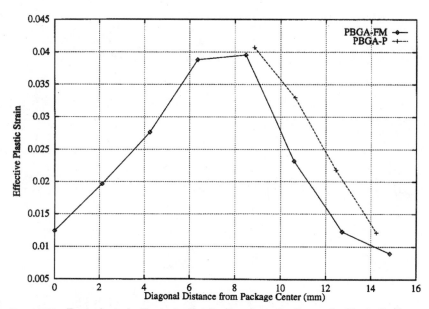

Figure 5.26   Equivalent plastic strain distribution along the diagonal solder balls.[111]

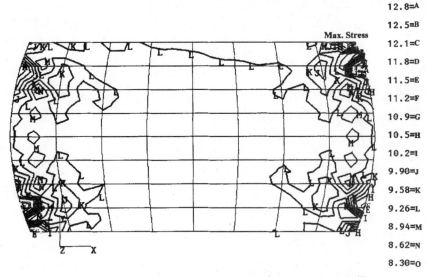

Max. Stress

```
12.8=A
12.5=B
12.1=C
11.8=D
11.5=E
11.2=F
10.9=G
10.5=H
10.2=I
9.90=J
9.58=K
9.26=L
8.94=M
8.62=N
8.30=O
```

Figure 5.27 von Mises stress (MPa) distribution in the critical solder ball.[111]

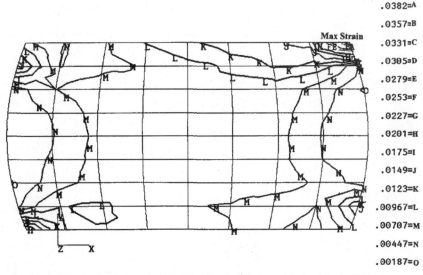

Max Strain

```
.0382=A
.0357=B
.0331=C
.0305=D
.0279=E
.0253=F
.0227=G
.0201=H
.0175=I
.0149=J
.0123=K
.00967=L
.00707=M
.00447=N
.00187=O
```

Figure 5.28 Equivalent plastic strain distribution in the critical solder ball.[111]

modulus for the solder are given in Figs. 5.21 and 5.22. The temperature boundary condition, partially shown in Fig. 5.23, started at 25°C, ramped-up to 125°C at a rate of 16.5°C/min, held at 125°C for 20 minutes, ramped-down to –40°C at a rate of 16.5°C/min, held at –40°C for 20 minutes, ramped-up to 125°C again at a rate of 16.5°C/min, and then repeated the cycling. Two complete cycles were executed.

Similar to the temperature-dependent elastoplastic analysis, the critical solder balls which have the maximum stress and strain hysteresis responses are those right underneath the edges of the chip (Fig. 5.29). The hysteresis loops of the shear stress and shear strain at the inner critical solder ball, i.e., the fourth solder ball from the package center (NODES A, B, C, D, E, F, G, H, and I on the upper interface 1) and at the outer critical solder ball, i.e., the fifth solder ball from the package center (NODES A, B, C, D, E, F, G, H, and I on the lower interface 2) are shown in Figs. 5.30 and 5.31, respectively. It can be seen from Fig. 5.30 that the maximum shear stress and shear strain range are 35 MPa and 3.5 percent, respectively, for NODE I near interface 1 of the fourth solder ball from the package center. Also, from Fig. 5.31, it can be seen that the maximum shear stress and shear strain range are 30 MPa and 3.0 percent, respectively, for NODE A near interface 2 of the fifth solder ball from the package center. These locations are susceptible to crack initiation and propagation. It is interesting to note that the maximum value of the creep shear strain range (3.5 percent) is very close to the maximum value of the plastic shear strain range (3.8 percent) predicted by the temperature-dependent elastoplastic method.

5.2.4 Temperature-dependent elastoplastic analysis of perimeter-arrayed PBGA solder joints

Figures 5.18b and 5.32 show a P-PBGA (perimeter-arrayed PBGA). The only difference between the 225-pin FM-PBGA and the 256-pin P-PBGA is that the 256 solder balls are on a 1.27-mm pitch and are arranged in four rows and columns near the perimeter-arrayed area of the package. The finite element model is shown in Fig. 5.33. There are 2492 plane strain elements and 7978 nodes. The temperature condition is from −40 to 125°C.

The deformation of the P-PBGA assembly is shown in Figs. 5.34 and 5.35. Again, the maximum deflection is at the center of the assembly and is equal to 0.0581 mm. The von Mises stress and equivalent plastic shear strain range along the diagonal direction are shown in (dotted line) Figs. 5.25 and 5.26, respectively. It can be seen that the maximum values occur at the most inner solder ball. This is mainly caused by the local thermal expansion mismatch between the silicon chip, molding compound, and the BT substrate. Also, the maximum stress and strain distributions and values in the FM-PBGA and P-PBGA are about the same. (P-PBGA has a little higher value.)

Figures 5.36 and 5.37 show the von Mises stress and equivalent plastic shear strain range in the critical solder ball. It can be seen that the maximum values occur at the upper left-hand corner near the chip

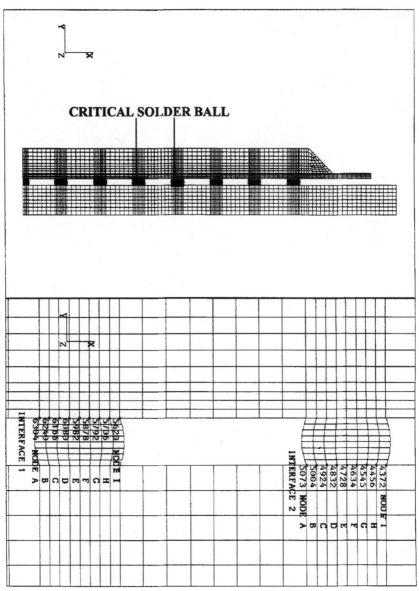

Figure 5.29 Temperature-dependent elasto-plastic-creep analysis of the full-matrix PBGA.[111]

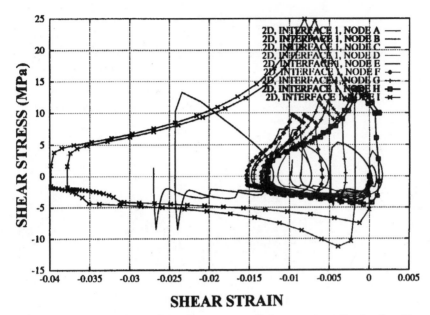

Figure 5.30 Hysteresis loops of the shear stress and shear strain at the fourth solder ball from the package center.[111]

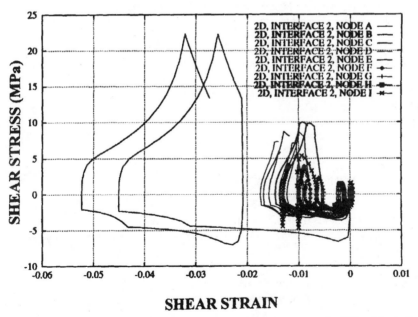

Figure 5.31 Hysteresis loops of the shear stress and shear strain at the fifth solder ball from the package center.[111]

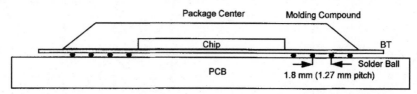

Figure 5.32 Schematic of the perimeter-arrayed PBGA along the diagonal direction.[111]

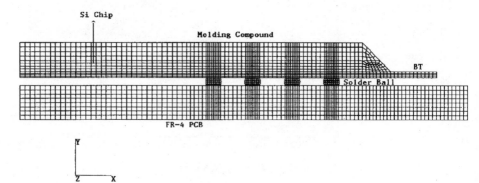

Figure 5.33 Finite element model of the perimeter-arrayed PBGA.[111]

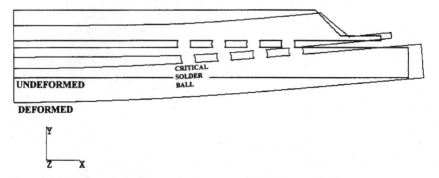

Figure 5.34 Deformation of the perimeter-arrayed PBGA assembly.[111]

direction. The maximum plastic shear strain range was 3.9 percent which is very close to that of the 225-pin FM-PBGA range (3.8 percent).

5.2.5 Temperature-dependent elasto-plastic-creep analysis of perimeter-arrayed PBGA solder joints

The same temperature boundary conditions applied to the FM-PBGAs were applied to the P-PBGAs. The critical location for the stress and strain hysteresis responses was, again, at the first ball from the package center (Fig. 5.38). The hysteresis loops of the shear stress and shear

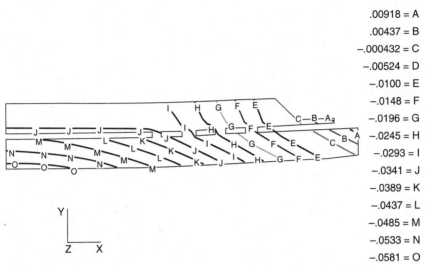

.00918 = A
.00437 = B
−.000432 = C
−.00524 = D
−.0100 = E
−.0148 = F
−.0196 = G
−.0245 = H
−.0293 = I
−.0341 = J
−.0389 = K
−.0437 = L
−.0485 = M
−.0533 = N
−.0581 = O

Figure 5.35   Deflection contours of the perimeter-arrayed PBGA assembly.[111]

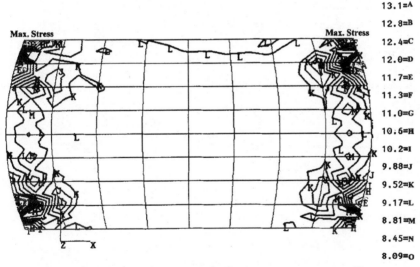

13.1=A
12.8=B
12.4=C
12.0=D
11.7=E
11.3=F
11.0=G
10.6=H
10.2=I
9.88=J
9.52=K
9.17=L
8.81=M
8.45=N
8.09=O

Figure 5.36   von Mises stress (MPa) distribution in the critical solder ball.[111]

strain at NODES A, B, C, D, E, F, G, H, and I on the upper interface 1
and on the lower interface 2 of the critical solder ball are shown in Figs.
5.39 and 5.40, respectively. It can be seen that the maximum shear
stress range (25 MPa) and shear strain range (2.0 percent) were at
NODE A near interfaces 1 and 2. This is because NODE A is the closest
to the chip and is subjected to the maximum local thermal expansion
mismatch between the chip, molding compound, and BT substrate. Any

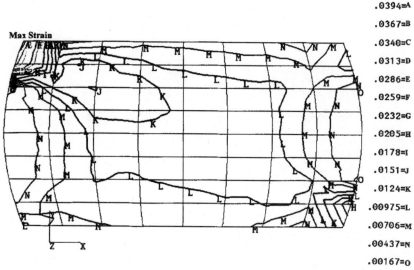

Figure 5.37   Equivalent plastic strain distribution in the critical solder ball.[111]

solder joint cracking should initiate at this location. Comparing the shear stress range and shear strain range between the P-PBGA and FM-PBGA (Sec. 5.2.3), it can be seen that the values of the P-PBGA solder joint are smaller. Thus, for optimum design of PBGA assemblies, the solder joints should be placed along the perimeters and not underneath the chip. Also, this will improve the signal routing on the PCB. If an FM-PBGA is used, then the solder joints underneath the chip can be used for mechanical support or thermal purposes.

5.3   Mechanical Reliability of PBGA Assemblies

A variety of environmental stress factors (e.g., temperature and humidity, shock and vibration, or thermal/power cycles) may lead to solder joint failure, and the most commonly observed failure modes in practice are *overload* and *fatigue*. Overload failure occurs whenever the stress in the solder joint, brought about by the imposed stress factors, is greater than the strength of the solder alloy. An example is the extensive bending and twisting of a PCB with PBGAs soldered to its surface. On the other hand, fatigue failure takes place via the initiation and slow propagation of a crack until the joint becomes unstable. The stress factors that typically cause failure by fatigue are far below the overload failure levels. Examples of fatigue failure include the vibration of a PCB with PBGAs soldered to its surface and the power cycling of a PBGA which is soldered to the surface of a PCB.

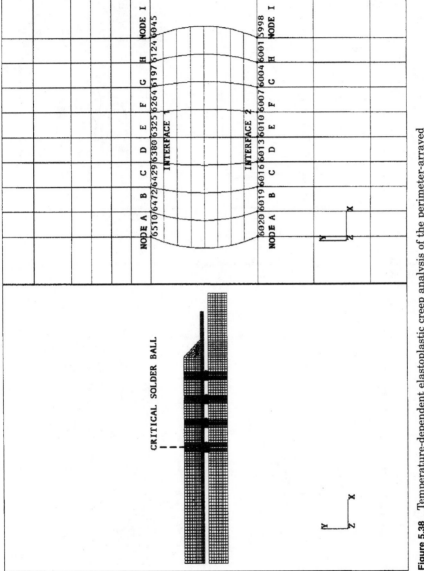

Figure 5.38 Temperature-dependent elastoplastic creep analysis of the perimeter-arrayed PBGA. [111]

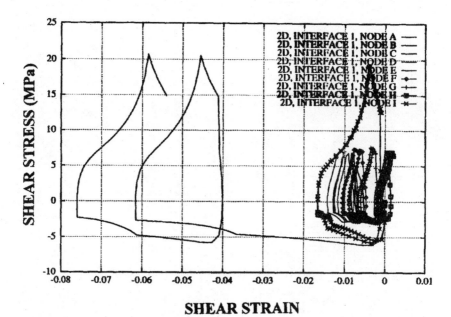

Figure 5.39 Hysteresis loops of the shear stress and shear strain at interface 1 (near the chip) of the first solder ball from the package center.[111]

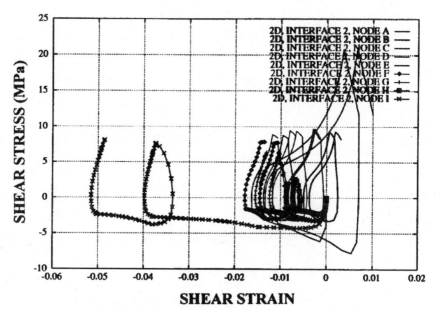

Figure 5.40 Hysteresis loops of the shear stress and shear strain at interface 2 (near the PCB) of the first solder ball from the package center.[111]

The thermal responses of PBGA solder joints have been discussed in previous sections. During manufacturing, handling, shipping, rework, etc., the PCB is subjected to bending, twisting, and vibration. The effects of these environmental stress factors on PBGA solder joints have been addressed through testing and analysis by Lau et al.[102–106] and are shown in the next two sections. Also, a no-clean mass reflow assembly of the PBGAs is discussed.

5.3.1 Test board assembly

The PBGAs, boards, and SMT assembly of the PBGA PCBs for the bending, twisting, and vibration tests will be presented.[102–104]

5.3.1.1 PBGA packages

Package molding. There are at least two forms of molding: one-side molding and overmold. One-side molding (Fig. 5.11) is a transfer molding method used for most of the PBGAs that are mass produced. Overmold for PBGAs was developed by Citizen (Fig. 5.41). It is mainly for PBGAs with very high pin counts and for multilayer PBGAs. In general, for both one-side and overmolding, the maximum wire loop above the silicon chip is 0.3 mm and the minimum molding thickness above the wire loop is 0.2 mm.

Package substrate. The substrate material used is BT resin (CCL-HL 832) manufactured by Mitsubishi Gas Chemical Company. For this material, the glass transition temperature is 170 to 215°C; the thermal expansion coefficient is 15 ppm/°C in the in-plane direction, and 52 ppm/°C in the vertical direction; the peel strength (35 µm) is 1.6 kgf/cm; the flexural strength is 52 kgf/mm; and the water absorption is 0.06 percent.

The hole diameters of the substrate are dependent on the substrate thickness. For 0.1–0.6-mm thick substrate the normal hole diameter is 0.4 mm, and for 0.8–1.0-mm thick substrate the normal hole diameter

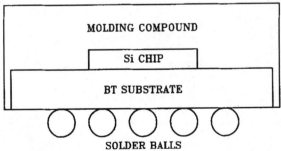

Figure 5.41 Schematic of overmold PBGA package.[103]

is 0.5 mm. In this study, the substrate thickness for the 324-pin and 396-pin PBGAs is 0.8 mm and 0.27 mm for the 225-pin PBGAs. The horizontal dimensions of the 396-pin, 324-pin, and 225-pin PBGAs are 35×35, 32×32, and 27×27 mm, respectively.

Based on selective plating processes, the PBGA substrates are subjected to two kinds of plating specifications. One is to plate soft gold (for wire bonding) on the top surface of the substrate. The other is to plate copper (for soldering) on the bottom surface of the substrate. Finally, photolithography with the subtractive method is used for pattern formation. In this study, the pads on the bottom surface of the PBGAs are interconnected via copper traces in an alternating pattern (Figs. 5.42 through 5.44) so as to provide daisy-chained connections when the PBGAs are soldered to the printed circuit board (PCB). In Figs. 5.42 through 5.44, the thin lines are the traces on the PBGAs. The wider lines are the traces on the PCB.

Solder balls. Figures 5.45*a* and *b* show a solder ball on the bottom surface of PBGAs. It can be seen that the thickness of the copper pattern is 38 µm, the diameter of the copper pattern is 889 µm, the thickness of dry film is 75 µm, the opening diameter of the copper pattern is 635 µm, the diameter of the solder ball is about 760 µm, and the height of the solder ball is about 625 µm.

The solder composition is 63wt%Sn/37wt%Pb. The Young's modulus of the solder is 1.5×10^6 psi (10,000 MN/m) and the Poisson's ratio is 0.4.

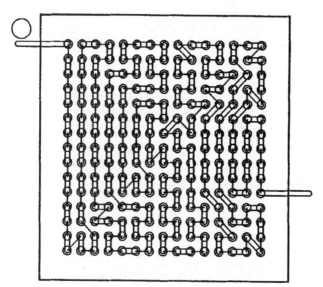

Figure 5.42 Schematic of the bottom pads of a 225-pin PBGA (solid, thin line between pads is on the PBGA, hollow bar between pads is on the PCB).[103,104]

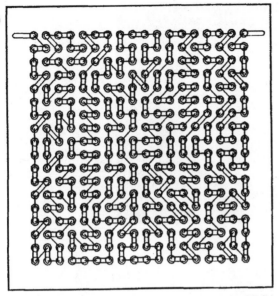

Figure 5.43 Schematic of the bottom pads of a 324-pin PBGA (solid, thin line between pads is on the PBGA, hollow bar between pads is on the PCB).[103,104]

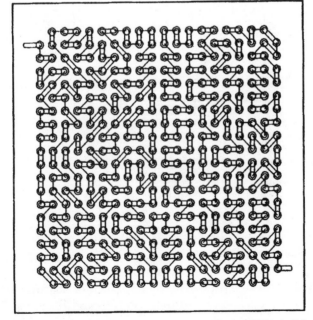

Figure 5.44 Schematic of the bottom pads of a 396-pin PBGA (solid, thin line between pads is on the PBGA, hollow bar between pads is on the PCB).[103,104]

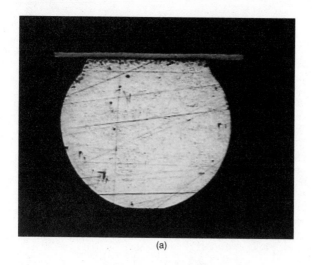

(a)

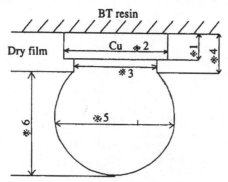

✖ 1 = thickness of copper pattern -- 38 μm
✖ 2 = diameter of copper pattern -- φ 889 μm
✖ 3 = opening diameter of copper pattern -- φ 635 μm
✖ 4 = thickness of dry film -- 75 μm
✖ 5 = diameter of solder ball -- φ 760 μm
✖ 6 = height of solder ball -- 600 μm ± 50 μm

(b)

Figure 5.45   (*a*) Cross-section view of a PBGA solder ball.
(*b*) Geometry of a PBGA solder ball.[103,104]

The thermal coefficient of linear expansion of the solder is 21 ppm/°C.
Solder balls are reflowed on the PBGAs, with an RMA-type of flux, in an
N_2 (100 ppm) gas infrared oven. The center-to-center spacing (pitch) of
the solder balls for the 225-pin, 324-pin, and 396-pin PBGAs is 1.5 mm.

5.3.1.2 PCB layout. Figure 5.46 schematically shows the test board
for this study. It is used for the three PBGAs and for all vibration, ther-

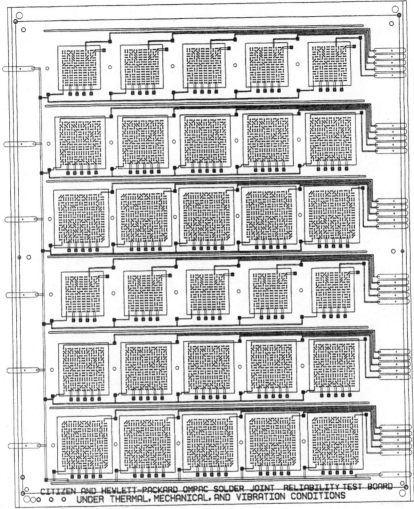

Figure 5.46 Test PCB design and layout.[103,104]

mal, and mechanical bending and twisting tests. Thus, only one set of layout, stencil, and board must be designed. It can be seen from Fig. 5.46 that there are 30 PBGA land patterns (10 for each size). Traces are routed on the test board to allow for daisy-chaining from the pad on the PCB surface through the PBGA solder balls, into the pad on the PBGA, and then over to the next pad and solder ball.

The PCB is made of FR-4 epoxy/glass and its dimensions are 305 × 255 × 1.5 mm. The detailed pad/mask/trace dimensions are shown in Fig. 5.47. It can be seen that (this is the case of NSMD solder joints) the diameter of the copper pad is 0.635 mm and the diameter of the solder

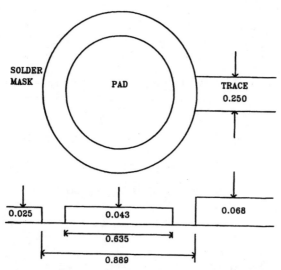

Figure 5.47   Details of a pad on the PCB (NSMD pad).[102–104]

mask opening is 0.889 mm (NSMD). The thickness of the copper pad is 0.043 mm and the width of the copper trace is 0.25 mm. The solder mask height is 0.025 mm and is 0.068 mm when it is covering the copper trace. Two different surface finishings were used; one is organic coated Entec Cu56 and the other is hot-air-leveled with an approximately 0.017 mm-thick layer of Sn/Pb solder.

The whole PCB was used intact for the thermal cycling test. However, for out-of-plane vibration, and mechanical bending and twisting tests, the board was routed into smaller test specimens (coupons) with only one PBGA at the center of each coupon.

5.3.1.3   PBGA test board assembly

Paste, flux, and stencil design and printing.   Stencil printing of solder paste was used before PBGA placements for solder ball attachments to the corresponding pads on the PCBs. Paste printing as opposed to flux-only techniques helps in at least two ways: (1) to further mitigate any ball nonplanarity that could potentially cause electrical opens, and (2) to ensure that the preliminary process-development efforts are toward making the PBGA assembly fully compatible with the standard surface-mount processes.

The paste used was a no-clean paste. It was a Sn63/Pb37 pre-alloyed powder. The flux vehicle used in the paste is a vendor-proprietary formulation which is comprised of natural and synthetic rosins with medium-boiling-point solvents, and additives to control viscosity and stabilize the paste. All processing guidelines (including tack time) of this paste were followed to ensure good paste deposition and good reflow of the joints.

An electropolished stencil was used in this study. Previous experiments with electropolished stencils and smaller (as low as three) aspect ratios (aperture diameter-to-thickness) and the same local thicknesses, rendered positive results with respect to sites' paste volume minimum variability from brick to brick, as well as from print to print. The standard SMT stencil printing equipment with standard parameter settings was used to print the PBGA sites.

Post-print inspection. As mentioned earlier, the PCB used in this study contained 10 sites of each of the three PBGA types of 225-pin, 324-pin, and 396-pin. Every printed pad on every site on every board was inspected after print using a Wild M-3 microscope to identify any defects (such as paste bridging, insufficiency, and potential smears). Although the volume of all the printed circular bricks visually appeared to have the same deviation from the theoretical goal, an acceptance criterion of 90 percent of the theoretical volume was set in advance for inspection. Over 100 PCBs were printed, half of which were hot-air-leveled (HAL), and the other half had organically coated copper pad surfaces. Of all the printed boards, no defects were observed.

Pick and place of PBGAs. All the PBGAs were placed on the test boards using a placement machine with look-up and look-down cameras.

Mass reflow. A 10-zone reflow forced-convection oven was used to reflow the parts under an inert N_2 environment. PCBs loaded with PBGAs and instrumented with thermocouples were used to develop a standard no-clean profile. One thermal profile accommodated all the different populations, as the numbers and locations of packages on the boards varied between boards assembled for thermal cycling, and for the different mechanical tests intended for solder joint reliability evaluations.

In this study, the reflow temperature profile is very similar to that of the fine-pitch technology. The peak temperatures of 229 and 221°C (corresponding to the corner and center, respectively) were achieved in the reflow zone with a profile PCB. Experience indicates that actual assemblies, which contain wet solder paste, attain temperatures very similar to the profile PCB (within 2 to 3°C). Also, a relatively high conveyor speed was used, which is desirable in production. The time above liquidus (103 seconds maximum) was optimized to maintain: (1) a small temperature gradient across the joints in one array and from part to part, and (2) a sufficiently high temperature (but still within desired specifications) to ensure proper reflow.

Post reflow inspection. Two methods of inspection were performed on each PBGA package after reflowing the solder joints in the N_2 convection oven. First an electrical continuity measurement was made to ensure that all of the solder joints were intact. Second, an X-ray microscope was used to inspect for solder bridges (which would not be detectable by the continuity measurements) and to detect any other anomalies in the solder joints, such as voids. These methods will be described in the following sections.

As mentioned earlier, each PBGA package circuit pattern and its corresponding PCB land pattern were designed to form a daisy chain that could be used to detect an open circuit in any solder joint. See Figs. 5.42 through 5.44 for layouts of the daisy-chain patterns used for the 225-pin, 324-pin, and 396-pin packages. Since the solder joints for area array packages including PBGAs are hidden under the package, conventional optical inspection of the solder joints is not feasible, except possibly for solder joints near the edges of the package. An X-ray inspection system proves useful in imaging the integrity and shape of these hidden solder joints.

A model FXS-160.30 X-ray microscope, manufactured by Fein Focus Roentgen-System Gmbh in Garbsen-Berenbostel, Germany, was chosen to inspect area array solder joints. A block diagram of this system is shown in Fig. 5.48. This machine has a number of attributes that make it a particularly useful tool for solder joint inspection. A primary feature of this system is its small focal spot size, which is achieved with a microfocus electron gun in combination with a thin transmission-type tungsten target that converts the focused electron beam to a less than 5-micron-diameter X-ray source. Figure 5.49 illustrates the improvement in feature edge definition with a smaller X-ray focal spot diameter.

This small focal spot, in conjunction with a 180-mm range of motion between the X-ray tube and the sample provides a geometric magnification range of 3.4 to 290 times. A 215-mm-diameter cesium iodide faceplate on a Thompson-CSF triple-field image intensifier converts the magnified X-ray image to an optical image for viewing via a CCD TV camera and display. This combination of high magnification and small focal spot size together with a 460×610 mm x-y positioning range allows high resolution inspection of the PBGA solder joints mounted on the 305×255 mm printed circuit test boards.

The noncentral solder balls appear distorted because object features farther away from the principal axis passing through the centerline of the X-ray source and image intensifier are magnified less than closer features. This effect is commonly known as *perspective distortion,* and is illustrated in Fig. 5.50 which shows an undistorted image of a solder ball viewed on axis, and a slightly distorted image viewed off axis. For this reason caution should be exercised in making shape judgments and measurements of off-axis features, especially, in X-ray images.

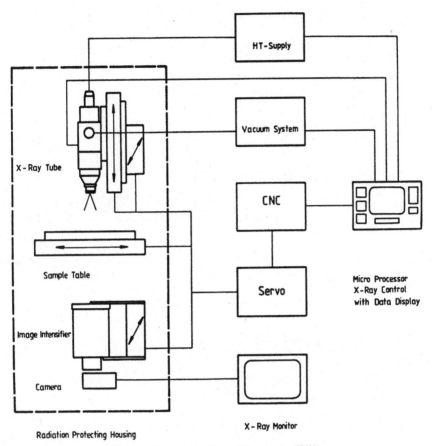

Figure 5.48   Block diagram of Fein Focus X-ray Microscope.[103,104]

Cross sections of PBGA assembly.   Figure 5.51 shows a typical cross section of the present PBGA assembly. It can be seen that all the solder joints look good. Because of the present PCB design (the solder mask opening is larger than the copper pad as seen in Fig. 5.47), the solder joint smoothly covers the copper pad, and the sharp corners of the solder joint are not apparent.

5.3.1.4 Self-alignment of PBGA assemblies.   The high accuracy of the placement machine allows a process robustness test by deliberately introducing placement offsets. PBGAs were placed with one-axis offsets of 5 to 45 mils (0.127 to 1.143 mm) in 5-mil (0.127-mm) steps. No tests were conducted with two axes simultaneously offset by introducing rotational errors. The solder balls on the PBGAs have a 60-mil (1.5-mm) pitch. With offsets at 25 mils (0.64 mm) and below, the parts move

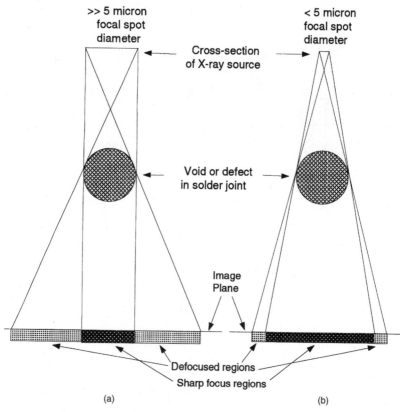

Figure 5.49   Improvement in X-ray image resolution due to small focal spot: (a) poor image resolution; (b) improved image resolution.[103,104]

during reflow and center themselves on the correct pads. At 35-mil (0.889-mm)-offset and above, the parts jog over a row and center themselves on the wrong pads. Finally, at 30-mil (0.762-mm)-offset the parts either center correctly (with occasional bridges) or jog over a row. A benefit of having such high placement error tolerance is that simple (nonvision) placement algorithms suffice in normal production.

5.3.1.5 PBGA moisture sensitivity evaluation.   PBGA packages were evaluated for robustness to SMT reflow stresses. A common package defect caused by SMT reflow is *popcorn cracking*. Moisture is absorbed by the package, then, during the rapid heating in a reflow oven, moisture in the package vaporizes. Interfacial delaminations and cracks through the bulk materials can occur. Scanning acoustic microscopy (SAM) is an excellent tool for imaging these internal defects. A Sonix Microscan IC, C-mode Scanning Acoustic Microscope operating at 35 MHz was used in this investigation.

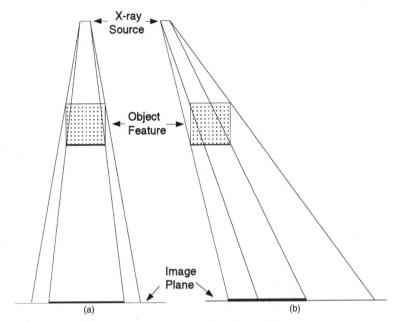

Figure 5.50 Off-axis perspective image distortion of X-ray image: (*a*) on-axis image; (*b*) off-axis image.[103,104]

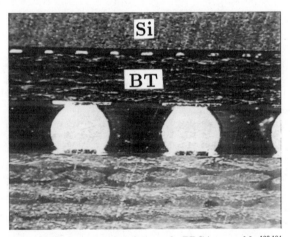

Figure 5.51 Cross-sectional view of a PBGA assembly.[103,104]

Eight each of the unstressed 225-pin, 324-pin, and 396-pin PBGA devices were acoustically imaged. All internal interfaces were well bonded. No delaminations or bulk material cracks were detected anywhere in the package. Devices were imaged from the top side only. The large number of solder balls on the backside obscured the internal features during backside imaging.

For each of the 225-pin, 324-pin, and 396-pin packages, five devices were subjected to 168 hours of 85°C/85%RH and three devices were baked at 125°C for 24 hours. The IPC (Institute for Interconnecting and Packaging Electronic Circuits) documents addressing IC package moisture sensitivity relate these conditions to an indefinite exposure to 30°C/60%RH and bake-before-use dry pack requirements, respectively. After the preconditioning exposure, the packages were reflowed in a forced convection oven with a peak temperature of approximately 225°C. The packages were not mounted to a PC board.

After acoustically imaging the stressed packages, all of the 85/85 preconditioned packages showed significant internal delaminations. The packages that were baked before reflow had no internal delaminations or cracks. Results are tabulated in Table 5.3 (reported as defective of samples inspected), and Figs. 5.52 and 5.53 show the PBGAs under the 85/85 and the baked condition, respectively.

Since the baked packages suffered no damage after the reflow stress and the moisture-saturated packages were highly damaged, the thermal mismatch contribution to interfacial delaminations appears insignificant. The delaminations resulted from the moisture adsorbed by the package. These packages should be considered moisture sensitive. Further evaluation will determine the acceptable level of moisture, or allowable time out of dry pack (shelf life), before reflow. The location and extent of delaminations appears to be dependent on the package geometry. Die surface delamination increased with the die/package surface area ratio. Solder mask to molding compound delamination increased with decreasing die/package surface area ratio.

5.3.1.6 PBGA assembly popcorn effect.   Two batches of PCBs were processed. The first batch with OCC pads were processed a week before the HAL boards. All the presealed desiccated packages containing the PBGA trays were opened for the first build. X-ray inspection of the OCC assemblies did not reveal any solder joint problems in terms of joint shape, alignment, or any characteristic examined with the X-ray. After the second build with HAL PCBs, the inspection showed a nonround effect of the central joints within the chip area on approximately 7 to 10 of the 225-pin PBGA sites out of all processed.

Figure 5.54*a* and *b* show *egg-shaped* solder balls that are not caused by perspective image distortion. Rather, the BT substrate delaminated

TABLE 5.3 PBGA Moisture Study Results[102,103,104]

No. of leads	Preconditioning stress	Die surface delamination	BT substrate to molding compound delamination	BT substrate to solder mask delamination	Solder mask to molding compound delamination
225	bake	0/3	0/3	0/3	0/3
324	bake	0/3	0/3	0/3	0/3
396	bake	0/3	0/3	0/3	0/3
225	85/85	5/5	5/5	5/5	5/5—least delam
324	85/85	5/5	5/5	5/5	5/5
396	85/85	3/5	5/5	5/5	5/5—most delam

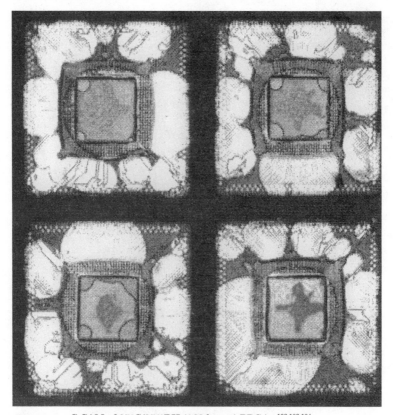

Figure 5.52   C-SAM of 85°C/85%RH (168 hours) PBGAs.[102,103,104]

from the chip and the bottom of this 225-pin PBGA bulged out due to *popcorning* of the package. This phenomenon is due to the expansion of moisture during reflow which exerts a downward force and deforms the substrate and solder joint in the central chip region. The deformation in the region apparently persists beyond the cooling and solidification of the joints to cause the nonround shape of the joint cross section. Cross sections of the tested samples of the PBGA assembly confirm this effect (Fig. 5.55).

In this study, the factors supporting the fact that popcorning phenomena caused the nonround joints are: (1) the PBGA packages were exposed to the lab environment for a week prior to the second build, and (2) the phenomena occurred only on the thinner-substrate 225-pin assembly in which the BT thickness was one-third that of the 324-pin and 396-pin.

5.3.1.7 PBGA test board assembly summary. The processing of the PBGAs involved assembling 240 of the three package types, amount-

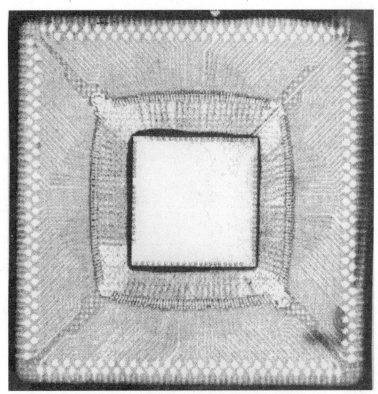

Figure 5.53   C-SAM of baked (125°C for 24 hours) PBGA.[102,103,104]

ing to grand total in excess of 222,000 solder joints. Thorough X-ray inspection of all the joints and electrical probing of the assembled sites showed: (1) no electrical opens, (2) excellent alignment of all the parts to the pads, and (3) a total of one solder bridge in a 225-pin assembly on a HAL board.

The popcorning phenomenon in the 225-pin parts was not unexpected since the parts were left exposed to the lab environment long enough to absorb moisture and were deliberately not baked. If the failures caused by the popcorning effect are disregarded, the measured defect rate in the assembly is 4.5 ppm with an upper bound value of ~20 ppm at a 95 percent statistical confidence level.

It was learned that: (1) PBGA is an SMT-compatible component, (2) PBGA is moisture sensitive, (3) the X-ray inspection machine is an excellent tool for process development, (4) the PBGA process is very forgiving due to the self-aligning characteristic of the molten solder during reflow, and (5) PBGA on PCB is a very high yield assembly process.

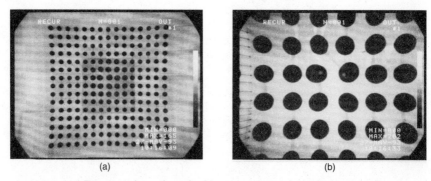

(a)   (b)

Figure 5.54 (*a*) Top view of egg-shaped solder joints due to delaminations. (*b*) Top view of egg-shaped solder joints due to delaminations.[103,104]

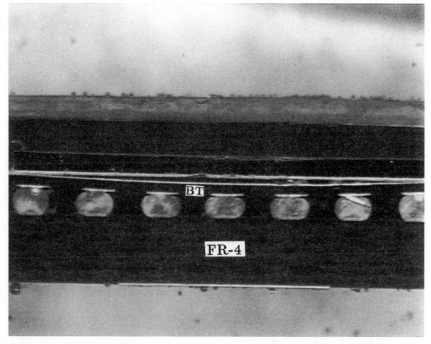

Figure 5.55 Cross-sectional view of egg-shaped solder joints due to delaminations.[103,104]

5.3.2 Three-point bending of PBGA assemblies' solder joints

The setup of the three-point bending test of the PBGA assemblies is shown in Fig. 5.56. It can be seen that the test board had a clear span of 152.4 mm and was supported at both ends. The load was applied at the center of the test board with a crosshead speed equal to 3.81 mm

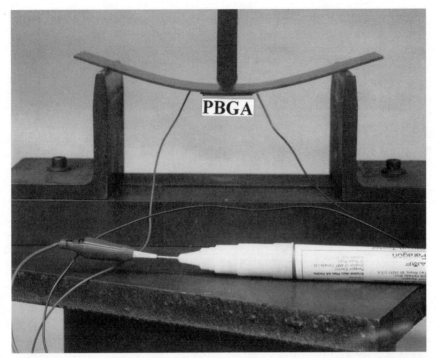

Figure 5.56   Three-points bending of a PBGA assembly.[105,106]

per min. The sample size was planned to be 10; however, because of the repeatability of the test results, most of the tests stopped at the fifth specimen.

Figures 5.57, 5.58, and 5.59 show the load-deflection curves of the test board with the 225-pin, 324-pin, and 396-pin PBGAs, respectively. It can be seen that for all the cases there was no solder joint failure even with a deflection greater than 12.7 mm. For this size board, this magnitude of deflection is beyond most of the manufacturing, shipping, handling, and rework conditions.

It can also be seen from Figs. 5.57 through 5.59 that failures did not happen until the applied load reached 125 N and the deflection reached 20 mm for the 396-pin PGA assembly. For this size board, this magnitude of deflection is far beyond all the manufacturing, handling, shipping, and rework conditions. Cross sections of the failure specimen showed that a few solder joints were broken on both sides (near the middle) of the PBGAs in the longer direction of the test board. The failure mode is solder cracking near the bottom surface of the PBGAs (Fig. 5.60).

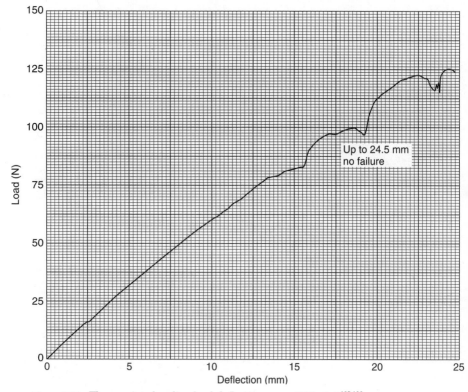

Figure 5.57 Three-points bending load-deflection curve (225-pin).[105,106]

5.3.3 Four-point twisting of PBGA assemblies' solder joints

The setup of the four-points twisting test of the PBGA assemblies is shown in Fig. 5.61. It can be seen that the overall dimensions of the board were 152.4 × 152.4 × 1.58 mm and the QFP was located at its center. The board was loaded by two forces at its two opposite corners (the other two corners were supported). These corners were not exactly at the corners of the board but 3.2 mm from the actual corners (for the purpose of applying the forces and supports). Thus, the clear distance between the forces and supports was 146 mm. In this study, the edges of the board were not reinforced (i.e., bending and shearing deformations in the edge planes were possible). The crosshead speed was 3.81 mm per min. Again, the sample size was planned to be 10; however, because of the repeatability of the test results, all the tests stopped at the fifth specimen.

Figures 5.62, 5.63, and 5.64 show the load-deflection curves of the test board with the 225-pin, 324-pin, and 396-pin PBGAs. It can be

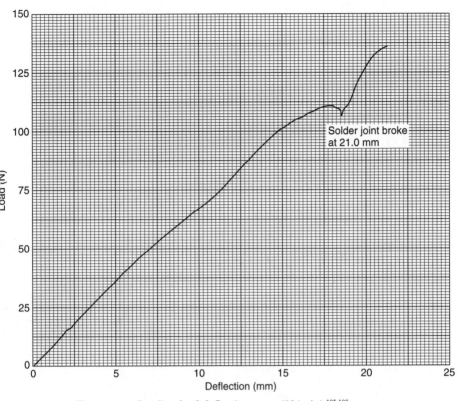

Figure 5.58   Three-points bending load-deflection curve (324-pin).[105,106]

seen that for all the cases there were no failures at a deflection equal to 12.7 mm. In fact, the solder joints were not cracked until the load reached 140 N and the deflection reached 17.5 mm for the 396-pin PBGA assembly. (These magnitudes are far beyond most manufacturing processes.) These broken solder joints were around the four corners of the PBGAs. The failure mechanism of each solder joint was close to the interface between the copper on the PBGA bottom surface and the solder joint (Fig. 5.65). The failure load and deflection of the 324-pin PBGA are larger than those of the 396-pin PBGA. There is no failure of the 225-pin solder joints up to 24.5 mm deflection.

5.4   Shock and Vibration Reliability of PBGA Assemblies

The shock and vibration tests of the PBGA assemblies built in the previous section are presented herein and the test results are discussed.

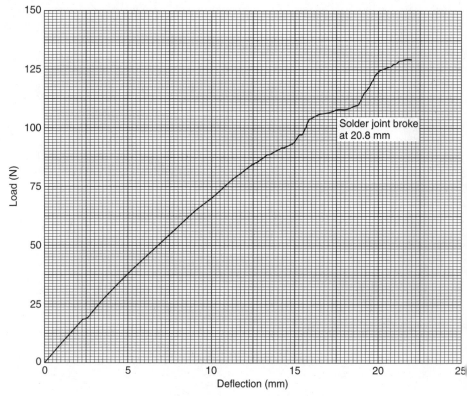

Figure 5.59 Three-points bending load-deflection curve (396-pin).[105,106]

5.4.1 Vibration of PBGA assemblies' solder joints

The test setup for the out-of-plane vibration of the PBGA assemblies is shown in Figs. 5.66 and 5.67. It consists of a table which was vertically driven by a shaker. A special test frame was designed and constructed to provide a two-point fatigue bending of the test board as shown schematically in Fig. 5.68. These two supports were made by clamping both ends of the test board with aluminum blocks. Tapes were used between the board and the blocks so that damage to the trace on the board could be avoided. The actual board-bending was caused by the movement of these supports which were bolted to the surface of the vertical shaker. The shaker was controlled by a data-acquisition unit and produced a sinusoidal excitation. There was only one PBGA on each test board. The sample size of each type of PBGA was 80. A total of 240 boards were tested.

In order to excite the test board, the excitation frequency had to include the natural frequency of the test board with these PBGAs. Con-

Figure 5.60 Solder joint cracking due to three-points bending.[105,106]

Figure 5.61 Four-points twisting of a PBGA assembly.[105,106]

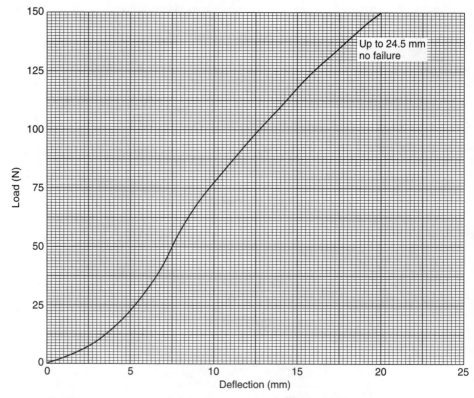

Figure 5.62 Four-points twisting load-deflection curve (225-pin).[105,106]

sequently, the next step was to estimate the natural frequencies of the test boards with these three different types of PBGAs.

Figure 5.68 shows the PCB with L as its largest dimension. It is supported by linear and rotational springs at both ends. The PBGA is attached at the center of the PCB. E is the Young's modulus, I is the area moment of inertia of the beam (PCB) cross section about the neutral axis, EI is the flexural rigidity of the PCB, M is the concentrated mass of the PBGA, and m is the mass per unit length of the PCB. K is the stiffness of the translational spring, and k is the stiffness of the rotational spring.

By assuming K and k approach infinity (the clamped condition), and using the data of the PBGAs and test board, we have the natural frequency f of 103 Hz for the 396-pin PBGA assembly, $f = 100$ Hz for the 324-pin PBGA assembly, and $f = 98$ Hz for the 225-pin PBGA assembly.

We also used the vibration table to sweep at very small magnitudes for a wide range of frequencies to determine the natural frequencies of our PBGA test boards. For example, Fig. 5.69 shows the measurement of the vibration result (four accelerometers on four different test

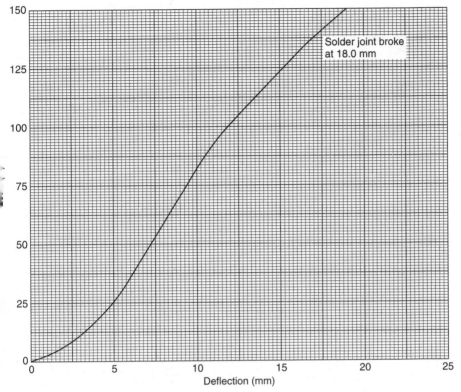

Figure 5.63 Four-points twisting load-deflection curve (324-pin).[105,106]

boards: one for 225-pin, one for 324-pin, and two for 396-pin). It can be seen that the natural frequencies range from 95 to 105 Hz and the average measured value of these four measurements is about 100 Hz. These values are very close to those predicted by the simple model.

Once the natural frequencies of the test boards were estimated, the excitation of the shaker was designed to perform a sinusoidal excitation with a frequency sweep between 80 and 120 Hz.

Once the excitation frequencies of the test boards were estimated, the shaker was designed to perform a sinusoidal excitation such that the average maximum (peak-to-peak) excitation magnitude (at the middle) of the PBGA PCB is 0.12 in (3.05 mm). The measured result (four accelerometers on four different PBGA test boards) is shown in Fig. 5.70. None of the solder joints of these 240 PBGAs failed up to 50 minutes of vibration. Figure 5.71 shows the cross section of a failed solder joint after three hours of vibration. It can be seen that the crack is near the interface between the solder joint and the copper on the bottom surface of the PBGA. Similar to the three-points bending case, sol-

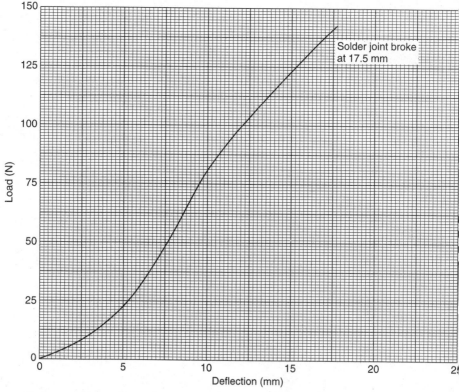

Figure 5.64 Four-points twisting load-deflection curve (396-pin).[105,106]

der joints were broken on both sides of the PBGAs in the longer direction of the test board.

5.4.2 Shock of PBGA assemblies' solder joints

The drop tests of the PBGA assemblies built in the previous section were performed. After three drops (both horizontal and vertical), there was no solder joint failure.

5.4.3 Summary of test results

The solder joints of the 225-pin, 324-pin, and 396-pin PBGA assemblies have been shown by mechanical bending and twisting tests to be reliable for use under normal manufacturing, handling, shipping, and rework conditions.

The natural frequencies, excitation frequencies, and excitation magnitude of the 225-pin, 324-pin, and 396-pin PBGA assemblies have

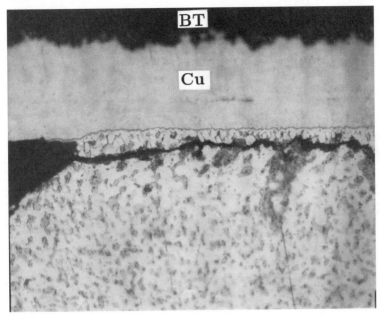

Figure 5.65   Solder joint cracking due to four-points twisting.[105,106]

Figure 5.66   Out-of-plane vibration test of PBGA assemblies.[105,106]

Figure 5.67   Out-of-plane vibration test of PBGA assemblies.[105,106]

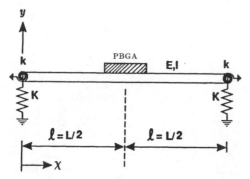

Figure 5.68   A model for the out-of-plane vibration
test of PBGA assemblies.[105,106]

been systematically and carefully determined by out-of-plane vibration
tests and a simple analysis. The results show that these PBGAs on
PCBs are reliable to use under shock and vibration conditions.

5.5   Thermal Reliability of TBGA Assemblies

In this section, the solder joint reliability of two different tape-
automated bonding BGA (TBGA) on PCB will be discussed. One (IBM's
TBGA) is subjected to power cycling and the other (LSI Logic's TBGA)
to thermal cycling.

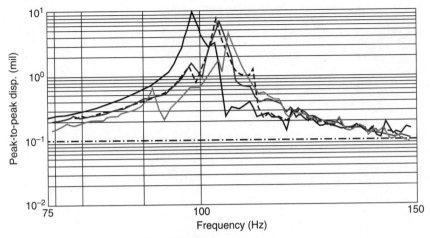

Figure 5.69   Free vibration of the PBGA assemblies.[105,106]

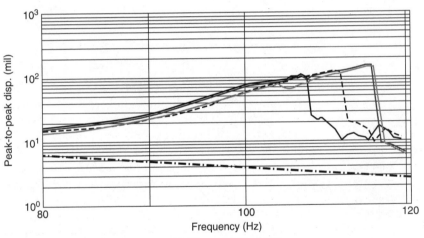

Figure 5.70   Forced vibration of the PBGA assemblies.[105,106]

5.5.1   Power cycling versus thermal cycling test

Power cycling is different from thermal cycling. Power cycling tests are defined as thermally stressing the system by alternately applying and removing power to the device (chip). The most common way to run a power cycling test is to supply power to a functional electronics package or to a package that includes circuitry designed to allow a controlled heat dissipation within the chip. Power cycling tests are more difficult and expensive to run than thermal cycling tests which are simply performed in a chamber. However, power cycling tests are more realistic.

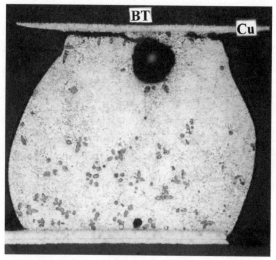

Figure 5.71 Solder joint cracking due to out-of-plane vibration.[105,106]

5.5.2 Power cycling of IBM's TBGA assemblies

Figure 5.72 schematically shows a cross section of IBM's TBGA.[20,24,63] A two-metal tape (a 2-mil-thick polyimide called Upilex sandwiched between two layers of copper) for ground and signal was used for the substrate. The chip is 95wt%Pb/5wt%tSn solder bumped by C4 technology. The inner lead bonding (i.e., bonding the solder-bumped chip on the Cu-Ni-Au tape) may be done by either pulse-thermode or hot-air-thermode reflow soldering. Underfill encapsulant is needed to overcome the thermal expansion mismatch between the chip and the copper tape. A copper plate with a center opening (for the chip and the underfill encapsulant) is attached (by a stiffener adhesive) to the two-metal tape to provide planarity and rigidity (as a stiffener) to the package. The bottom side of the substrate is attached with an area array 50-mil (1.27-mm)-pitch 90wt%Pb/10wt%Sn solder balls (25 mils or 0.65 mm in diameter). The back of the chip is attached to a heat sink to improve package thermal performance. TBGA can be used for ASIC and microprocessor applications. TBGA is SMT compatible, which can be assembled on the PCB by mass reflowing the eutectic solder. A typical cross section of the TBGA solder is shown in Fig. 5.73.[63]

The test vehicle used for the power cycle test[63] contained a 40 × 40 mm stitched 736 I/O IBM TBGA. The chip size was 14 × 14 mm. The PCB had four signal layers and two power planes. The test was run to 12,000 cycles of 25 to 140°C junction temperature with fails on approximately 10 percent of the eutectic solder joints. The solder joint failures occurred not only at the exterior joints due to PCB deflection and tem-

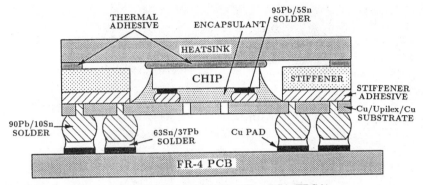

Figure 5.72   IBM's area array tape-automated bonding BGA (TBGA).

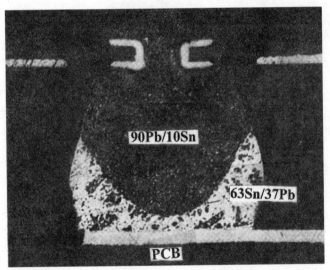

Figure 5.73   A cross section of IBM's TBGA on FR-4 PCB.[63]

perature gradient, but also at the interior joints due to local thermal expansion mismatch between the chip and the surrounding materials. A typical failure mode of the TBGA solder joint is shown in Fig. 5.74. It can be seen that it failed at the 63Sn/37Pb solder.

5.5.3   Thermal cycling of LSI Logic's TBGA assemblies

Figure 5.75 shows a cross section of LSI Logic's electrical- and thermal-enhanced TBGA-T2.[138] It integrates the heatspreader and groundplane together with the novel use of downbonds to groundplane providing a low-inductance ground path. Also, the solder balls are 63wt%Sn/

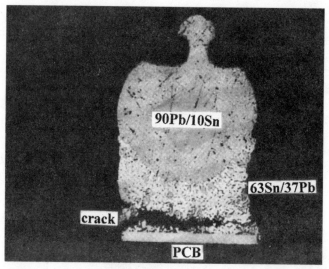

Figure 5.74 A typical failure mode (subjected to power cycling) for IBM's TBGA on FR-4 PCB.[63]

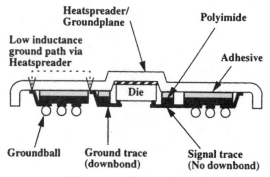

Figure 5.75 Schematic of LSI Logic's TBGA assembly.[138]

37wt%Pb and are on 1.27 mm pitch. The ball diameter is 0.76 mm (Table 5.4).

The 342-pin TBGA-T2s were mounted on FR-4 PCBs with a 0.2-mm-thick layer of eutectic solder paste. Table 5.5 shows the humidity test results. It can be seen that after 500 hours of 85°C/85%RH and 500 thermal cycles, there is no failure.

Tables 5.6 and 5.7 show the mechanical and thermal tests of the as-built assembly and the repair (up to two package replacements) assembly, respectively. In all these tests, a mechanical heat sink was attached to the package by using a leaf spring riveted to the PCB on two diagonal corners. It can be seen from Tables 5.6 and 5.7 that after

TABLE 5.4   LSI Logic's TBGA Material Properties and Descriptions[138]

Material property	Description
Die bump material	25-μm-thick Au
Die barrier metal	Ti/W
Bonding method (inner and downbonds)	Au-Au thermosonic
Polyimide thickness	125 μm
Tape interconnect	25-μm-thick Cu
Tape lead finish (inner and downbonds)	1.27-μm-thick Au
Heat sink material	635-μm Cu, Ni/Au-plated
Heat sink attach adhesive	Pressure-sensitive acrylic
Encapsulants (die and downbond slots)	Dispensed epoxy resins
Solder ball composition	63/37 Sn/Pb
Solder ball size	760 μm
Solder ball pitch	1.27 mm

TABLE 5.5   LSI Logic's TBGA Assemblies (Humidity Test Results) Without Heat Sink[138]

Test	Test conditions	Result
Humidity storage	85°C, 85%RH, 500 hours	
Thermal shocks with permanent continuity measurement	0 to 100°C; 500 cycles	0/20

TABLE 5.6   Mechanical and Thermal Tests of LSI Logic's TBGA Assemblies with Heat Sinks[138]

Test	Test conditions	Result
Vibration with permanent continuity monitor	0.25 g; 15 minutes	
Torque with permanent continuity monitor	2%; 50 cycles	
Aging	125°C; 168 hours	
Thermal shocks with permanent continuity measurement	0 to 100°C; 1000 cycles	0/20

168 hours of aging (125°C), 15 minutes of vibration (0.25 g), 50 cycles of twisting (2 percent), and 1000 thermal cycles (0 to 100°C), none of the solder joints failed.

Acknowledgments

The principal author (Lau) would like to thank J. Miremadi, J. Gleason, R. Haven, and S. Ottoboni of Hewlett-Packard and S. Mimura of Citizen for their contributions on no-clean mass reflow of large PBGAs,[103,104]

TABLE 5.7   Repair Evaluation (Up to Two Package Replacements) of LSI Logic's TBGA Assemblies with Heat Sinks

Test	Test conditions	Result
Vibration with permanent continuity monitor	0.25 g; 15 minutes	
Torque with permanent continuity monitor	2%; 50 cycles	
Aging	125°C; 168 hours	
Thermal shocks with permanent continuity measurement	0 to 100°C; 1000 cycles	0/8

K. Gratalo, E. Schneider, T. Marcotte, and T. Baker of Hewlett-Packard for their contributions on solder joint reliability of large PBGAs.[106] The authors want to thank Wen-Je Jung of Ford Research Laboratory for his contribution on nonlinear finite element modeling of PBGA-P and PBGA-FM.[111] It has been a great pleasure and fruitful experience to work with them.

References

1. Master, R. N., M. Cole, G. Martin, and A. Caron, "Ceramic Column Grid Array for Flip Chip Applications," *Proceedings of IEEE Electronic Components & Technology Conference,* May 1995, pp. 925–929.
2. Caulfield, T., J. Benenati, and J. Acocella, "Surface Mount Array Interconnections for High I/O MCM-C to Card Assembles," *Proceedings of the 2nd International Conference and Exhibition on Multichip Modules,* April 1993, pp. 320–325.
3. Banks, D., C. Heim, R. Lewis, A. Caron, and M. Cole, "Second-Level Assembly of Column Grid Array Packages," *Proceedings of SMI Conference,* August 1993, pp. 92–98.
4. Banks, D. R., T. Burnette, R. Gerke, E. Mammo, and S. Mattay, "Reliability Comparison of Two Metallurgies for Ceramic Ball Grid Arrays," *Proceedings of the 3rd International Conference and Exhibition on Multichip Modules,* April 1994, pp. 529–534.
5. Gerke, R. D., "Ceramic Solder Ball Grid Array Interconnection Reliability Over a Wide Temperature Range," *Proceedings of NEPCON West,* February 1994, pp. 1087–1094.
6. Sigliano, R., "Ceramic Substrates for Ball Grid Array Packages," in *Ball Grid Array Technology,* J. H. Lau, ed., McGraw-Hill, New York, 1995, pp. 65–92.
7. Fujikawa, O., and M. Kato, "Plastic Substrates for Ball Grid Array Package," in *Ball Grid Array Technology,* J. H. Lau, ed., McGraw-Hill, New York, 1995, pp. 93–113.
8. Hession, P., "Printed Circuit Board Routing Considerations for Ball Grid Array Packages," in *Ball Grid Array Technology,* J. H. Lau, ed., McGraw-Hill, New York, 1995, pp. 115–129.
9. Caulfield, T., M. Cole, F. Cappo, J. Zitz, and J. Benenati, "An Overwiew of Ceramic Ball and Column Grid Array Packaging," in *Ball Grid Array Technology,* J. H. Lau, ed., McGraw-Hill, New York, 1995, pp. 131–169.
10. Banks, D., K. Hoebener, and P. Viswanadham, "Ceramic Ball Grid Array Assembly," in *Ball Grid Array Technology,* J. H. Lau, ed., McGraw-Hill, New York, 1995, pp. 171–192.

11. Lee, Y. C., J. Liu, C. Tsai, and J. Zitz, "Thermal and Electrical Management of Ceramic Ball Grid Array Assembly," in *Ball Grid Array Technology*, J. H. Lau, ed., McGraw-Hill, New York, 1995, pp. 193–221.
12. Guo, Y., and J. Corbin, "Reliability of Ceramic Ball Grid Array Assembly," in *Ball Grid Array Technology*, J. H. Lau, ed., McGraw-Hill, New York, 1995, pp. 223–265.
13. Marrs, R. C., "Plastic Ball Grid Array Packaging Technology," in *Ball Grid Array Technology*, J. H. Lau, ed., McGraw-Hill, New York, 1995, pp. 267–294.
14. Mullen, B., A. Hertz, B. Miles, and R. Darveaux, "Plastic Ball Grid Array Assembly," in *Ball Grid Array Technology*, J. H. Lau, ed., McGraw-Hill, New York, 1995, pp. 295–330.
15. Rogren, P., "Thermal and Electrical Performance Management in Plastic Ball Grid Array Packages from the Vendor's Perspective," in *Ball Grid Array Technology*, J. H. Lau, ed., McGraw-Hill, New York, 1995, pp. 331–357.
16. Walshak, D. B., Jr., and H. Hashemi, "Thermal and Electrical Management of Plastic BGA Packages—A User's Perspective," in *Ball Grid Array Technology*, J. H. Lau, ed., McGraw-Hill, New York, 1995, pp. 359–377.
17. Darveaux, R., K. Banerji, A. Mawer, and G. Dody, "Reliability of Plastic Ball Grid Array Assembly," in *Ball Grid Array Technology*, J. H. Lau, ed., McGraw-Hill, New York, 1995, pp. 379–442.
18. Huang, C. C., and A. Hamzehdoost, "Area Tape Automated Bonding Ball Grid Array Technology," in *Ball Grid Array Technology*, J. H. Lau, ed., McGraw-Hill, New York, 1995, pp. 443–464.
19. Adams, J. A., "Inspection of Ball Grid Array Assembly," in *Ball Grid Array Technology*, J. H. Lau, ed., McGraw-Hill, New York, 1995, pp. 465–489.
20. Chung, T. C., and P. Mescher, "Rework of Ball Grid Array Assemblies," in *Ball Grid Array Technology*, J. H. Lau, ed., McGraw-Hill, New York, 1995, pp. 491–516.
21. Schmolze, C. A., "Burn-In Sockets for Ball Grid Arrays (BGAs)," in *Ball Grid Array Technology*, J. H. Lau, ed., McGraw-Hill, New York, 1995, pp. 517–541.
22. Vardaman, E. J., "BGA Infrastructure," in *Ball Grid Array Technology*, J. H. Lau, ed., McGraw-Hill, New York, 1995, pp. 543–554.
23. Gedney, R. W., "Packaging Glossary," in *Ball Grid Array Technology*, J. H. Lau, ed., McGraw-Hill, New York, 1995, pp. 555–612.
24. Lau, J. H., "A Brief Introduction to Ball Grid Array Technologies," in *Ball Grid Array Technology*, J. H. Lau, ed., McGraw-Hill, New York, 1995, pp. 1–64.
25. Kromann, G., D. Gerke, and W. Huang, "Motorola's PowerPC 603 and PowerPC 604 RISC Microprocessor: The C4/Ceramic-Ball-Grid Array Interconnect Technology," *Proceedings of the IEEE Electronic Components & Technology Conference*, May 1995, pp. 1–9.
26. Cohn, C., R. Richman, L. Saxena, and M. Shih, "High I/O Plastic Ball Grid Array Packages—AT&T Microelectronics Experience," *Proceedings of the IEEE Electronic Components & Technology Conference*, May 1995, pp. 10–20.
27. Carichner, K., "A Comparison of a Single Tier and Double Tier Enhanced PBGA Package Design," *Proceedings of the IEEE Electronic Components & Technology Conference*, May 1995, pp. 21–27.
28. Mulgaonker, S., and H. Berg, "Thermal Sensitivity Analysis for the 119 PBGA—A Framework for Rapid Prototyping," *Proceedings of the IEEE Electronic Components & Technology Conference*, May 1995, pp. 28–37.
29. Hasegawa, S., T. Suyama, Y. Shimada, K. Tamura, and Y. Tanada, "Hole Grid Array (HGA)—A New Tape Carrier Package with High Count I/O," *Proceedings of the IEEE Electronic Components & Technology Conference*, May 1995, pp. 38–41.
30. Wilson, J., S. Moore, and E. Laine, "A Low-Cost Metal Ball Grid Array for Flip Chip Die," *Proceedings of the IEEE Electronic Components & Technology Conference*, May 1995, pp. 42–45.
31. Master, R. N., R. Jackson, S. Ray, and A. Ingraham, "Ceramic Mini-Ball Grid Array Package for High Speed Device," *Proceedings of the IEEE Electronic Components & Technology Conference*, May 1995, pp. 46–50.
32. Oota, K., and K. Shigeno, "Development of Molding Compounds for BGA," *Proceedings of the IEEE Electronic Components & Technology Conference*, May 1995, pp. 78–85.

33. Beder, B., G. Cokkinides, and A. Agrawal, "Electrical Modeling of CBGA Packages," *Proceedings of the IEEE Electronic Components & Technology Conference,* May 1995, pp. 251–254.
34. Edwards, D., B. Stearns, and M. Helmick, "The Effect of Internal Package Delaminations on the Thermal Performance of PQFP, Thermally Enhanced PQFP, LOC, and BGA Packages," *Proceedings of the IEEE Electronic Components & Technology Conference,* May 1995, pp. 285–292.
35. Poborets, B., Q. Ilyas, M. Potter, and J. Argyle, "Reliability and Moisture Sensitivity Evaluation of 225-pin, 2 Layered Overmolded (OMPAC) Ball Grid Array Package," *Proceedings of the IEEE Electronic Components & Technology Conference,* May 1995, pp. 434–439.
36. Heinrich, S., S. Shakya, Y. Wang, P. Lee, and S. Schroeder, "Improved Yield and Performance of Ball Grid Array Packages: Design and Processing Guidelines for Uniform and Non-Uniform Arrays," *Proceedings of the IEEE Electronic Components & Technology Conference,* May 1995, pp. 793–803.
37. Ghahghahi, F., and T. Chung, "A Case Study of High Pin Count Area Array Ceramic Package Crack," *Proceedings of the IEEE Electronic Components & Technology Conference,* May 1995, pp. 917–921.
38. Caggiano, M., "A PC Program That Generates a Model of the Parasitics for BGA Packages," *Proceedings of the IEEE Electronic Components & Technology Conference,* May 1995, pp. 959–963.
39. Puttlitz, K., T. Caulfield, and M. Cole, "Effect of Material Properties on the Fatigue Life of Dual Solder (DS) Ceramic Ball Grid Array (CBGA) Solder Joints," *Proceedings of the IEEE Electronic Components & Technology Conference,* May 1995, pp. 1005–1010.
40. Solomon, D., P. Hoffman, G. Brathwaite, P. Robinson, and T. Madelung, "Thermal and Electrical Characterization of the Metal Ball Grid Array (MBGA)," *Proceedings of the IEEE Electronic Components & Technology Conference,* May 1995, pp. 1011–1015.
41. Bradley, E., and K. Banerji, "Effect of PCB Finish on the Reliability and Wettability of Ball Grid Array Packages," *Proceedings of the IEEE Electronic Components & Technology Conference,* May 1995, pp. 1028–1038.
42. Miremadi, J., "Impact of PBGA-Ball-Coplanarity on Formation of Solder Joints," *Proceedings of the IEEE Electronic Components & Technology Conference,* May 1995, pp. 1039–1050.
43. Cole, M., and T. Caulfield, "A Review of Available Ball Grid Array (BGA) Packages," *Proceedings of SMI Conference,* August 1995, pp. 207–213.
44. Marrs, R., R. Molnar, and B. Lynch, "An Enhanced Performance Low Cost BGA Package," *Proceedings of SMI Conference,* August 1995, pp. 214–225.
45. Chopra, M. A., "Fabrication Issue in the Development of Low Cost Flip Chip Plastic BGA Assemblies," *Proceedings of SMI Conference,* August 1995, pp. 226–237.
46. Ross, R., "Unique Thermal Management System for BGAs and Other Power Packages," *Proceedings of SMI Conference,* August 1995, pp. 238–242.
47. Tarter, T., M. Goetz, and M. Papageorge, "BGA Performance Characteristics: A User's Design Guide," *Proceedings of SMI Conference,* August 1995, pp. 245–254.
48. Johnston, P., "Printed Circuit Board Design Guidelines for Ball Grid Array Packages," *Proceedings of SMI Conference,* August 1995, pp. 255–260.
49. Schueller, R., and A. Plepys, "Design of a Low Cost Wire Bond Tape Ball Grid Array (TBGA) Package," *Proceedings of SMI Conference,* August 1995, pp. 261–269.
50. Cole, M., and K. Hoebener, "Planarity and Centrality Requirements for Ceramic Ball Grid Array Packaging," *Proceedings of SMI Conference,* August 1995, pp. 273–278.
51. O'Hara, W., and N. Lee, "Voiding in BGA," *Proceedings of SMI Conference,* August 1995, pp. 279–285.
52. Rorgren, R., P. Carlsson, and J. Liu, "A Comparative Study of Ball Grid Array and Ultra Fine-Pitch QFP Technologies Using Solder Paste Stencil Printing," *Proceedings of SMI Conference,* August 1995, pp. 286–294.
53. Petrucci, M., C. Ramirez, and J. Brown, "High Volume SMT Assembly of High Pin Count PBGA Devices," *Proceedings of SMI Conference,* August 1995, pp. 297–304.

54. Holliday, A., T. Ejim, G. Munie, R. Lockwood, and G. Tashjian, "Designed Experiment to Evaluate Assembly Defect Drivers for Plastic BGA Packages," *Proceedings of SMI Conference,* August 1995, pp. 305–312.
55. Darveaux, R., and A. Mawer, "Thermal and Power Cycling Limits of Plastic Ball Grid Array (PBGA) Assemblies," *Proceedings of SMI Conference,* August 1995, pp. 315–326.
56. Albrecht, H., J. Gamalski, and G. Petzold, "Metallurgical, Interfacial and Constructive Aspects on the Reliability of BGA Solder Joints," *Proceedings of SMI Conference,* August 1995, pp. 327–343.
57. Dody, G., and T. Burnette, "BGA Assembly Process and Rework," *Proceedings of SMI Conference,* August 1995, pp. 361–366.
58. Corbin, J. S., "Finite Element Analysis of Solder Ball Connect Structural Design Optimization," *IBM Journal of Research and Development,* 37:585–596, 1993.
59. Thomas, S., and C. Thornton, "BGA Process Development and SPC Implementation Using In-Line X-ray Laminography Measurements," *Proceedings of SMI Conference,* August 1995, pp. 367–372.
60. Freyman, B., M. Kim, and Y. Heo, "The Move to Perimeter-Arrayed Plastic BGAs," *Proceedings of SMI Conference,* August 1995, pp. 373–382.
61. Ejim, T., A. Holliday, F. Bader, and S. Gahr, "Design Experiment to Determine Attachment Reliability Drivers for PBGA Packages," *Proceedings of SMI Conference,* August 1995, pp. 385–392.
62. Banks, D., B. Carpenter, W. DeMarco, M. Govindasamy, E. Mace, and A. Mawer, "A Reliability Evaluation of Molded and Glob-Top Plastic Ball Grid Array," *Proceedings of SMI Conference,* August 1995, pp. 393–401.
63. Knadle, K., J. Perkins, and J. Potenza, "Verifying a TBGA Card Assembly Process Using Innovative Reliability Tests and DOE," *Proceedings of SMI Conference,* August 1995, pp. 402–410.
64. Mearig, J., "An Overview of Manufacturing BGA Technology," *Proceedings of NEPCON West,* February 1995, pp. 295–299.
65. Mammo, E., A. Mawer, A. Srikantappa, S. Vasan, G. Dody, and T. Burnette, "Solder Joint Reliability Study on Area Array and Peripheral Leaded Packages," *Proceedings of the SMTA National Symposium,* October 1995, pp. 43–58.
66. Huang, W., and J. Ricks, "Electrical Characterization of PBGA for Communication Application by Simulation and Measurement," *Proceedings of NEPCON West,* February 1995, pp. 300–307.
67. Knight, J., E. Dibble, and D. Trevitt, "Reliability Data for the IBM PBGA and PPGA Chip Carriers," *Proceedings of NEPCON West,* February 1995, pp. 308–326.
68. Di Stefano, T., "The μBGA as a Chip Size Package," *Proceedings of NEPCON West,* February 1995, pp. 327–333.
69. Chung, T., and P. Mescher, "Rework of Ball Grid Array Assemblies," *Proceedings of NEPCON West,* February 1995, pp. 334–345.
70. Abbagnaro, L., "Repairing BGA Components," *Proceedings of NEPCON East,* June 1995, pp. 321–338.
71. Economou, M., S. Sato, G. Vial-David, and L. Repellin, "The Challenges of Area Array Rework," *Proceedings of NEPCON West,* February 1995, pp. 1633–1645.
72. Sanyal, A., and R. Chase, "Considerations for Ceramic Ball Grid Array Package Removal/Re-Attach Process Design," *Proceedings of NEPCON West,* February 1995, pp. 1670–1679.
73. Schiesser, T., E. Menard, T. Smith and J. Akin, "Micro Dynamic Solder Pump: An Innovative Liquid Solder Dispense Solution to FCA and BGA Challenges," *Proceedings of NEPCON West,* February 1995, pp. 1680–1687.
74. Goers, W., "Rework of BGAs: A Comparative Study," *Proceedings of NEPCON West,* February 1995, pp. 360–364.
75. Heck, L., "Card Assembly Rework for Ceramic Column Grid Array Packages," *Proceedings of NEPCON West,* February 1995, pp. 1971–1975.
76. Abbagnaro, L., "Repairing BGA Components," *Proceedings of NEPCON West,* February 1995, pp. 1017–1034.
77. Crawshaw, R., "Cross-Sectional X-Ray Measurements Test Ball Grid Array Connections," *Proceedings of NEPCON West,* February 1995, pp. 1035–1047.

78. Phelan, G., M. Welch, S. Wand, and M. Cole, "Card Assembly & Reliability of 44mm Ceramic Solder Column Array Modules," *Proceedings of NEPCON West,* February 1995, pp. 1048–1058.
79. Kromann, G., D. Gerke, and W. Huang, "A Hi-Density C4/CBGA Interconnect Technology for a CMOS Microprocessor," *Proceedings of NEPCON West,* February 1995, pp. 1523–1529.
80. Miks, J., and R. Lawrence, "Assembly Process Issues of Hermetic BGA's," *Proceedings of NEPCON East,* June 1995, pp. 339–343.
81. Yip, W., A. Chandra, P. Celaya, and C. Tsai, "Package Characterization of a 313 Pin BGA," *Proceedings of NEPCON West,* February 1995, pp. 1530–1541.
82. Langston, K., and J. Maloney, "Heatsink Attach—The Agony and the Ecstasy," *Proceedings of NEPCON West,* February 1995, pp. 1542–1553.
83. Yacovitch, W., G. Bartlett, and J. Posivik, "PWB Optimization and Control for CBGA Assemblies," *Proceedings of NEPCON West,* February 1995, pp. 1847–1854.
84. Hattas, D., and A. Wakigawa, "Mounting Technology of BGA-P and BGA-T," *Proceedings of IEEE International Electronics Manufacturing Technology Symposium,* October 1995, pp. 417–421.
85. Bolger, J., and J. Czarnowski, "Area Bonding Conductive Epoxy Adhesives for Low Cost Grid Array Chip Carriers," *Proceedings of IEEE International Electronics Manufacturing Technology Symposium,* October 1995, pp. 422–427.
86. Mita, M., G. Murakami, T. Kumakura, N. Okabe, and S. Shinzawa, "Advanced Interconnect and Low Cost µ Stud BGA," *Proceedings of IEEE International Electronics Manufacturing Technology Symposium,* October 1995, pp. 428–433.
87. Mearig, J., and B. Goers, "An Overview of Manufacturing BGA Technology," *Proceedings of IEEE International Electronics Manufacturing Technology Symposium,* October 1995, pp. 434–437.
88. Lanzone, R., "Dimple Ball Grid Array (D-BGA): A New CBGA Package," *Proceedings of NEPCON West,* February 1996, pp. 657–677.
89. Matsuda, S., K. Kawabata, and N. Itoh, "High Reliability Ceramic BGA," *Proceedings of International Symposium on Microelectronics,* October 1995, pp. 13–17.
90. Brofman, P., K. Puttlitz, and R. Master, "Hard Ball Grid Arrays for Pluggable BGA Connector," *Proceedings of International Symposium on Microelectronics,* October 1995, pp. 1–6.
91. Ahn, S., H. Yoon, and S. Oh, "Reliability Improvement of Plastic Grid Array Package by UV/Ozone Cleaning," *Proceedings of International Symposium on Microelectronics,* October 1995, pp. 7–12.
92. Quintal, J., "100 Micron Fine Pitch on IBM µLaminate PBGA," *Proceedings of International Symposium on Microelectronics,* October 1995, pp. 253–256.
93. Kamath, S., "Ceramic Ball and Column Grid Array for ASIC Applications," *Proceedings of NEPCON West,* February 1996, pp. 678–681.
94. Kato, M., "Plastic Substrate for Ball Grid Array Package," *Proceedings of NEPCON West,* February 1996, pp. 682–686.
95. Mattei, C., and R. Marrs, "SuperBGA, Designing for Enhanced Performance," *Proceedings of NEPCON West,* February 1996, pp. 687–698.
96. Vardaman, E., and R. Crowley, "Worldwide Trends in Ball Grid Array Developments," *Proceedings of NEPCON West,* February 1996, pp. 699–701.
97. Hundt, M., A. Saboui, J. Exposito, and V. Motta, "Thermal Enhancements of Ball Grid Arrays," *Proceedings of NEPCON West,* February 1996, pp. 702–711.
98. Verma, A., and I. Sterian, "Manufacturing Experience with PBGA, CBGA, and TBGA," *Proceedings of NEPCON West,* February 1996, pp. 989–993.
99. Shu, W., "PBGA Wire Bonding Development Using Statistical Design of Experiment," *Proceedings of NEPCON West,* February 1996, pp. 994–1004.
100. Windsor, C., S. Pienimaa, and T. Chung, "Chip Scale Packaging Technologies," *Proceedings of NEPCON West,* February 1996, pp. 1005–1012.
101. Raad, P., S. Chen, and R. Keith, "BGA Package Design: A Characterization of the Effects of Molten Solder Sphere Parameters on Critical Package Design," *Proceedings of NEPCON West,* February 1996, pp. 1013–1019.
102. Lau, J. H., *Ball Grid Array (BGA) Technology,* McGraw-Hill, New York, November 1994.

103. Lau, J. H., J. Miremadi, J. Gleason, R. Haven, S. Ottoboni, and S. Mimura, "No Clean Mass Reflow of Large Over Molded Plastic Pad Array Carriers (OMPAC)," *Proceedings of IEEE International Manufacturing Technology,* October 1993, pp. 63–75.
104. Lau, J. H., J. Miremadi, J. Gleason, R. Haven, S. Ottoboni, and S. Mimura, "No Clean Mass Reflow of Large Plastic Ball Grid Array Carriers," *Circuit World,* **20:**15–22, March 1994.
105. Lau, J. H., "Solder Joint Reliability of Flip Chip and Plastic Ball Grid Array Assemblies Under Thermal, Mechanical, and Vibration Conditions," *Proceedings of Japan International Electronic Manufacturing Technology Symposium,* December 1995, pp. 13–19.
106. Lau, J., K. Gratalo, E. Schneider, T. Marcotte, and T. Baker, "Solder Joint Reliability of Large Plastic Ball Grid Array Assemblies Under Bending, Twisting, and Vibration Conditions," *Circuit World,* **22:**27–32, October 1995.
107. Teuber, K., J. Self, and D. Rose, "Large Void Phenomenon in PBGA Solder Joints," *Proceedings of NEPCON East,* June 1995, pp. 47–55.
108. Boutin, L., "Mold Compound Study for Plastic Ball Grid Array Applications," *Proceedings of NEPCON East,* June 1995, pp. 279–290.
109. Glenn, Y., "Ball Grid Array (BGA) Packaging for High Performance Applications," *Proceedings of NEPCON East,* June 1995, pp. 291–320.
110. Moore, K., Machuga, S., Bosserman, S., and Stafford, J., "Solder Joint Reliability of Fine Pitch Solder Bumped Pad Array Carriers," *Proceedings of NEPCON West,* Feb. 1990, pp. 264–274.
111. Jung, W., J. H. Lau, and Y.-H. Pao, "Nonlinear Analysis of Full-Matrix and Perimeter-Arrayed Plastic Ball Grid Array Solder Joints," *ASME Paper, ASME Winter Annual Meeting,* November 1966.
112. Lee, S., and J. H. Lau, "Effect of Chip Dimension and Substrate Thickness on Plastic Ball Grid Array Solder Joint Reliability," to be published in the Proceedings of SMI Conference, September 1996.
113. Freyman, B., and R. Pennisi, "Overmolded Plastic Pad Array Carriers (OMPAC): A Low Cost, High Interconnect Density IC Packaging Solution for Consumer and Industrial Electronics," *Proceedings of the 41st IEEE Electronic Components and Technology Conference,* May 1991, pp. 176–182.
114. Vardaman, E.J., "Ball Grid Array Packaging," Consulting Report, January 1993.
115. Johnson, R., A. Mawer, T. McGuiggan, B. Nelson, M. Petrucci, and D. Rosckes, "A Feasibility Study of Ball Grid Array Packaging," *Proceedings of NEPCON East,* June 1993, pp. 413–422.
116. Johnson, R., and D. Cawthon, "Thermal Characterization of 140 and 225 Pin Ball Grid Array Packages," *Proceedings of NEPCON East,* June 1993, pp. 423–430.
117. Ho, T., J. Lee, R. Lee, and A. Lin, "Linear Finite Element Stress Simulation of Solder Joints on 225 I/O Plastic BGA Package Under Thermal Cycling," *Proceedings of IEEE Electronic Components & Technology Conference,* May 1995, pp. 930–936.
118. Ju, T., Y. Lee, S. Hareb, and Y. Chan, "An Integrated Model for Ball Grid Array Solder Joint Reliability," *Proceedings of the ISHM/IEEE ICEMCM,* April 1995, pp. 308–313.
119. Nagaraj, B., and M. Mahalingam, "Package-to-Board Attach Reliability—Methodology and Case Study on OMPAC Package," *ASME Paper EEP-4-1,* November 1993.
120. Parry, J., H. Rosten, and G. Kromann, "The Development of Component-Level Thermal Compact Models of a C4/CBGA Interconnect Technology: The Motorola PowerPC 630 and PowerPC 604 RISC Microprocessors," *Proceedings of IEEE Electronic Components & Technology Conference,* May 1996, pp. 195–204.
121. Shu, W., "PBGA Wire Bonding Development," *Proceedings of IEEE Electronic Components & Technology Conference,* May 1996, pp. 219–225.
122. Chan, B., and P. Singh, "BGA Socket—A Dendritic Solution," *Proceedings of IEEE Electronic Components & Technology Conference,* May 1996, pp. 460–466.
123. Shid, D., P. Lauro, K. Fogel, B. Beaman, Y. Liao, and J. Hedrick, "New Ball Grid Array Module Test Sockets," *Proceedings of IEEE Electronic Components & Technology Conference,* May 1996, pp. 467–470.

124. Kromann, G., "Thermal Management of a C4/CBGA Interconnect Technology for a High-Performance RISC Microprocessor: The Motorola PowerPC 620 Microprocessor," *Proceedings of IEEE Electronic Components & Technology Conference,* May 1996, pp. 652–659.

125. Tan, G., C. Hoo, G. Chew, J. Low, N. Tay, K. Chakravorty, and T. Lim, "Reliability Assessment of BGA Packages," *Proceedings of IEEE Electronic Components & Technology Conference,* May 1996, pp. 687–693.

126. Liang, D., "Warpage Study of Glob Top Cavity-up EPBGA Packages," *Proceedings of IEEE Electronic Components & Technology Conference,* May 1996, pp. 694–701.

127. Carichner, K., and S. Dandia, "Enhanced PBGA—The Next Generation," *Proceedings of IEEE Electronic Components & Technology Conference,* May 1996, pp. 702–706.

128. Domadia, A., and D. Mendoza, "TBGA Bond Process for Ground and Power Plane Connections," *Proceedings of IEEE Electronic Components & Technology Conference,* May 1996, pp. 707–712.

129. Mulgaonker, S., G. Hawkins, K. Ramakrishna, A. Mawer, and E. Winkler, "PBGA for High Power: Extending the Thermal Envelope," *Proceedings of IEEE Electronic Components & Technology Conference,* May 1996, pp. 713–718.

130. Abbott, J., G. Hamilton, N. Kalidas, M. Murtuza, C. Thornton, S. Thomas, Y. Umeda, D. Malladi, D. Towne, and S. Chao, "Design and Development of a High Performance PBGA Package for the UltraSPARC-I Processor," *Proceedings of IEEE Electronic Components & Technology Conference,* May 1996, pp. 719–726.

131. Matsuda, S., K. Kata, H. Nakajima, and E. Hagimoto, "Development of Molded Fine-Pitch Ball Grid Array (FPBGA Using Through-Hole Bonding Process," *Proceedings of IEEE Electronic Components & Technology Conference,* May 1996, pp. 727–732.

132. Yip, L., T. Massingill, and H. Naini, "Moisture Sensitivity Evaluation of Ball Grid Array Packages," *Proceedings of IEEE Electronic Components & Technology Conference,* May 1996, pp. 829–835.

133. Evans, T., "Practical Considerations for the Design, Performance, and Application of Plastic BGA Packages," *Proceedings of IEEE Electronic Components & Technology Conference,* May 1996, pp. 875–883.

134. Chung, T., F. Ghaghahi, B. Oberlin, D. Carey, and D. Nelson, "Area Array Packaging Technologies for High-Performance Computer Workstations and Multiprocessors," *Proceedings of IEEE Electronic Components & Technology Conference,* May 1996, pp. 902–910.

135. Tsai, C., "Inductance and SSN Performance Comparison of a 225 Plastic BGA," *Proceedings of IEEE Electronic Components & Technology Conference,* May 1996, pp. 918–924.

136. Lall, P., G. Gold, B. Miles, K. Banerji, P. Thompson, C. Koehler, and I. Adhihetty, "Reliability Characterization of the SLICC Package," *Proceedings of IEEE Electronic Components & Technology Conference,* May 1996, pp. 1202–1210.

137. Syed, A., "Thermal Fatigue Reliability Enhancement of Plastic Ball Grid Array (PBGA) Packages," *Proceedings of IEEE Electronic Components & Technology Conference,* May 1996, pp. 1211–1216.

138. Gainey, T., M. Stover, and M. Auray, "Reliability Evaluations on a New Tape Ball Grid Array (TBGA)," *Proceedings of IEEE Electronic Components & Technology Conference,* May 1996, pp. 1217–1221.

139. Delisle, G., and E. Dibble, "Transition: Lessons Learned in the Development and Production of IBM's μLaminate PBGA," *Proceedings of IEEE Electronic Components & Technology Conference,* May 1996, pp. 1244–1250.

140. Tostado, S., and J. Chow, "Assembly Process and Solder Joint Integrity of the Metal Ball Grid Array (MBGA) Package," *Proceedings of IEEE Electronic Components & Technology Conference,* May 1996, pp. 1265–1270.

141. Karnezos, M., M. Goetz, F. Dong, A. Ciaschi, and N. Chidambaram, "Flex Tape Ball Grid Array," *Proceedings of IEEE Electronic Components & Technology Conference,* May 1996, pp. 1271–1277.

6

Solder Joint Reliability of Flip Chip Assemblies

The solder-bumped flip chip technology was introduced by IBM in the early 1960s for its solid logic technology (SLT), which became the foundation of the IBM System/360 computer line. The so-called C4 (controlled-collapse chip connection) technology utilizes solder bumps deposited on wettable metal terminals on the chip and a matching footprint of solder wettable terminals on the substrate. The solder-bumped flip chip is aligned to the substrate, and all solder joints are made simultaneously by reflowing the solder.[1–168] In this chapter, solder joint reliability of different kinds of high and low melting point solders on different types of substrates (e.g., ceramic and FR-4 PCB) will be considered.

6.1 Solder-Bumped Flip Chips on Ceramic Substrates

In this section, the responses of a self-stretching flip chip solder joint under thermal fatigue crack propagation are studied. Emphasis is placed on the determination of the J-integral and stress intensity factor around the crack tips of the solder joint by finite element methods. A fifth order polynomial is used to best-fit the J-integral and stress intensity factor as a function of the crack length in the solder joint. Thermal fatigue life of the solder joint is then estimated based on the calculated stress intensity factor, Paris's law, and fatigue crack growth rate data on solders. Furthermore, a correlation between the analytical and experimental results is also made.

Figure 6.1a shows a set of self-stretching flip chip (SSFC) solder joints connecting an Si chip and an alumina substrate.[32] Figure 6.1b

220   Chapter Six

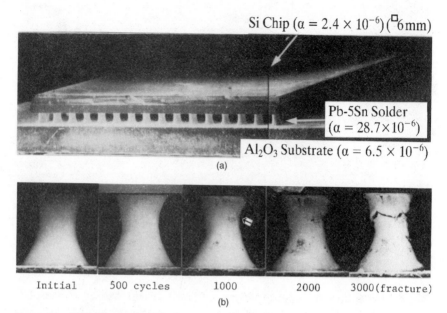

Si Chip ($\alpha = 2.4 \times 10^{-6}$) (□6mm)

Pb-5Sn Solder
($\alpha = 28.7 \times 10^{-6}$)

Al$_2$O$_3$ Substrate ($\alpha = 6.5 \times 10^{-6}$)

(a)

Initial   500 cycles   1000   2000   3000(fracture)

(b)

Figure 6.1 (a) Self-stretching flip chip solder joints. (b) Self-stretching flip chip solder joints subjected to thermal cycling.[32]

shows a corner SSFC solder joint subjected to 3000 thermal cycles (–50 to +150°C at 1 cycle/h).[32] It can be seen that the corner SSFC solder joint cracked through its cross section. The purpose of this section is to model (by means of fracture mechanics) the thermal fatigue crack propagation of the corner SSFC solder joint based on the plane strain theory.[37] Similar work based on the plane stress theory has been reported by Lau.[36]

6.1.1 Boundary-value problem

The horizontal dimensions of the Si chip are about 6 by 6 mm (Fig. 6.1). The coefficient of thermal expansion is 2.4 ppm/°C for the Si chip and 6.5×10^{-6}/°C for the Al$_2$O$_3$ substrate. The geometry of the SSFC solder joint is shown in Figs. 6.1, 6.2, and 6.3. Figure 6.3 shows that the joint is 0.32 mm high and the diameters of the bottom and top surfaces are 0.38 and 0.29 mm, respectively. The solder joint is made of 5wt%Sn/95wt%Pb and has a Young's modulus $E = 2600$ MPa, yield stress $\sigma_y = 18$ MPa, Poisson's ratio $\nu = 0.3$, and coefficient of thermal expansion $\alpha = 28.7 \times 10^{-6}$/°C.

Because of the assembly's global thermal expansion mismatch between the Si chip ($4.24 \times 200 \times 0.0000024 = 0.002$ mm) and the Al$_2$O$_3$ substrate ($4.24 \times 200 \times 0.0000065 = 0.0055$ mm), and the thermal

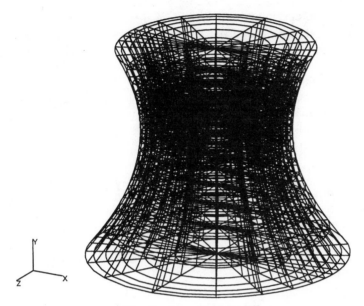

Figure 6.2   Geometry of the flip chip solder joint.[36,37]

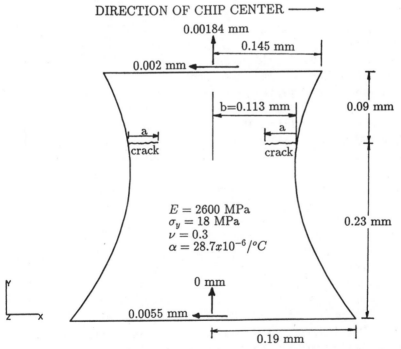

Figure 6.3   Material properties, boundary conditions, and geometry of the flip chip solder joint.[36,37]

expansion of the solder joint in the vertical direction ($0.32 \times 200 \times 0.0000287 = 0.00184$ mm), the corner solder joint is subjected to a complex state of stress and strain during the temperature cycling. (The local thermal expansion mismatch between the Si chip and solder and the Al_2O_3 substrate and solder are not considered in this study.) These stresses and strains produce the driving force for solder joint failures. Since most of the thermal fatigue life of ductile materials is spent in propagating the crack (i.e., fatigue crack growth),[44] the strains and stresses around the crack tip (J-integral and stress intensity factor) for different crack lengths in the solder joint are of the utmost interest. In this study, seven different crack lengths were investigated. The boundary conditions imposed on the corner SSFC solder joint are shown in Fig. 6.3, where the bottom surface of the joint is subjected to a 0.0055 mm displacement moving to the left, and the top surface of the joint is subjected to a 0.002 mm displacement moving to the left and a 0.00184 mm displacement moving in the upward direction. Only linear analysis is presented.

6.1.2   Analysis and results

The finite element model (dotted lines) and deformation (solid lines) of the corner SSFC solder joint with no crack, subjected to a $\Delta T = 200°C$, are shown in Fig. 6.4a. Because of the symmetry of the problem, only half of the solder joint was modeled. A high-order 3D solid element was used for the model. Each element had 20 nodes. Each node had 3 degrees of freedom. It can be seen from Fig. 6.4a that the corner SSFC solder joint moves away from the center of the chip because of the thermal expansion of the Si chip and Al_2O_3 substrate. It can also be seen that the lower part of the solder joint moves more than the upper part.

The von Mises stress (MPa) contours in the SSFC solder joint are shown in Fig. 6.4b. It can be seen that the maximum stress (42.1 MPa) occurs near the throat of the solder joint (see also Fig. 6.1). Thus, any solder joint cracks should begin at this location. In this study, we modeled the crack propagation near the throat of the solder joint.

The 3D calculation of the stress intensity factor and J-integral for various crack lengths of the SSFC solder joint (Figs. 6.2 and 6.3) is very time consuming. Thus, in the present study, only 2D plane strain theory is used. Figure 6.5 shows the von Mises stress (MPa) contours in the solder joint. A high-order 2D plane strain element was used for the model. Each element has eight nodes. Each node has 2 degrees of freedom. It can be seen from Figs. 6.4b and 6.5 that the stress distribution for the 3D and 2D cases is very similar except for the stress magnitude (the 2D results are 15 percent less than the 3D results).

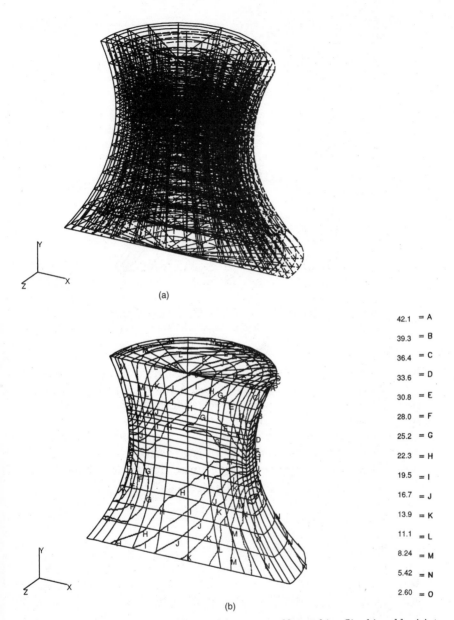

42.1	= A
39.3	= B
36.4	= C
33.6	= D
30.8	= E
28.0	= F
25.2	= G
22.3	= H
19.5	= I
16.7	= J
13.9	= K
11.1	= L
8.24	= M
5.42	= N
2.60	= O

(a)

(b)

Figure 6.4 (*a*) Deformation (solid lines) of the corner self-stretching flip chip solder joint (no crack). (*b*) 3D von Mises stress distribution (MPa) in the self-stretching flip chip solder joint (no crack).[36,37]

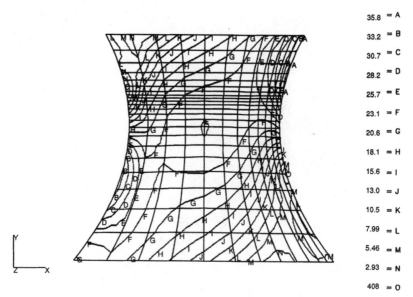

35.8	= A
33.2	= B
30.7	= C
28.2	= D
25.7	= E
23.1	= F
20.6	= G
18.1	= H
15.6	= I
13.0	= J
10.5	= K
7.99	= L
5.46	= M
2.93	= N
408	= O

Figure 6.5 2D von Mises stress distribution (MPa) in the self-stretching flip chip solder joint (no crack).[36,37]

Figure 6.6 shows the deformed (solid lines) and undeformed (dotted lines) shapes of the corner SSFC solder joint with a crack near the throat. The crack length is $a = 0.00565$ mm ($\eta = a/b = 0.05$). The element type and boundary conditions are exactly the same as those of the case with no crack. In this analysis, ΔJ was calculated at three contours around the crack tip. The average value is shown in Fig. 6.6 ($\Delta J = 0.0093$ N/mm). Once ΔJ is determined, the stress intensity factor ΔK for a plane strain condition can be determined by

$$\Delta J = (1 - \nu^2)\frac{\Delta K_I^2 + \Delta K_{II}^2}{E} + \frac{\Delta K_{III}^2}{2G} \qquad (6.1)$$

where ΔK_I is the stress intensity factor for a mode I (opening)–type of fracture, ΔK_{II} is the stress intensity factor for a mode II (in-plane shear)–type of fracture, ΔK_{III} is the stress intensity factor for a mode III (transverse shear)–type of fracture, E is the Young's modulus, ν is the Poisson's ratio, and G is the shear modulus. In this study, $\Delta K_{II} \approx 0$, and $\Delta K_{III} = 0$, then

$$\Delta K_I = \Delta K = \sqrt{\frac{E\Delta J}{(1 - \nu^2)}} \qquad (6.2)$$

Thus, for a crack length $a = 0.00565$ mm ($\Delta J = 0.0093$ N/mm), $\Delta K = 5.16$ MPa$\sqrt{\text{mm}}$ (Fig. 6.6).

$$\eta = a/b = 0.05$$
$$\Delta J = 0.0093 \text{ N/mm}$$
$$\Delta K = 5.16 \text{ MPa}\sqrt{mm}$$

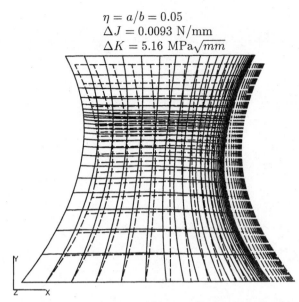

Figure 6.6 Deformation of the corner self-stretching flip chip solder joint with a crack length $a = 0.00565$ mm.[36,37]

Figures 6.7 through 6.12 show the deformed (solid lines) and undeformed (dotted lines) shapes of the corner SSFC solder joint with crack lengths a = 0.0113, 0.0226, 0.0339, 0.0565, 0.0791, and 0.1017 mm, respectively. It can be seen that for a = 0.0113 mm, ΔJ = 0.0162 N/mm and ΔK = 6.81 MPa$\sqrt{mm}$; for a = 0.0226 mm, ΔJ = 0.0245 N/mm and ΔK = 8.37 MPa$\sqrt{mm}$; for a = 0.0339 mm, ΔJ = 0.0284 N/mm and ΔK = 9.01 MPa$\sqrt{mm}$; for a = 0.0565 mm, ΔJ = 0.0285 N/mm and ΔK = 9.02 MPa$\sqrt{mm}$; for a = 0.0791 mm, ΔJ = 0.0286 N/mm and ΔK = 9.04 MPa$\sqrt{mm}$; and for a = 0.1017 mm, ΔJ = 0.0287 N/mm and ΔK = 9.06 MPa$\sqrt{mm}$. These values are plotted in Figs. 6.13 and 6.14 for ΔJ and ΔK, respectively. The best-fit equations for ΔJ and ΔK are, respectively,

$$\Delta J = 0.01(-0.02 + 21.97\eta - 62.80\eta^2 + 84.28\eta^3 - 52.85\eta^4 + 12.20\eta^5) \quad (6.3)$$

and

$$\Delta K = 2.87 + 54.97\eta - 189.18\eta^2 + 314.22\eta^3 - 252.81\eta^4 + 79.17\eta^5 \quad (6.4)$$

where

$$\eta = \frac{a}{b} \quad (6.5)$$

Figure 6.15a shows the von Mises stress distribution (MPa) in the SSFC solder joint with a crack length a = 0.0339 mm (Fig. 6.9). A very

$$\eta = a/b = 0.1$$
$$\Delta J = 0.0162 \text{ N/mm}$$
$$\Delta K = 6.81 \text{ MPa}\sqrt{mm}$$

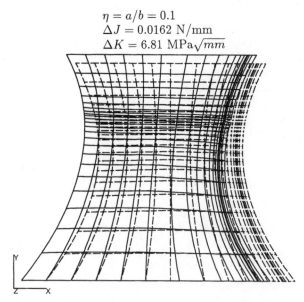

Figure 6.7 Deformation of the corner self-stretching flip chip solder joint with a crack length $a = 0.0113$ mm.[36,37]

$$\eta = a/b = 0.2$$
$$\Delta J = 0.0245 \text{ N/mm}$$
$$\Delta K = 8.37 \text{ MPa}\sqrt{mm}$$

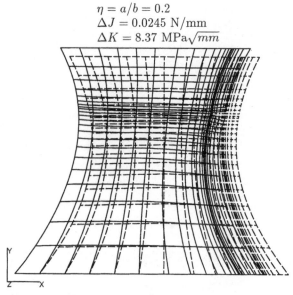

Figure 6.8 Deformation of the corner self-stretching flip chip solder joint with a crack length $a = 0.0226$ mm.[36,37]

$$\eta = a/b = 0.3$$
$$\Delta J = 0.0284 \text{ N/mm}$$
$$\Delta K = 9.01 \text{ MPa}\sqrt{mm}$$

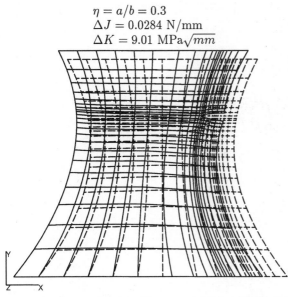

Figure 6.9 Deformation of the corner self-stretching flip chip solder joint with a crack length a = 0.0339 mm.[36,37]

$$\eta = a/b = 0.5$$
$$\Delta J = 0.0285 \text{ N/mm}$$
$$\Delta K = 9.02 \text{ MPa}\sqrt{mm}$$

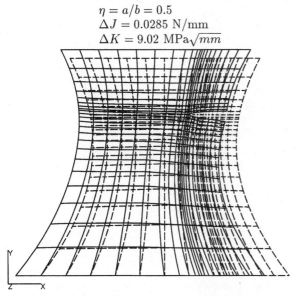

Figure 6.10 Deformation of the corner self-stretching flip chip solder joint with a crack length a = 0.0565 mm.[36,37]

$$\eta = a/b = 0.7$$
$$\Delta J = 0.0286 \text{ N/mm}$$
$$\Delta K = 9.04 \text{ MPa}\sqrt{mm}$$

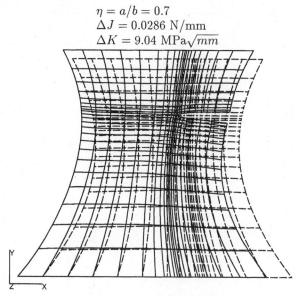

Figure 6.11 Deformation of the corner self-stretching flip chip solder joint with a crack length $a = 0.0791$ mm.[36,37]

$$\eta = a/b = 0.9$$
$$\Delta J = 0.0287 \text{ N/mm}$$
$$\Delta K = 9.06 \text{ MPa}\sqrt{mm}$$

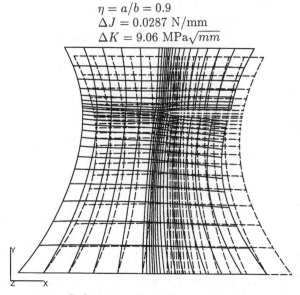

Figure 6.12 Deformation of the corner self-stretching flip chip solder joint with a crack length $a = 0.1017$ mm.[36,37]

$$\Delta J = 0.01(-.02 + 21.97\eta - 62.80\eta^2 + 84.28\eta^3 - 52.85\eta^4 + 12.20\eta^5)$$

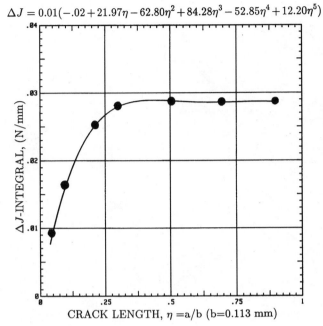

Figure 6.13 J-integral versus crack length.[36,37]

$$\Delta K = 2.87 + 54.97\eta - 189.18\eta^2 + 314.22\eta^3 - 252.81\eta^4 + 79.17\eta^5$$

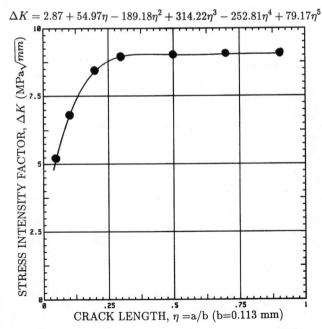

Figure 6.14 Stress intensity factor K versus crack length.[36,37]

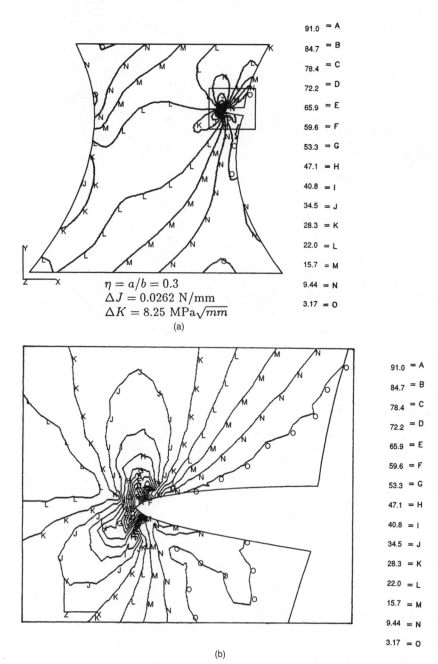

$$\eta = a/b = 0.3$$
$$\Delta J = 0.0262 \text{ N/mm}$$
$$\Delta K = 8.25 \text{ MPa}\sqrt{mm}$$

(a)

91.0	= A
84.7	= B
78.4	= C
72.2	= D
65.9	= E
59.6	= F
53.3	= G
47.1	= H
40.8	= I
34.5	= J
28.3	= K
22.0	= L
15.7	= M
9.44	= N
3.17	= O

(b)

Figure 6.15 (a) von Mises stress distribution (MPa) in the flip chip solder joint with a crack length a = 0.0339 mm. (b) von Mises stress distribution (MPa) around the crack tip with a crack length a = 0.0339 mm.[36,37]

large stress concentrates around the crack tip shown in Fig. 6.15b. This stress, however, becomes much smaller as soon as it moves away from the crack tip. In fact, except near the crack tip, the stresses in the cracked solder joint (Figs. 6.15a and b) are smaller than those in the uncracked solder joint (Fig. 6.5).

6.1.3 Thermal fatigue life by fracture mechanics

Paris's law is expressed as[36,37]

$$\frac{da}{dN} = \gamma \Delta K^{\beta} \tag{6.6}$$

where a is crack length, N is number of cycles to failure, and γ and β are material constants. For 63wt%Sn/37wt%Pb solder, γ and β have been determined by Logsdon, Liaw, and Burke[33] at 24°C and 10 cycle/sec. Substituting their values for $\gamma = 2.77 \times 10^{-7}$ and $\beta = 3.26$ into Paris's equation, we have

$$N = \int \frac{10^7 \, da}{2.77(\Delta K)^{3.26}} \tag{6.7}$$

In this equation, a is in cm, N is in cycles, and ΔK is in MPa$\sqrt{\text{m}}$. Thus, fatigue life of a solder joint can be estimated as soon as ΔK is a function of crack length a.

Substituting Eq. (6.4) for Eq. (6.7) and noting the units of the crack length a and ΔK, we have

$$N = \int_{0.05}^{0.9} \frac{0.113(10^7) \, d\eta}{(10)2.77(0.03162)^{3.26}(\Delta K)^{3.26}} \tag{6.8}$$

or

$$N = \int_{0.05}^{0.9} \frac{3.1675(10^9) \, d\eta}{[2.87 + 54.97\eta - 189.18\eta^2 + 314.22\eta^3 - 252.81\eta^4 + 79.17\eta^5]^{3.26}} \tag{6.9}$$

After integration, we have $N = 2,620,000$ cycles.

As mentioned earlier, the solder material constants determined by Logsdon et al.[33] are at 10 cycles/sec and 24°C. However, the thermal cycling results obtained by Satoh et al.[32] are at 1 cycle/h and from −50 to +150°C. Thus, a factor which captures the frequency and temperature effects is needed to transfer the lower stress condition to the higher stress condition. In this study, we assume the thermal fatigue life N_f of the corner SSFC solder joint is

$$N_f = \lambda N \tag{6.10}$$

where
$$\lambda = \left(\frac{f_o}{f_t}\right)^{1/3} \left(\frac{\text{Max } T_t}{\text{Max } T_o}\right)^2 \tag{6.11}$$

For the present study, $f_o = 1$ cycle/h, $f_t = 10$ cycle/sec, Max $T_t = 24°C$, Max $T_o = 150°C$. Then,

$$\lambda = (1/36000)^{1/3}(24/150)^2 = 0.00078 \tag{6.12}$$

and
$$N_f = (0.00078)(2620000) = 2030 \tag{6.13}$$

This value is 32 percent lower than that (3000) of Satoh et al.[32] This may be caused by the 2D plane strain theory (for the 3D boundary-value problem) and the fatigue crack growth data of the 63wt%Sn/37wt%Pb solder (for the 5wt%Sn/95wt%Pb solder).

6.1.4 Recommendations

1. The finite element analysis showed that the maximum stress occurred near the throat (or neck) of the corner SSFC solder joint. This result agreed very well with the observed failure location.

2. Based on the plane strain theory, the ΔJ and ΔK have been calculated for seven different crack lengths of the corner SSFC solder joint. Also, the ΔJ and ΔK have been best-fitted into functions of the crack length.

3. The thermal fatigue life ($\Delta T = 200°C$, 1 cycle/h) of the corner SSFC solder joint has been estimated by the calculated ΔK, Paris's law, and solder fatigue crack growth rate data to be 2030 cycles. This compared well with the experimental results (3000 cycles).

4. The present fracture mechanics approach to predict the thermal fatigue life of solder joints was just a small step beyond the Coffin-Manson approach. However, we hope that this has opened the door for advanced solder joint fatigue analysis with the fracture mechanics method.

5. The fatigue of materials is not fully understood. This is especially true for solder alloys due to the small amount of experimental data. In order to better understand and predict the solder joint thermal fatigue life, additional mechanical properties of bulk solders and joints are needed.[44] These are: the true stress-strain-temperature-frequency-time relation, fracture toughness, and fatigue crack growth rate at various temperatures and frequencies, and the peeling and shearing fracture initiation strength at various temperatures.

6.2   Solder-Bumped Flip Chips on Low-Cost Substrates

IBM in Yasu, Japan, has been assembling solder-bumped flip chips on organic PCBs since 1990.[14–16] Applications for these flips chips include personal computers, Personal Computer Memory Card International Association (PCMCIA) cards, and token-ring local area network (LAN) adapter cards. IBM's results show that solder-bumped flip chip assembly technology is applicable for low-cost PCBs as well as high-cost substrates, such as ceramics. In this section the design, material, process, and reliability of different kinds of solder-bumped flip chips on low-cost substrates will be discussed.

6.2.1   Design and procurement of solder-bumped flip chips and organic substrates[40]

The infrastructural challenges associated with the design and procurement of solder-bumped flip chips and their corresponding organic substrates will be explained with an example from Lau et al.[40] The issues associated with transferring integrated circuit data to printed circuit design software are identified. The bare and bumped wafers are characterized to identify possible criteria for lot qualification.

6.2.1.1 Design and data transfer.   The data flow is shown in Fig. 6.16. A schematic of the desired designs was provided to both the substrate and wafer designers. The final wafer design data were used for several steps: wafer fabrication, bumping mask fabrication, and wafer dicing. The final substrate design data were transmitted as usual to the substrate supplier.

Wafer design.   The electronic layout of the wafer was carried out by the IC supplier that fabricated the wafers, based on schematic information provided to them. This conceptual design information was provided in the form of mechanical drawings that specified part locations and feature dimensions. The wafer data was output in GDS-II format (note: GDS-III format is emerging), which is a common format accepted by many wafer designers and IC fabricators. Because this format is widely used in the integrated circuit industry, we were able to transfer the data to the bump mask manufacturer without incident. There is potential for errors, however, in the step-and-repeat distance and other features if the data formats and software revisions are not identical.

Substrate design.   The substrate was designed on a CAD (computer-aided design) workstation using a standard printed circuit board

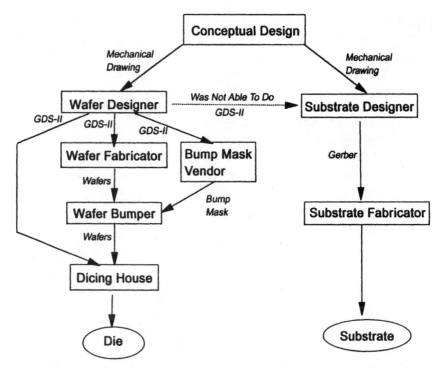

Figure 6.16 Bare die and substrate procurement flow. (Italics indicate data or objects transferred between suppliers.)[40]

design package. Again, the design was based on mechanical drawings that specified part locations and feature dimensions. The firm that did the layout would have preferred electronic data since the board designer would have been able to read-in the GDS-II design of the wafer to match the bumps on the die electronically with the pads on the board. Unfortunately, the software to do this is relatively uncommon, and drawings from the wafer designer had to be substituted.

Wafer design information could not be used in the board design because board design systems typically accept and output data in the Gerber format. Gerber is a standard format commonly used in the printed circuit board industry because it is compatible with large-format photoplotters. For a more complex die, the manual process used would be prohibitively time-consuming and likely to introduce errors.

Since the pad designs for real integrated circuit die are not symmetric, flipping the die introduces additional complexity. If pad layouts for bare die were standardized, then the bare ICs could be treated as any other component (i.e., a standard land pattern could be inserted). In the absence of standardization, this gap in data formats between the IC world, which uses GDS-II, and the world of large-scale photoplotters and organic substrates, which uses Gerber data, is a significant barrier to design with flip chip-on-board.

The output format from the substrate design package was Gerber. The difficulties associated with getting the data to substrate suppliers are not unique to flip chip designs. Different suppliers use different computer operating systems, data transfer media (tape, floppy disk, e-mail, or modem), and data compression programs. Identifying and resolving these incompatibilities caused significant time delays. Thus, it is important to set up a method to transfer data accurately with a new supplier for the first time.

6.2.1.2 The wafer

Wafer design. Figure 6.17 shows the 6-in wafer containing seven different chip designs: 4-mil square pads and 7-mil pitch, 4-mil octagon pads and 7-mil pitch, 6-mil square pads and 10-mil pitch, 6-mil octagon pads and 10-mil pitch, 8-mil square pads and 14-mil pitch, 8-mil octagon pads and 14-mil pitch, and a double row of 6-mil octagon pads and 10-mil pitch (see Fig. 6.18). All of the pads are arranged symmetrically around the perimeter of the chip. One chip design has a 5-watt heater. The chip size for all of the designs is 0.5 inch square and 25 mil

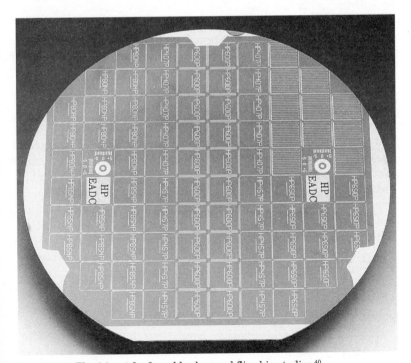

Figure 6.17 The 6-in wafer for solder-bumped flip chip studies.[40]

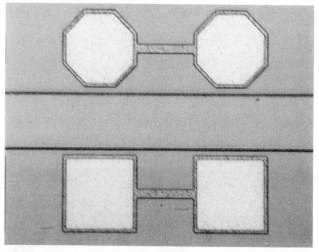

Figure 6.18   Scribe street separating chips with square and octagonal pads.[40]

thick. The street width between all the chips is 6 mil. Because the real estate of silicon is expensive, the distance between the chip pads and the passivation edge is only 12 μm (a very aggressive design). All of the pads are interconnected via traces on the chip in an alternating pattern so as to provide daisy-chained connections when the chip is soldered to the organic substrate.

Wafer fabrication. The silicon wafers consist of a patterned aluminum layer on a layer of silicon dioxide, which was covered with a patterned silicon nitride passivation layer. The wafer fabrication process flow starts with a 2500-Å layer of silicon dioxide deposited by plasma enhanced chemical vapor deposition (PECVD) on a <111> silicon substrate. For the metal layer, an 8500-Å layer of Al/1%Si/0.1%Ti alloy is sputtered over the silicon dioxide (Fig. 6.19). The metal pattern is then defined by coating with positive resist, exposed by projection alignment, developed, and wet-chemical-etched. After the photoresist is removed by plasma stripping, the metal pattern is sintered at 450°C to remove film stresses. For the passivation layer, a 7500-Å layer of silicon nitride is deposited over the entire surface of the wafer by PECVD. The pad opening in the passivation layer is then defined by coating with positive resist, exposed by projection alignment, developed, and plasma-etched. Finally, the resist is removed by plasma stripping, leaving the silicon nitride passivation layer to overlap the perimeter of the Al pads by 10 μm.

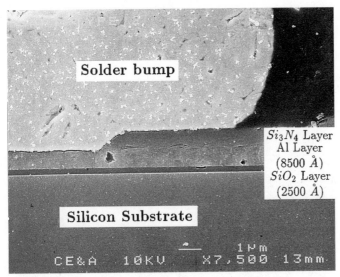

Figure 6.19   Terminal metals and solder bump on silicon.[40]

6.2.1.3   Wafer bumping with eutectic Sn/Pb solder[40]

Terminal metals and solder—materials and processes. The terminal metals of the wafers were Ti/Cu. They were sputtered over the entire surface of the wafer—1000 to 2000 Å of Ti first, followed by 5000 to 8000 Å of Cu. A 40-μm layer of resist was then overlaid on the Ti/Cu and a solder bump mask was used to define the bump pattern (Fig. 6.20a). The openings in the resist were 7 to 10 μm wider than the pad openings in the passivation layer. A 6- to 8-μm layer of Cu was then plated over the Ti/Cu, followed by electroplated 60wt%Sn/40wt%Pb solder. In order to plate enough solder to achieve a final solder ball height of 100 μm, the solder was plated over the resist coating by about 15 μm to form a *mushroom bump*. The resist was then removed and the Ti/Cu was stripped off with a hydrogen peroxide etch. The wafers were then reflowed at 215°C, which created smooth, spherical solder balls due to surface tension (Fig. 6.20b).

Wafer dicing and handling. The wafers were diced using a standard dicing process. A thin circular blade made of a nickel matrix with 35-mil-diameter synthetic diamonds was used. The blade was 1.5 mils thick and rotated at 28,000 rpm. Approximately, 2 mils of the street was removed.

After wafer dicing, die were manually removed from the backing tape and placed in carriers. Since silicon is extremely brittle, this manual process caused cracks in some die, particularly at the corners. Also,

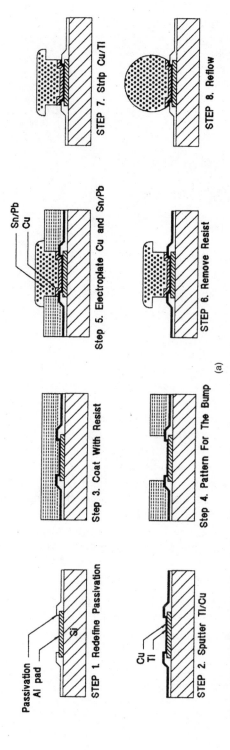

Figure 6.20 (a) Electroplated wafer bumping process flow. (b) Solder bump.[40]

238

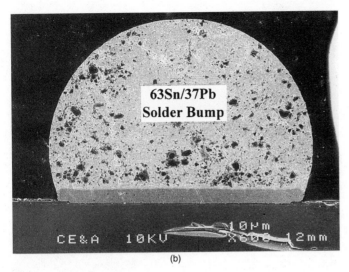

Figure 6.20   (*Continued*)

because we were unaccustomed to working with solder-bumped die, some of the bumps were squashed by the tweezers used to move them. Typical die with Al bond pads would not be damaged by this process.

6.2.1.4 Solder bump characterization.

Since one purpose of this evaluation was to exercise the procurement process, the as-received wafers were extensively characterized. One goal was to better understand which characteristics must be specified to qualify supplier lots.

Solder bump height measurement.   The solder bump height distribution on the wafers was characterized because large variations can cause problems during the assembly process. Solder bump heights were measured on several wafers; their average height was about 100 μm. For example, Fig. 6.21*a* and *b* shows the distribution of solder bump height measurements of balls on 4-mil square and 4-mil octagon pads, respectively. The pad diameter had the greatest impact on ball height (Fig. 6.22). The average ball height on the 8-mil-diameter pads was 0.32 mil, which is greater than that for the 6-mil-diameter pads and 1.04 mils greater than balls on the 4-mil-diameter pads. The pad location on the wafer had a smaller effect (Fig. 6.23). The average ball height at the edge of the wafer was 0.53 mil greater than balls at the center of the wafer. This difference was caused by current density variations across the wafer during the electroplating process. There was no statistical difference between ball heights on the octagon and the square pads.

Solder bump strength measurement.   The shear (push) strength of the solder balls was measured on a Dage automated test machine at a speed

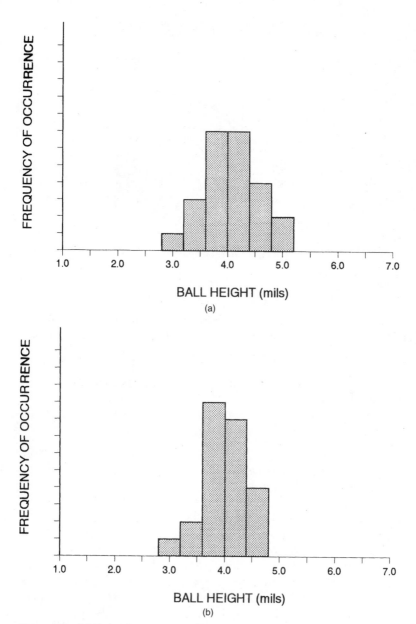

Figure 6.21 (*a*) Bump height of 4-mil octagon pads. (*b*) Bump height of 4-mil square pads.[40]

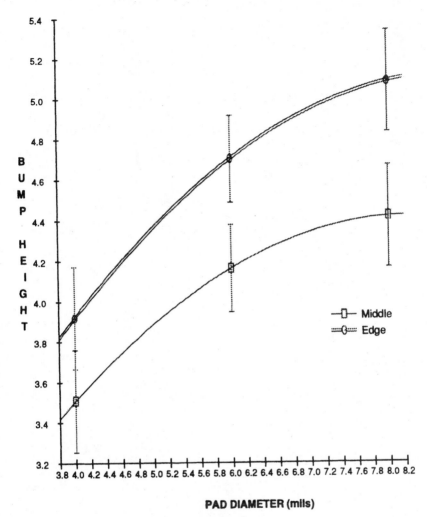

Figure 6.22   Effect of pad size on bump height.[40]

of 100 μm/sec with the shear blade tip 25 μm from the wafer surface (i.e., about one-fourth of the solder bump height). The shear strength increased with increasing pad size (Table 6.1). Again, there was no statistical difference between the octagon and square pads. Figure 6.24 shows the fractured surface of a sheared solder bump. The failure location was in the solder and not at the terminal metals. It should be noted that if the failure location is at the under bump metal, that means the wafer bumping process is poor and the chip may fall off during reflow.

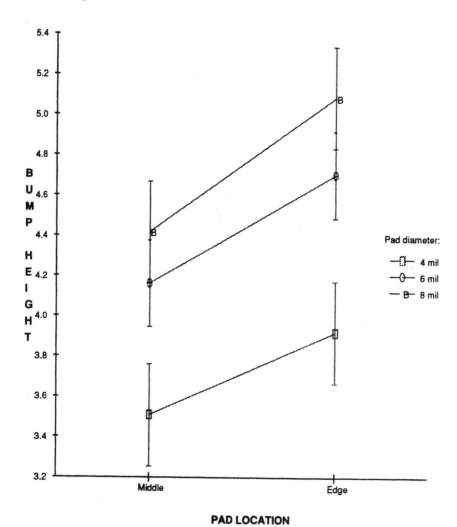

Figure 6.23 Bump height versus pad location on the wafer.[40]

SEM of solder bumps. Four important sources of solder bump defects were characterized by SEM (scanning electron microscopy) of the cross sections: (1) undercutting of the pad or the scribe edge, (2) integrity of the terminal metal and underlying Al pad, (3) adhesion of the metallic interfaces, and (4) distribution of Sn and Pb within the solder bump.

The electroplating process for wafer bumping uses the terminal metal layer as a conductive film during solder plating. Excessive overhang at either the pad or the edge of the scribe line causes step coverage problems during sputter or evaporative deposition of the terminal metal layer. Figure 6.25 shows excessive overhang at the edge of the

TABLE 6.1   Shear Force Measurement
Results for Square and Octagon Pads[40]

Pad size	Shear force (gm)	
	Square	Octagon
4 mil	35	34
6 mil	64	60
8 mil	96	95

scribe. Inadequate step coverage causes reduced or no solder plating at all. This illustrates the importance of communication between the wafer fabricator and the wafer bumper on the requirements of the wafer-bumping process.

The terminal metal adheres to the Al pad and forms a wettable surface for solder bumps. Thus, an unbroken terminal metal layer is required between the solder and the Al pad. Figure 6.26 shows a broken terminal metal layer.

The Al pad, terminal metal, and solder bump must form an integrated structure as indicated by the intermetallic growth and edge wetting shown in Fig. 6.27. A gap or crack at the interfaces between the Al, terminal metal, or the solder bump may indicate residual oxides at the interface of metal layers, possibly caused by vacuum leaks in the deposition system or inadequate metal surface preparation.

The Sn- and Pb-rich regions should be uniformly distributed within the solder ball as seen in Fig. 6.28. The cooling rate after the bumps are reflowed determines the scale of the microstructure and the extent to which dendrites are present.

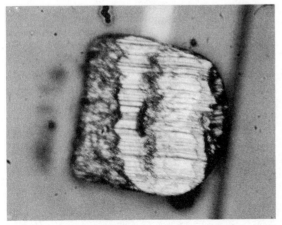

Figure 6.24   Fractured surface of a sheared solder bump.[40]

OVERHANG AT THE EDGE OF SCRIBE LINE

Figure 6.25   Overhang at the edge of scribe line.[40]

6.2.1.5 Procurement of solder-bumped chips.   There are at least four different ways to obtain solder-bumped chips and they are shown in Figs. 6.29 through 6.32. In the first method (Fig. 6.29), the first step is to place the order for the wafers with the IC vendor. It tests and sends the wafers to its preferred bumper for solder-bumping. After the wafers are solder-bumped, they are sent back to the IC vendor for testing, dicing, and sorting. Finally, the IC vendor ships the known good die (KGD) to the end user.

BROKEN TERMINAL METAL

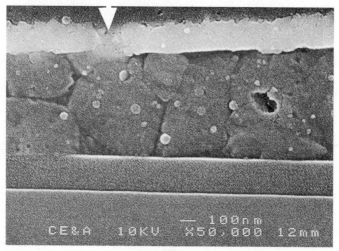

Figure 6.26   A broken terminal metal layer.[40]

Figure 6.27   Intermetallic growth at interfaces.[40]

Figure 6.30 shows the second method of obtaining solder-bumped chips. After the IC vendor receives the order, it fabricates and tests the wafers and returns them to the end user. The end user sends the chips to a bumper for solder-bumping and then receives the bumped wafers. The bumped wafers are then sent back to the IC vendor for testing, dicing, and sorting. Finally, the KGD is received from the IC vendor.

The third method is shown in Fig. 6.31, where the end user is provided with the wafers and a corresponding wafer map to indicate the locations of good and bad IC chips from the IC vendor. Once the wafers get out of the door, the IC vendor does not want to hear anything about these wafers again. In this case, the end user manages the bumping,

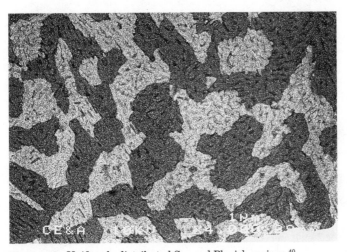

Figure 6.28   Uniformly distributed Sn- and Pb-rich regions.[40]

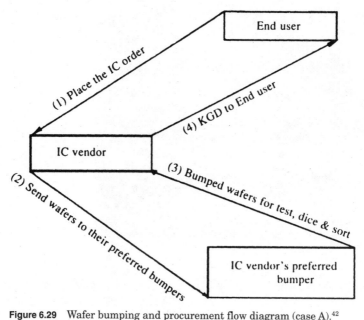

Figure 6.29   Wafer bumping and procurement flow diagram (case A).[42]

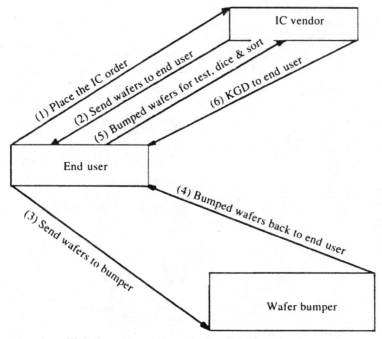

Figure 6.30   Wafer bumping and procurement flow diagram (case B).[42]

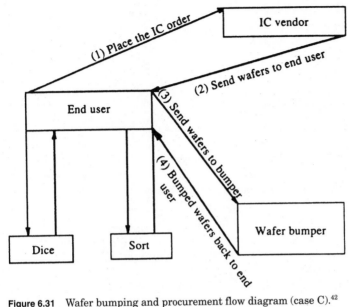

Figure 6.31   Wafer bumping and procurement flow diagram (case C).[42]

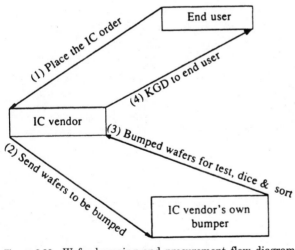

Figure 6.32   Wafer bumping and procurement flow diagram (case D).[42]

dicing, and sorting, and the wafers or the chips are neither tested nor burned-in after bumping for KGDs.

The last case is the most ideal method for solder-bumped flip chip on board applications (i.e., place the order with the IC vendor and receive the KGD from it; see Fig. 6.32). It should be noted that this is more desirable than the procurement flow in the first case where, even though the IC vendor is willing to accept responsibility for bump quality from its preferred bumper, communication problems existed with the bump vendor as well as extra shipping and handling requirements. Furthermore, such an arrangement has the disadvantage that people other than the IC vendor see the wafers.

6.2.1.6 Test board

Board design. A matching test board was designed along with the wafer. It was designed to accommodate 30 die on several different land patterns with various electrical trace layouts and solder mask designs.

Figure 6.33 shows the test board layout. Since one of the die designs has a 7-mil pitch, the 4-mil-diameter pads have 3-mil spacing and the traces exiting those pads were specified to be 3 mils wide. Such features are beyond the production capabilities of many substrate vendors.

The board has five columns of different land patterns. The pad diameters and pitches match the dimensions on the die, except that the pads are round, not square or octagonal.

The board has two basic electrical trace layouts: a single daisy chain and a double daisy chain. The former connects all of the solder joints in series for opening measurements; the latter has outer and inner daisy chains connected to every other two solder joints. The double daisy chain further provides detection of solder bridging and current leakage across the chains. Because the wafer has only one metal layer, a maximum of 50 percent of the solder connections can be tested for bridging. There are also electrical inputs for four-terminal resistance measurements on some of the solder joints and power connections for the heater chip.

Test board fabrication and characterization. The substrates were fabricated by several different suppliers. The features of the test boards characterized included pad width, height, and substrate flatness.

6.2.1.7 Summary. Solder-bumped wafers and matching substrates were designed and fabricated using external suppliers to test the industry infrastructure. Difficulties were encountered in moving data between suppliers tied to the integrated circuit industry and those tied to large-format organic substrates. Incompatibilities in computer operating systems, data transfer media, and data compression software caused further delays.

Although fabrication of the die and substrates was successful, the industry infrastructure was still quite immature. Despite predictions that flip chip technology will be widely used by the end of the decade, improvements in the following areas are urgently needed: standardization of data formats and transfer methods, pad layouts, shipping and handling processes for bumped bare die, and criteria for qualifying

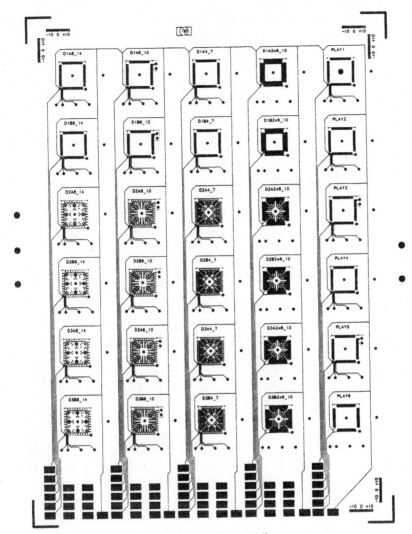

Figure 6.33   Top signal layer of the FR-4 PCB board.[40]

incoming lots of bumped wafers. Because this evaluation did not use functional die, we did not experience other well-documented problems such as burn-in and test to provide known good die (KGD); see Ref. 34.

6.2.2   Thermal reliability of high-temperature solder-bumped flip chips on boards

IBM Yasu's direct chip attach (DCA) technology (Fig. 6.34) is based on IBM's C4 technology, namely, the use of the 5wt%Sn/95wt%Pb (liq-

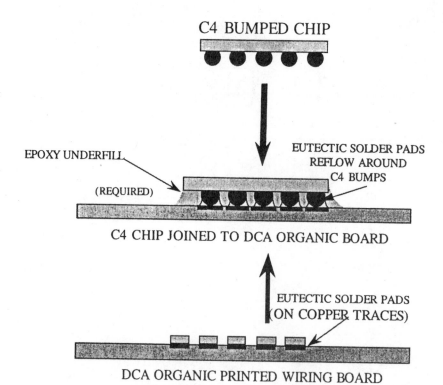

Figure 6.34   Low-cost C4 bumped chip on eutectic-solder-coated PCB.

uidus temperature = 312°C) or 3wt%Sn/97wt%Pb (liquidus temperature = 320°C) solders.[14–16] In order to be SMT-compatible (at a maximum reflow temperature of about 240°C), the Cu pads on the PCB are coated with 63wt%Sn/37wt%Pb solder using either IBM's own molten-solder injection machine (Fig. 6.35) or conventional electroplating methods. However, both methods are quite expensive compared to Motorola's E-3 (evaporated, extended, eutectic) approach.[57,65]

The E-3 method takes advantage of existing C4 technology to create a new, on-chip solder bump structure[57,65] (Fig. 6.36a, b, and c). These E-3 solder-bumped flip chips can be attached to the PCB without intermediate eutectic deposits on the board. Also, the E-3 technology may be used to join chips to a variety of low-cost substrates, such as OCC (organic coated copper) boards (Fig. 6.37).

The problem with a large thermal expansion mismatch between the silicon chip and the PCB is reduced by filling the gap between the chip and PCB with a liquid epoxy resin. Due to capillary action, this resin fills the space beneath the chip. The chip and PCB are then firmly bonded by curing the resin (Figs. 6.34 and 6.37). The underfill encap-

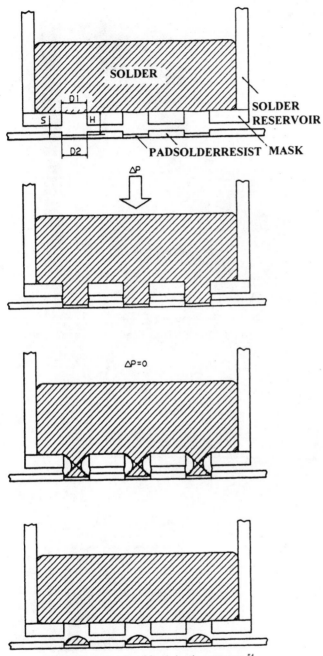

Figure 6.35 IBM at Yasu's solder injection process.[54]

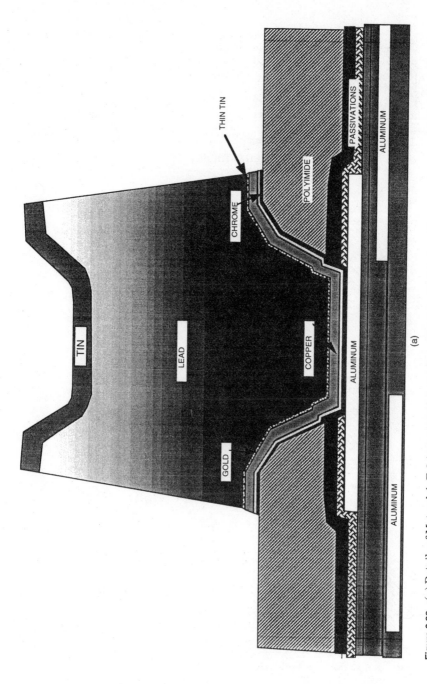

TIN

THIN TIN

CHROME

LEAD

COPPER

GOLD

POLYIMIDE

PASSIVATIONS

ALUMINUM

ALUMINUM

ALUMINUM

(a)

Figure 6.36 (a) Details of Motorola's E-3 (evaporated, extended, eutectic) solder bump; (b) a cross section of E-3 solder bump; (c) 3-D view of the E-3 bumps. After Greer.[57]

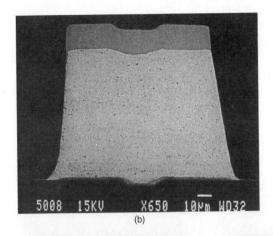
(b)

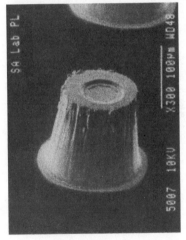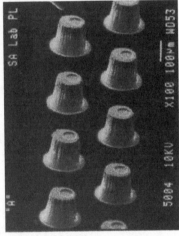
(c)

Figure 6.36   (Continued)

sulant is one key to the solder-bumped flip chip on PCB technology; it
will be presented in more detail in Sec. 6.2.4.

6.2.2.1   Thermal cycling test of high-temperature solder-bumped flip chips on PCB.

The temperature cycling test results of a high-temperature
solder-bumped flip chip on organic PCB are shown in Fig. 6.38.[138,139]
The silicon chip was 7.12×7.12 mm and it had perimeter pads with
95Pb/5Sn solder bumps. The substrate was a conventional FR-4 PCB
of 0.51-mm thickness. The copper pads on the PCB were coated with
63wt%Sn/37wt%Pb solder (Fig. 6.34). The thermal coefficient of expansion (TCE) of the PCB was found to be 16 ppm/°C. The underfill material's TCE, Young's modulus, and glass transition temperature were 24
ppm/°C, 0.9 Mpsi, and 150°C, respectively, when it was cured.

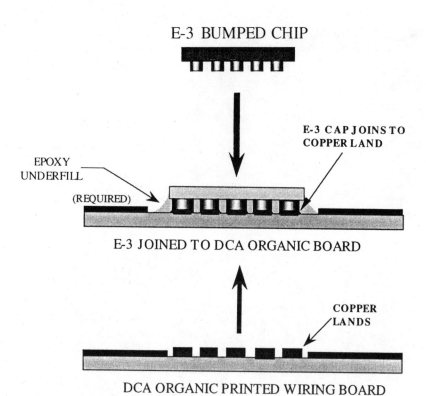

Figure 6.37   Low-cost E-3 bumped chip on bare Cu PCB.

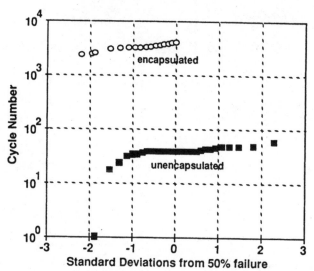

Figure 6.38   Cumulative failure thermal cycling data for C4 bumped chip on eutectic-coated PCB.[138,139]

The samples were in a completely automated air-to-air thermal shock chamber (–55 and 125°C) with a dwell time of 10 minutes. The two-point resistance of each chip interconnection was measured continuously at the hot and cold temperatures. The failure criterion for the air-to-air thermal cycling tests consisted of an increase in the total daisy-chain resistance of 5 ohms or a complete open for five consecutive readings at a given dwell temperature during a given cycle.

Figure 6.38 shows the failure distribution (log-normal plot) of the flip chip assemblies with and without underfill encapsulant. The abscissa shows the standard deviations from the MTTF. The MTTF of a flip chip assembly without encapsulant was 38 cycles and that all the solder joints failed before 60 cycles. With the underfill encapsulant, however, the MTTF of the flip chip solder joints was 3900 cycles and the failure mode was at the eutectic solder interconnects.

It is worthwhile to point out that the solder joints behaved quite differently when tested at liquid-to-liquid thermal shock conditions (–55 and 125°C). For the flip chip assemblies with underfill epoxy, the MTTF was less than 600 cycles and the failure mode was exclusively encapsulant failure.[138,139]

The reliability assessments on E-3 solder-bumped flip chip-on-board have been reported by Motorola[57,65] and are shown in Table 6.2. It can be seen that the E-3 bumps are very reliable under thermal cycling, thermal shock, 85°C/85%RH, and shock and vibration tests. More on E-3 bumped flip chip-on-board solder joint reliability is shown in Sec. 6.2.5.2.

6.2.2.2 Thermal fatigue modeling of high-temperature solder-bumped flip chips on PCB. Figure 6.39 shows the finite element model of one-quarter of a high-temperature solder-bumped flip chip on PCB[54] (Fig.

TABLE 6.2 Reliability Assessments on E-3 Bumps[57]

Assembly stress tests (with underfill)	
Thermal cycle, –20–110°C:	1,800 cycles, no fails (1CPH)
Thermal cycle, –45–100°C:	2,000 cycles, no fails (3CPH)
Liq/Liq shock, –55–125°C:	800 cycles, no fails (3CPH)
THB, 85°C/85%RH/15v bias:	1,000 cycles, no fails
THB, 110°C/85%RH/3v bias:	200 cycles, no fails
Assembly stress tests (without underfill)	
Vibration, 6 hours:	no fails
Shock: 35″ drops:	no fails
Bump shear:	all fails in ductile Pb
3- and 4-point bending:	no fails
Electrical test probe stress:	no fails
Multiple reflow:	no fails
(std. joining profile, 6 reflows)	

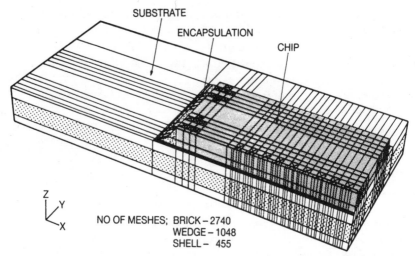

SUBSTRATE

ENCAPSULATION

CHIP

Z
Y
X

NO OF MESHES; BRICK – 2740
WEDGE – 1048
SHELL – 455

Figure 6.39   Finite element model of the C4 bumped chip on eutectic-coated PCB.[54]

6.34). A detailed model for a solder bump is shown in Fig. 6.40. The chip dimensions were $10.7 \times 5.7 \times 0.6$ mm. The solder joint was 180 μm in diameter with a 100-μm height. The PCB thickness was 1.6 mm. The material properties for the isotropic linear analysis are shown in Table 6.3.

The equivalent plastic strain versus solder bump location is shown in Fig. 6.41a. It can be seen that the solder joint strain is strongly

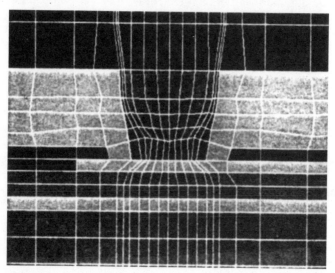

Figure 6.40   A close look at the composite solder bump (high-lead and eutectic).[54]

TABLE 6.3   Material Properties for Modeling IBM at Yasu's Solder-Bumped Flip Chip-on-Board[54]

Material	Modulus (MPa)	CTE (ppm/°C)	Poisson ratio
Silicon (chip)	169540	3.5	0.07
Glass epoxy substrate	21560	X,Y-14	0.20
Photo-sensitive epoxy	4724	69	0.39
Solder (63/37)	6270	24	0.40
(5/95)	2940	29	
Encapsulant	4900–6370 (0–100°C)	26–60 (0–100°C)	0.36

affected by the underfill epoxy. The ones with underfill have a much smaller strain than those without underfill. This is especially true for solder joints located near the corners. The underfill encapsulant cements the chip on the PCB[54] (Fig. 6.34), i.e., the shear strain in the encapsulated solder joint is reduced because the chip-joint-PCB bends like a single unit (Fig. 6.41b).

6.2.3   Thermal reliability of eutectic solder-bumped flip chips on boards

Unlike IBM and Motorola's composite solder interconnects (high-lead solder bumps with eutectic solder) using the evaporation method, Oki,[132] Sharp,[50,51] Hewlett-Packard,[40–42] and Toshiba[96] use the electro-plating method to bump wafers with eutectic solder. The substrates for such eutectic solder-bumped flip chips can be standard FR-4 or BT (bis-maleimide triazine) with a CuNiAu metal finish or OCC PCBs. They do not use the E-3 process or coat the PCB with eutectic solder. They treat the eutectic solder-bumped chips like other surface-mount components (SMCs), which can be mass-reflowed on the low-cost PCBs with all the other SMCs.

6.2.3.1   Thermal cycling test of eutectic solder-bumped flip chips on boards.   Some of the thermal cycling test results of eutectic solder-bumped flip chips on FR-4 PCB have been reported.[34,50,51] In this section we will discuss Toshiba's test results.[96]

Figure 6.42 shows a cross section of Toshiba's FR-4 flip chip assembly with eutectic solder bumps. The chip dimensions were 17.26×8.13 mm. The minimum pitch of the periphery pads on the chip was 200 μm. The bump height was variable (from 20 to 80 μm).

The resistance changes of the bumps during the thermal cycling test were monitored by the four-probe method. The number of cycles to reach 50 percent cumulative failure (N_f) for individual sample conditions was measured. Figure 6.43 shows the resistance change of the

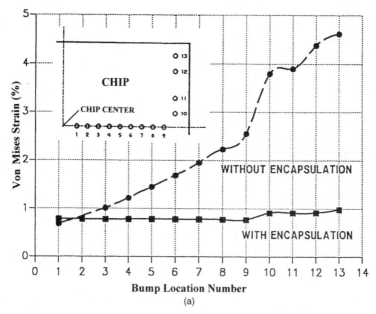

Figure 6.41 (*a*) von Mises (equivalent) strain distribution along the solder bumps. (*b*) Deflection of the C4 flip chip bumped assembly.[54]

solder bumps (80-μm-height bumps) with encapsulant (Young's modulus = 3.5 GPa, thermal coefficient of expansion = 74 ppm/°C) during the thermal cycling test. It can be seen that the initial resistance of the bump was under 1 mΩ (resistances under 1 mΩ have been plotted as 1 mΩ in Fig. 6.43). The resistance began to increase at 1000 cycles and reached 100 mΩ at 1500 cycles. Thus, the solder joints were considered failures when their bump resistance reached 100 mΩ. The thermal cycling test condition was –65 (30 min), 25 (5 min), and 100°C (30 min).

Figure 6.44 shows the life distribution of the eutectic solder-bumped flip chips on FR-4 PCB with three different bump heights (30-μm, 50-μm, and 80-μm). The open marks indicate the results for flip chips without underfill epoxy and the filled marks indicate the results for flip chips with underfill encapsulant. It can be seen that the ones without underfill failed very early compared to those with underfill. Also, the ones with taller bump heights last longer, i.e., are more reliable (Fig. 6.44).

6.2.3.2 Thermal fatigue modeling of eutectic solder-bumped flip chips on boards.
Finite element analysis and fatigue life prediction of eutectic solder-bumped flip chips on FR-4 PCB under thermal conditions have been reported.[36–38,43,50–51] In this section, Toshiba's results are discussed.

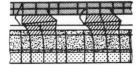

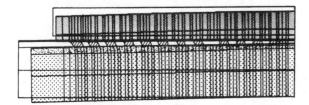

WITHOUT ENCAPSULATION

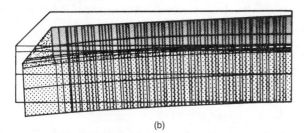

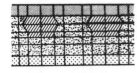

WITH ENCAPSULATION

(b)

Figure 6.41   (*Continued*)

Figure 6.42   Toshiba's cross-sectional view of eutectic solder bump on FR-4 PCB.[96]

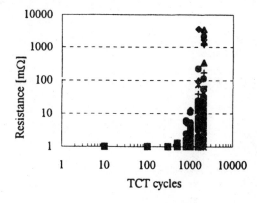

Figure 6.43 The resistance change of the eutectic bumps during thermal cycling test.[96]

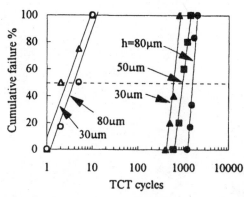

Figure 6.44 Life distribution (Weibull) of Toshiba's eutectic solder bumps on FR-4 substrate. (The open marks indicate the results for unencapsulated samples and the closed marks indicate the results for encapsulated samples.)[96]

Toshiba uses two different finite models to determine the plastic strain range.[96] The first is the quarter model shown in Fig. 6.45. It was used to determine the relative shear (horizontal) displacement and axial (vertical) displacement at the top and bottom faces of solder bumps under thermal cycling test conditions. The unit bump model (Fig. 6.46) is used to determine the cumulative equivalent plastic strain range distribution in the solder joint. The boundary conditions for the unit bump model were the relative shear and axial displacements obtained by the quarter model.

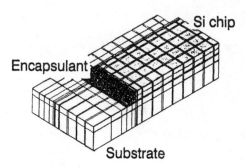

Figure 6.45 Finite element (overall) model for Toshiba's eutectic solder bumps on FR-4 substrate.[96]

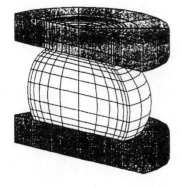

Figure 6.46 Finite element (local) model for Toshiba's eutectic solder bumps on FR-4 substrate.[96]

Figures 6.47 and 6.48 show the relative displacements for the unencapsulated and encapsulated flip chip solder joints, respectively. In both cases, the relative axial (vertical) displacement Δz is almost constant. The slightly larger (Δz) values for the encapsulated flip chip assembly are caused by the additional thermal expansion of the underfill encapsulant in the vertical direction.

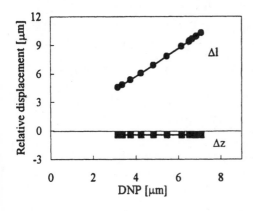

Figure 6.47 Relative displacements along the distance from neutral point (DNP) of Toshiba's eutectic solder bumps on FR-4 substrate (unencapsulated).[96]

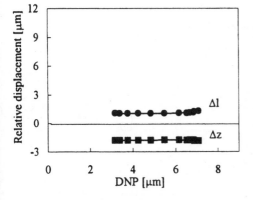

Figure 6.48 Relative displacements along the distance from neutral point (DNP) of Toshiba's eutectic solder bumps on FR-4 substrate (encapsulated).[96]

However, similar to the high-temperature solder-bumped flip chip assembly, the relative horizontal (shear) displacement Δl for the unencapsulated flip chip is much larger than for the encapsulated flip chip assembly. Again, this is due to the fact that the underfill encapsulant cements the chip on the FR-4 PCB and reduces the effect of the global thermal expansion mismatch between the silicon chip and the organic substrate, i.e., it reduces and redistributes the stresses and strains over the entire chip area that would otherwise be increasingly concentrated at the corner solder joints.

The effect of the Young's modulus and TCE of the underfill epoxy has also been determined by Toshiba with its finite element models. It found that both Young's modulus and TCE play important roles in the thermal fatigue life of solder joints (Fig. 6.49). In general, the larger the Young's modulus (up to 9 GPa), and the smaller the TCE, the better the solder joint thermal fatigue life. Tables 6.4 and 6.5 show the mechanical and plastic properties, respectively, of Toshiba's flip chip-on-board assembly.

6.2.4   Underfill encapsulants for solder-bumped flip chip assemblies

In flip chip applications, the underfill is expected to provide good adhesion, good matching of its thermal coefficient of expansion (TCE) with the solder bump, substrate material, and good moisture resistance. The ultimate measure is determined by how much it enhances the reliability of the solder-bumped flip chip assemblies. Table 6.6 provides a sample list of the different ingredients commonly found in current commercial products and their respective functions.

6.2.4.1   Physicochemical properties of the underfill polymers.   The basic requirements for the underfill materials, such as relatively low TCE,

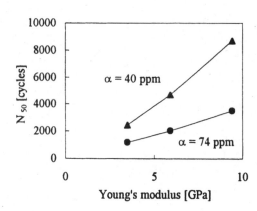

Figure 6.49   Dependence of thermal fatigue life calculated by Toshiba with finite element method on the physical (TCE) and mechanical (Young's modulus) properties of encapsulant.[96]

TABLE 6.4   Mechanical Properties of Toshiba's
Flip Chip-on-Board Assembly[96]

Material	Young's modulus (GPa)	α (ppm/°C)
Si	127	2.5
FR-4	11	12.1
Encapsulant	3.5	74

TABLE 6.5   Plastic Properties of
Toshiba's Eutectic Solder[96]

Temperature (°C)	Yield stress (MPa)
−65	40
0	26.9
25	21.5
50	16.5
75	11.7
100	5.3

TABLE 6.6   Underfill Epoxy Ingredients[41]

Chemical ingredient	Function
Resins: epoxy, polyesters	Polymer network with good surface adhesion
Fillers: fused silica; alumina; calcium carbonate	Mechanical strength and TCE modifier; affects viscosity/flow
Crosslinking agents: acid anhydrides; amines; peroxides	Material hardening; determines the gel and cure kinetics
Latent curing agent: dicyandiamide	Prolongs open time for one-part system
Diluents/flexibilizers: butyl glycidyl ether; polysulfides	Lower viscosity; improves adhesion and electrical properties

low ionics, and low alpha particle emission have been well documented. Some additional constraints imposed by flip chip assembly in a production environment are[41]: one-part system with at least a few hours of open time, fast flow, fast cure, and chemical compatibility with the flux used to promote solder wetting. Low-chlorine epoxies with silica filler are by far the most common formulation used commercially. The use of cycloaliphatics to minimize chlorine contamination imposes the preference for anhydrides as the curing agent, which, in turn, requires storage of the one-part system at −40°C. From a formulation standpoint, fast cure and fast flow are usually not problematic to obtain, but the post-cure requirements, such as high glass transition temperature T_g, high modulus, and high moisture resistance, make the balancing act so critical. Since most of the PCB assemblies have been converted to a no-clean process, another concern is whether the flux residues impede the underfill flow or create undesirable chemical by-products.

6.2.4.2 Underfill flow evaluation.   While the data sheets from underfill manufacturers usually provide the viscosity data at room temperature, the actual flow rate at an elevated temperature and at a given vertical clearance can be determined empirically using heated parallel glass plates separated by shims of known thicknesses. Figure 6.50 shows how the flow rates of different underfill materials compared at a known gap and at a fixed temperature. Underfills A and B were unable to flow through a 0.05-mm gap. Using a fineness-of-grind gage, they were both found to have particle sizes ranging from 0.075 to 0.1 mm.

It is important to note that a fast-flow underfill is good only if there is no significant settling of the filler as it traverses a distance equal to the die length. Using a top glass plate for this experiment allows us to observe any settling effects of the underfills. Indeed, while underfills D and E flow the fastest (see Fig. 6.50), they also show the largest

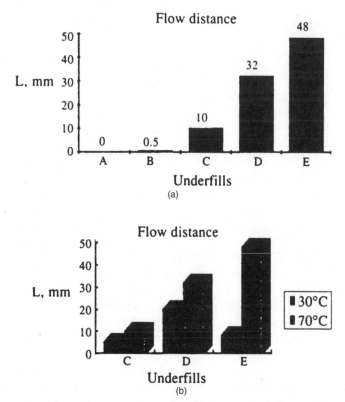

Figure 6.50   (*a*) Flow distance of underfills at 70°C and 0.05 mm gap for nine minutes; (*b*) increase in flow distance from 30 to 70°C for nine minutes at 0.05 mm gap.[41]

amount of settling, as observed by the lightening of the streak as they moved forward. This filler-settling effect may also contribute to the much greater temperature sensitivity of underfill E because, as more filler gradually separates out, the liquid portion simulates the Newtonian flow to an increasingly greater extent.

The next step is to verify how well the materials underfill a solder-bumped flip chip with a vertical clearance of between 0.02 and 0.0375 mm. The underfill was dispensed as a string along one (if the chip edge is less than 12.5 mm) or two edges (if the chip edge is equal to or larger than 12.5 mm) of the reflowed chip and allowed to flow in an oven at 75°C for a short period. Given a sufficient time, the underfill could be observed at the opposite side of the chip (a distance of about 12 mm).

In order to check the uniformity of the underfill spreading beneath the chip and whether there were a large number of voids (especially for the fast flow/fast cure processes), scanning acoustic microscopy (SAM) was used to sample chips underfilled with different materials and at different process conditions. Since micrographs from SAM show mainly contrasting shades reflected from different densities of material, it was difficult to distinguish between different grey levels. Therefore, the silicon was lapped down to expose the underfill where voids could be directly observed.

In general, it was found that, with the fast-flow process, microvoids were present in most samples. The extent of voiding depends on three factors: the cure schedule, the quantity and chemistry of the flux residues, and the method of underfilling. It was also observed that, as long as there was sufficient underfill, vias under the chip did not seem to create more voids. However, if there was photoresist under the chip, the underfill process was not as effective. Finally, with regard to using SAM to verify underfilling, further investigation is required to better interpret the micrographs. Rutherford backscattering micrographs showed platelet formation as the filler of the underfill strings along its path. This mechanism is, of course, highly dependent on the filler size distribution and, in the worst case, may create blockage of further underfill flow. Using very small fillers, e.g., around 0.01 mm, may minimize the platelet effect, but smaller fillers tend to settle more easily at the beginning of the flow. Therefore, a delicate balance between the platelet and the settling effects is very important for a complete and even distribution of the underfill.

6.2.4.3 Underfill thermal and mechanical analysis.   The underfill thermal and mechanical analysis was performed using the Perkin-Elmer Series 7 TA System, which consists of a differential scanning calorimeter (DSC) unit, a dynamic mechanical analyzer/temperature mechanical analyzer (DMA/TMA) unit, a thermal gravimetric analyzer (TGA), and software for kinetics and viscoelasticity analysis. Figures 6.51, 6.52, 6.53, and 6.54 show examples of these measurements. Table 6.7 indicates the strong dependence of the material parameters on the cure schedule.

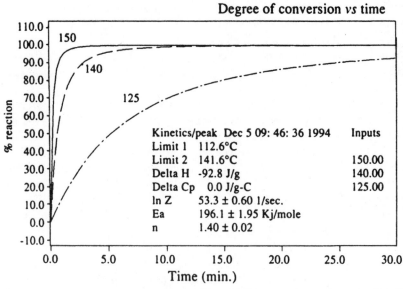

Figure 6.51   Percent cure as a function of time at different isotherms for underfill E.[41]

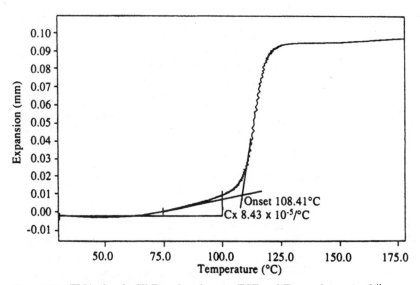

Figure 6.52   TMA of underfill E to show how its TCE and T were determined.[41]

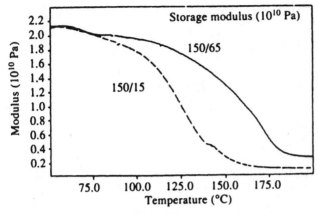

Figure 6.53   DMA of underfill C to show the storage modulus and how the T changes as the cure schedule changes.[41]

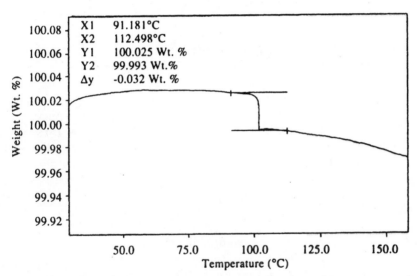

Figure 6.54   TGA of underfill C (150°C and 15 minutes cure) to show the extent of moisture adsorption (absorption after storage at ambient for 30 days).[41]

If based on only the thermal and mechanical parameters of the underfills shown in Table 6.7, underfill B, which was cured at 150°C for an hour, seems to have the most desirable properties. However, because underfill B took a long time to fill the low-clearance gap (Fig. 6.50), it is not a good choice for this process. Underfills D and E showed insufficient cure at 150°C for 15 min in addition to the filler-settling problem discussed earlier. Underfill C seems to fit the requirements of fast flow and fast cure best. It also meets the generic requirements for low ionics, low moisture absorption, and low alpha-particle emission. It

TABLE 6.7   Physical and Mechanical Properties of Some Underfill Epoxies

Glass transition temperature (T_g), shear modulus (S. modul.), and thermal coefficient of expansion (TCE) of some underfills. (Epoxy-glass laminate (FR-4) was quoted for comparison.)

Cure at 150°C	T_g (°C)		S. modul. (GPa)		TCE (ppm/°C)	
	15 min	60 min	15 min	60 min	15 min	60 min
E	108		0.8		84	
		137		11		31
D	112		0.3		234	
		133		11		153
C	138		21		40	
		149		21		32
B	103		0.6		19	
		160		26		16
FR-4 (Ref. 48)	115 (x/y plane)		9 (x/y plane)		18 (x/y plane)	

NOTE: The modulus and TCE were measured at below T_g.

is also interesting to note that underfill C is a *snap cure material* because the differences between the measured parameters in Table 6.7 are small from 15 to 60 minutes of cure at 150°C.

The fast flow–fast cure process may increase the risk of voiding, but how that affects the ongoing temperature cycling results remains to be verified. The shear strength of the underfilled flip chip assembled with this fast process seems to be adequate. On the other hand, the properties of the underfill can be further improved for a high-throughput, high-efficiency assembly process such as better underfilling with even shorter times and preferably much longer gel times at ambient temperatures.

6.2.4.4   Effects of flux on underfill flow and adhesion.

The SIR (surface insulation resistance) and mechanical shear tests were applied to three different fluxed, eutectic, solder-bumped flip chip assemblies. A low SIR reading indicates possible surface contamination from the underfill and flux residues, and/or surface corrosion due to underfill/flux/substrate interactions. Table 6.8 also includes a sample of shear data as the indicator of underfill adhesion strength.

The three fluxes tested were essentially concentrated flux extracts with added tack promoter. Flux I was a nonrosin, low-solids flux extract for no-clean soldering under nitrogen. Fluxes II and III were conventional RMA fluxes with decreased solids contents intended for no-clean soldering in air. It was observed that, for the same process with the same amount of underfill, underfill C seemed to show less consistency in underfilling in the presence of flux III residues. The SIR

TABLE 6.8   SIR and Mechanical Shear Data of Fluxes
with Different Flux Solids Contents[41]

Flux	Underfill Flow	SIR (ohms) (85/85/100V/7d) No UF	UF	Shear (to failure) (kg) No UF	UF
I	Very good	$3E10$	5E8	14–18	~180
II	Very good	$2E10$	4E8	~16	15–126
III	Good	n/a	5E8	~9	85–250

NOTE: In the SIR column, 'E-x' stands for 10^x. Both SIR and shear data reported are raw data without converting to the proper normalizing factors. Shear rate was 0.05 mm/min. 'No UF' stands for no underfill applied to the reflowed die.

showed a large drop in insulation resistance for all three fluxes with added underfill after seven days of 85°C/85%RH/100 V bias. This was probably caused by an increase in moisture absorption by the underfill under the stress conditions. However, the SIR recovered to a reading of about 10^{11} ohm when the underfilled samples were brought back to ambient conditions. This was an indication that no permanent degradation or ionic contamination existed. How moisture absorption by the underfill affects the device's reliability and is a major objective in reliability testing.

The shear test data shown in Table 6.8 are sketchy at best, but they serve as an indication of whether the underfills have increased the bonding strength of the bare die to the substrate. The data with flux II seemed to show some inconsistencies. This may be caused by experimental variations or be an indication of poor adhesion due to interference between flux II and underfill C. The inconsistencies may also be explained by the observation that some of the test chips appeared to have had an under-bump metallization problem. All in all, it can be concluded that the assembly process with flux I and underfill C has demonstrated a reasonably robust and fast production process, albeit with lots of room for improvement.

6.2.5   Mechanical reliability of high-temperature solder-bumped flip chips on boards

Mechanical shearing tests of solder-bumped flip chips on PCB have been determined by Lau et al.[46] In this section, the twisting and pulling (Fig. 6.55a) of solder-bumped flip chips on FR-4 PCB with and without underfill encapsulant are presented.

6.2.5.1   Twisting of solder-bumped flip chips on boards.   The assembly process of the test samples was reported.[46] The chip has 351 95wt%Pb/

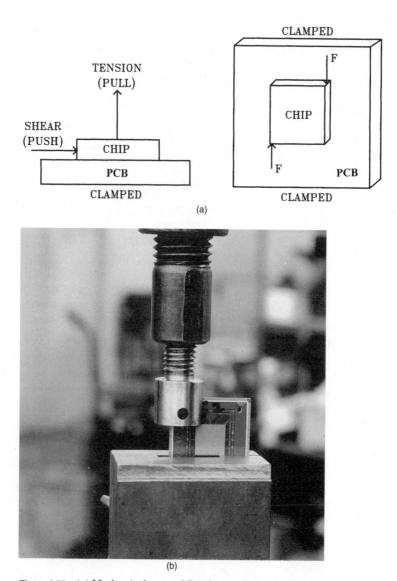

Figure 6.55 (a) Mechanical tests of flip chips on boards. (b) Twisting of a flip chip-on-board.[46]

5wt%Sn solder bumps patterned in three rows. The bump diameter was approximately 0.13 mm and the bumps had a 0.254-mm pitch. The chip dimensions were $12.7 \times 12.7 \times 0.7$ mm. The PCB thickness was 1.7 mm and its copper pads were coated with 63wt%Sn/37wt%Pb solder. The underfill material's TCE = 30 ppm/°C, Young's modulus = 6 GPa, and Poisson's ratio = 0.35.

The setup of the torsion test is shown in Fig. 6.55*b*, where the PCB was clamped and two equal but opposing forces were applied at the corners of the chip. This setup allowed the chip to rotate freely while subjected to a twist (= applied force × moment arm). For flip chip solder bumps without encapsulant, the average applied force was 27 lb (120 N) (Fig. 6.56). The failure mode was at the solder joint near the chip but not on the chip pad.

For the flip chip assemblies with underfill encapsulant, the tests became more difficult because most of the chips cracked near the corners. Figure 6.57 shows the average load-deflection curve for those successful twisting tests (i.e., with no broken chips). It can be seen that the failure load (342 lb or 1522 N) of the encapsulated interconnects was more than 10 times that of those without the encapsulant. Also, the failure mode was quite different. In this case, the solder joint cracked near the chip and on the chip pad. The encapsulant broke at the interface of the chip and the chip surface was free of encapsulant. The broken encapsulant contained some replication of the silicon chip's top-level metal pattern, which indicates excellent encapsulant-adhesion results.

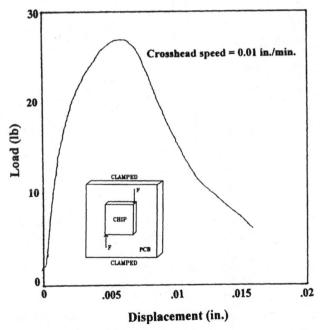

Figure 6.56 Load-deflection curve of a C4 solder-bumped flip chip-on-board (unencapsulated).

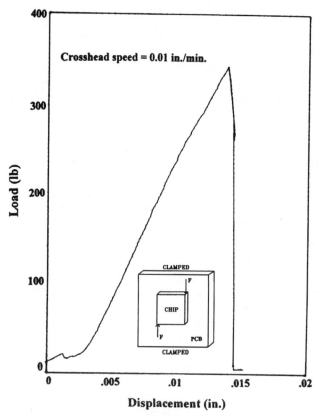

Figure 6.57 Load-deflection curve of a C4 solder-bumped flip chip-on-board (encapsulated).

6.2.5.2 Pulling of high-temperature solder-bumped flip chips on boards. Figure 6.58 shows the daisy-chained L8AA test chip with 6.7 by 6.5 mm dimensions. There were 84 bumps. The bump standoff height was 90.9 to 97 µm and the bump pitches were 0.25 to 0.36 mm. The Young's modulus of the silicon was 131 GPa and the Poisson's ratio was 0.3. There were two kinds of solder bumps,[65] the standard C4 (97Pb/3Sn) bump and the E-3 bump. The board was constructed with three layers of $T_g = 175°C$ photo-via dielectric and copper planes on both surfaces of a BT-glass core. The dimensions were $21 \times 25 \times 1$ mm and the copper pads were Ni-plated prior to solder cladding.[65]

The test setup was prepared by mounting an aluminum stud to the backside of the chip. The assembly was fixed and an increasing tensile load was applied to the stud until the assembly was pulled apart. Figure 6.59 shows the test results[65] and Table 6.9 lists the codes for various results. The *pull strength* (gram/bump) was defined as the

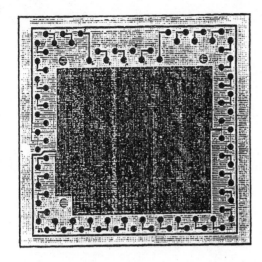

Figure 6.58   L8AA test chip.[65]

breaking pull force divided by the number of solder joints (84) whether the part was underfilled or not. Thus, for the unencapsulated case, the pull strength reflected the reflowed solder joint's integrity (for both C4 and E-3 bumps). For the encapsulated case, the unit of the pull strength was insignificant but makes logical sense for the purpose of a comparative study.

It can be seen from Fig. 6.59 and Table 6.9 that: (1) the E-3 solder joint pull strength is slightly stronger than the C4's, (2) the pull strength of the underfilled assemblies (both C4 and E-3 bumps) is much stronger than that of unencapsulated assemblies, (3) the pull strength of the underfilled assemblies (both C4 and E-3 bumps) without three times rework and reflow are stronger than those with either three times rework or three times reflow or both, and (4) the under-

🔲 C4   ☐ E-3

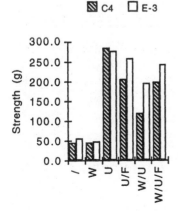

Figure 6.59   Pull test results for C4 and E-3 solder-bumped flip chip-on-board with and without underfill encapsulant.[65]

TABLE 6.9   Notations (Codes) for Pull Test Results of C4 and E-3 Solder-Bumped Flip Chips on FR-4 PCB[65]

Code	3× Rework	Underfill	3× Reflow
/			
W	✓		
U		✓	
U/F		✓	✓
W/U	✓	✓	
W/U/F	✓	✓	✓

filled E-3 solder joint pull strength is stronger than the C4's after three times of rework and reflow.

6.2.6 Vibrational reliability of high-temperature solder-bumped flip chips on boards

In this section, the dynamic effects of solder-bumped flip chips on PCB are studied. The engineering data, such as the natural frequencies, excitation frequencies, excitation magnitude, velocity, and acceleration, of the solder-bumped flip chips on OCC PCB under shock and vibration conditions are provided by Lau et al.[127]

6.2.6.1   Vibration test and results

Test board. Figure 6.60 shows the test board for dynamics testing. It is made of FR-4 epoxy/glass and is $10.25 \times 6.25 \times 0.062$ in ($260 \times 159 \times 1.57$ mm). The thermal coefficient of linear expansion of the FR-4 material is 18 ppm/°C in the x and y (i.e., horizontal) directions and is 70 ppm/°C in the z (i.e., vertical) direction. The Young's modulus of the PCB is 22 GPa and the Poisson's ratio is 0.28. There are five chips on the center portion of the PCB.

Test chip. The test chip (L8AA) with E-3 solder bumps is shown in Fig. 6.58. In this study, a single L8AA chip is called *single,* and a group of four single L8AA chips ($\approx$14 by 14 mm) is called *quad.*

Sample size. The sample sizes for the vibration test are 20 single and 17 quad. The underfill encapsulants are: encapsulant I for 10 single, encapsulant II for 10 single, and encapsulant I for 17 quad.

Vibration test setup. The test setup for the out-of-plane vibration of the high-lead solder-bumped flip chip PCB assembly consisted of a table which was vertically driven by a shaker. A special test frame was designed and constructed to provide a two-point fatigue bending of the test board as shown schematically in Fig. 6.61. These two supports were made by clamping both ends of the test board with aluminum

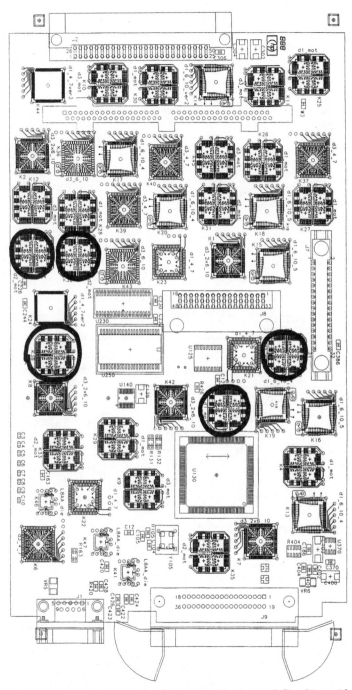

Figure 6.60 Vibration test board for E-3 solder-bumped flip chips with and without underfill encapsulant (the ones with circle have chips).[127]

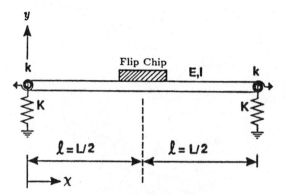

Figure 6.61 Model for vibration test and analysis.[127]

blocks. Tapes were used between the board and the blocks so that damages to the trace and the board would be avoided. The actual board bending was caused by the movement of these supports which were bolted to the surface of the vertical shaker. The shaker was controlled by a data acquisition unit and produced a sinusoidal excitation.

Natural frequencies of the flip chip systems. In order to excite the test board, the excitation frequencies had to include the natural frequency of the test board. Consequently, the next step was to estimate the natural frequencies of the test boards with these chips.

Figure 6.61 shows the PCB with L as its largest dimension. It is supported by linear and rotational springs at both ends. The flip chips are attached near the center of the PCB. E is the Young's modulus, I is the area moment of inertia of the beam (PCB) cross section about the neutral axis, EI is the flexural rigidity of the PCB, M is the concentrated mass of the flip chips, and m is the mass per unit length of the PCB. K is the stiffness of the translational spring, and k is the stiffness of the rotational spring.

By assuming K and k approach infinity, i.e., the clamped condition, and by using the data of the test chip and test board, it can be shown that the natural frequency[127] f: $f = 130.2$ Hz for the single and $f = 120.1$ Hz for the quad. We also used the vibration table to sweep at very small magnitudes for a wide range of frequencies to determine the natural frequencies of our flip chip test boards. For example, in the case of the L8AA quad flip chip system, the measurement of the vibration result (four accelerometers on four different flip chip test boards) is shown in Fig. 6.62. It can be seen that the average measured value of these four measurements is about 110 Hz. Similarly, the average value of the L8AA single is about 120 Hz. These values are very close to those predicted by the simple model (Table 6.10).

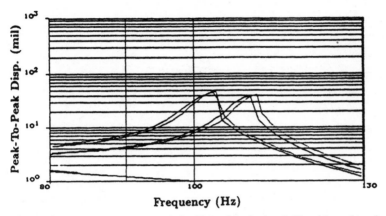

Figure 6.62 Natural frequency of the E-3 solder-bumped flip chip-on-board (free vibration).[127]

Excitation frequencies for the flip chip systems. Once the natural frequencies of the test boards were estimated, the excitation of the shaker was designed to perform a sinusoidal excitation with a frequency sweep between 100 and 150 Hz for the L8AA single, and 90 and 140 Hz for the L8AA quad.

Excitation magnitudes for the flip chip systems. Once the excitation frequencies of the test boards were estimated, the shaker was designed to perform a sinusoidal excitation such that the maximum (peak-to-peak) excitation magnitude (at the middle) of the flip chip PCB is 0.12 in (3 mm) for the L8AA single, and is 0.15 in (3.8 mm) for the L8AA quad. The measured result (four accelerometers on four different flip chip test boards) for the L8AA quad is shown in Fig. 6.63.

Vibration test results. The test results are summarized in Table 6.10. It can be seen that with the severe test conditions described, failure was not observed after six hours of vibration for the L8AA quad and after seven hours of vibration for the L8AA single solder-bumped flip chip assemblies. One explanation could be that because of the small mass of the solder-bumped flip chip, the resultant dynamic force is too small to crack the joint.

TABLE 6.10 Vibration Test Results of E-3 Solder-Bumped Flip Chips on PCB[127]

Chip	Natural freq. (Hz)		Excitation freq. (Hz)	Excitation magnitude (pp-mm)	Test time (hour)	No. of failures
	Calculated	Measured				
L8AA Single	130.2	120	100–150	3	7	0
L8AA Quad	120.1	110	90–140	3.8	6	0

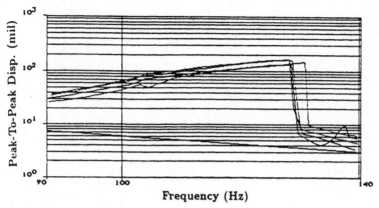

Figure 6.63   Forced vibration of the E-3 solder-bumped flip chip-on-board.[127]

6.2.6.2 Shock test and results. The test board and test chip for the shock tests were the same as for the vibration tests. For the shock tests, the sample sizes were 10 single with encapsulant II, 10 single with encapsulant I, 20 single without underfill, 10 quad with encapsulant I, and 10 quad without underfill.

In-plane shock test and results. The test setup for the in-plane (vertical) shock test consisted of a frame with a guide cylinder. The frame was resting on absorption springs. The test board was bolted vertically to a fixture which was bolted to the platform. The platform was able to drop on the absorption springs along the guide cylinder. An accelerometer was attached to the bottom of the platform to record the acceleration time-history $g(t)$ (Fig. 6.64). An integration of the acceleration time-history with respect to time led to the velocity time-history $v(t)$ (Fig. 6.65). It can be seen that the maximum acceleration was 665 g and the maximum velocity was 113 in/s (2.87 m/s). After three drops, no solder joint failures for any of the E-3 solder-bumped flip chip PCBs were

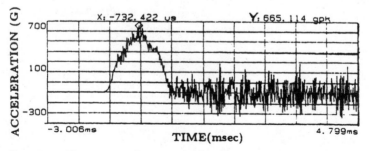

Figure 6.64   Drop test acceleration of the E-3 solder-bumped flip chip-on-board.[127]

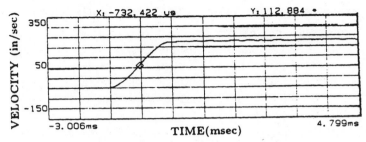

Figure 6.65 Drop test velocity of the E-3 solder-bumped flip chip-on-board.[127]

observed. In addition to the small mass of the chip, the natural frequencies of the flip chip assemblies in the in-plane direction were very high and, in general, they will not be excited by shock and vibrations occurring under normal operating conditions.

Out-of-plane shock test and results. The test setup of the out-of-plane (horizontal) drop test was very similar to that of the in-plane shock test, except the flip chip PCB was laid flat on the table. All the test boards that went through the in-plane shock test were tested again in the horizontal direction with the same magnitudes. After three drops, there were no failures (Table 6.11).

6.2.6.3 Summary. The natural frequencies, excitation frequencies, excitation magnitudes, velocity, and acceleration of the E-3 solder-bumped flip chip vibration system have been systematically and carefully determined. The shock (drop) test results for flip chip assemblies with and without underfill encapsulants have also been provided. The results show that E-3 solder-bumped flip chips on OCC PCB are reliable to use under shock and vibration conditions.

It is worthwhile to mention that the shock and vibration test results of the solder-bumped flip chip assemblies are much better than those of other SMT assemblies, such as plastic quad flat pack and plastic leaded chip carrier,[169] and SMT connectors,[170] even though the test conditions for flip chips are more severe.

TABLE 6.11   **Shock Test Results of E-3 Solder-Bumped Flip Chips on PCB**[127]

Chip	1st Drop	2nd Drop	3rd Drop
L8AA Single (encapsulant I)			
L8AA Single (encapsulant II)		NO	
L8AA Single (no underfill)			
L8AA Quad (encapsulant I)		FAILURES	
L8AA Quad (no underfill)			

6.2.7 Low-cost solder-bumped flip chip MCM-L demonstration

A low-cost multichip module employing solder-bumped flip chips mounted on an organic substrate was demonstrated by Kelly and Lau.[42] This functional prototype was used to help assess the preliminary feasibility of low-temperature (eutectic) solder-bumped flip chip applications, from wafer design, sourcing, bumping, and substrate design and fabrication to MCM-L assembly.

The demonstration vehicle utilized the digital portion of the electronics from an existing HP product. The digital core was designed into a small MCM-L and separately attached to the main board using peripheral solder balls, similar to a ball grid array (BGA)[35] (see Fig. 6.66). There were six decoupling capacitors (0805 body) and five IC chips, namely, one application specific integrated circuit (ASIC), one microprocessor (μP), two random access memory (RAM), and one read only memory (ROM), on the MCM-L (Fig. 6.67). These four different chip sets were supplied by four different IC vendors (Table 6.12).

6.2.7.1 MCM-L substrate description.   The MCM-L used was an organic laminate substrate. Figure 6.67 shows the front side of the module layout. The overall size of the module was 2.0 × 1.9 in (50 × 48 mm). The module has been designed in four layers with 5-mil traces and spaces globally, and 4-mil traces and spaces used locally to access the bond pads around two edges of the ASIC. Vias were 12 mils (0.3 mm) in diameter with 22-mil (0.56 mm) via pad diameters. Signals and power were routed using the top side of the board as well as layers 3 and 4. Layer 2 is for the ground plane only. Typical construction of the substrate is shown in Fig. 6.68. The top side of the board was populated with five IC chips and six decoupling capacitors located on the power lines of each chip.

Two types of flip chip land area designs were used. One design used a rectangular solder mask opening at the die site with exposed traces exiting the solder mask and terminating at locations which matched

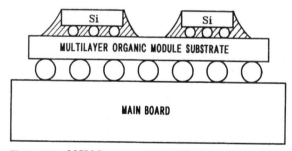

Figure 6.66   MCM-L on a main PCB.[42]

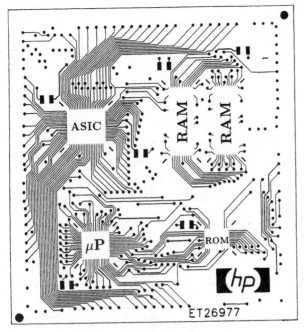

Figure 6.67   MCM-L front-side layout.[42]

the bond pad footprint of the flipped chips. The second design used solder masks underneath the die and solder mask openings to define the bonding areas used on each trace. These designs are shown schematically in Figs. 6.69a and b.

6.2.7.2 Wafer/IC description.

The product used for this demonstration was selected for the following reasons: (1) the ICs were in production, (2) the design was stable, and (3) the minimum pitch of 200 μm on any of the ICs was considered a good starting point for the first demonstration of this technology.

All of the selected IC chips used peripheral bond pads originally designed for wire bonding. Current wire bonding capability ranges

TABLE 6.12   MCM-L IC Chips' Pin Counts, Size, Pitch, and Passivation Openings[42]

IC Chip	Pin counts	Size (mm)	Min. pitch (μm)	Sq. passi. opening (μm)
Chip A (μP)	68	5.40 × 5.20	213	96
Chip B (ASIC)	88	7.02 × 6.10	200	100
Chip C (RAM)	32	11.4 × 4.49	298	100
Chip D (ROM)	28	4.50 × 3.69	232	108

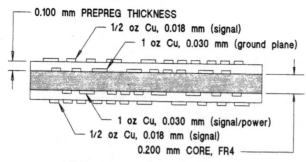

Figure 6.68   MCM-L substrate construction.[42]

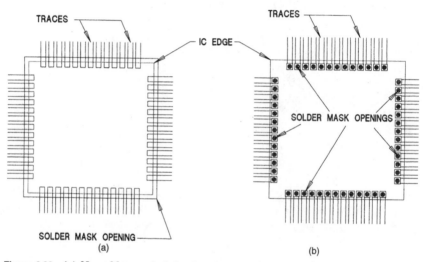

Figure 6.69   (*a*) Nonsolder-mask-defined (NSMD) pad; (*b*) solder-mask-defined (SMD) pad.[42]

from 75- to 100-µm minimum pad pitch. As such, we were fortunate that none of the IC chips used utilized pad pitches in this range. This is an obvious concern for future implementation of flip chip technology.

The digital ICs which are being attached to the MCM-L are the microprocessor (chip A), ASIC (chip B), RAM (chip C), and ROM (chip D). Each came from a different supplier (Table 6.12). The passivation layer in all cases was a silicon dioxide and silicon nitride. Chip A and chip C used an additional polyimide layer.

6.2.7.3 Wafer/IC sourcing. Of the four bare IC chip sets that we needed to build the demonstration vehicle, we were assured of bare die availability for only the ASIC (chip B), which is fabricated internally by

HP. We communicated with the other three vendors regarding availability of bare IC chips. Responses ranged from tentative to hostile. With this backdrop there was a good deal of uncertainty surrounding the availability issue, especially since we would need full wafers, which are required for the bumping process.

We formally requested a quotation for 20 wafers from each of the three remaining vendors. With each of these vendors, there appeared to be three particularly important items that, when made clear, enabled them to respond positively to our needs:

The vendor for chip A was quite enthusiastic about helping our engineering investigation after we told it that we were not planning an immediate conversion to bare chip-on-board.

Vendors for chip C and chip D were much more comfortable with providing bare die once they fully understood that this was an engineering experiment. Many phone calls were exchanged before they could be assured that they were not in any way committing to production volumes.

Naturally, all of our IC vendors were concerned about sharing knowledge of their yield on the wafer. For vendors C and D, an agreement to fix the number of tested good dies which we would receive per wafer greatly reduced the apprehension surrounding this issue for the suppliers. The number agreed upon was believed to be considerably less than actual yield. This arrangement was satisfactory for our engineering demonstrations, but was only an interim step toward a full production volume scenario.

6.2.7.4 Wafer/IC bumping. Getting wafers from different vendors was not easy. Getting the wafers solder-bumped, however, was much more difficult due in part to a lack of infrastructure to support this activity.

Wafer bumping of chip A. The wafer bumping and procurement flow diagram for case A is shown in Fig. 6.29. For solder bumping, the normally square passivation (for wire bonding) usually requires modification. The change is required to provide a solder bump which has the required height after reflow, to minimize stress concentrations and to ensure that there is an adequate adhesion and seal between the passivation and underbump metal (UBM) structure. Fortunately, passivation is the last masking step in most of the IC processes and, as such, it is not a large burden on the vendor to make changes to it. A typical passivation opening design is shown in Fig. 6.70. The square passivation opening was changed to a circular passivation opening. It should be pointed out that the octagonal passivation openings work as well as the circular passivation openings but with much less computer aid design (CAD) effort.

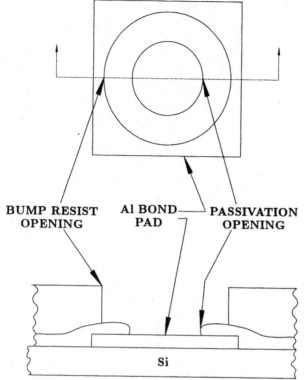

**BUMP RESIST   Al BOND   PASSIVATION
  OPENING   PAD   OPENING**

Si

Figure 6.70   A typical passivation and bump mask openings.[42]

In this study, we were able to arrange for passivation design changes for chip A. The passivation and bump mask openings were octagonal and were modified and fabricated by the IC vendor.

Wafer bumping for chip B.   The wafer bumping and procurement flow diagram for case B is shown in Fig. 6.30. Again, the IC vendor was able to arrange for passivation design changes, and the final result was very similar to that shown in Fig. 6.70.

In this case, the passivation and bump mask openings were circular. Circular, rather than octagonal, openings were used because chip B had two pad pairs that were 170 μm in pitch. In order to keep the 200-μm-pitch-minimum rule in this study, the passivation centers were moved relative to the pad centers by the IC vendor. The bump mask with circular openings was fabricated by the bumper.

Wafer bumping for chip C.   The wafer bumping and procurement flow diagram for case C is shown in Fig. 6.31. In this case, the vendor was unable to make changes to the passivation design and did not want to do anything with wafer bumping. Thus, we worked with the bumper to redefine the passivation openings. The bumper fabricated an octagonal bump mask and used it to define openings in a thin layer of polyimide, which was spun onto the original passivation. In this case, we managed the dicing and sorting, and the wafers/ICs were not tested or burned-in after bumping.

Wafer bumping for chip D. The wafer bumping and procurement flow diagram for case D is shown in Fig. 6.32. It should be emphasized that this is the most ideal case for flip chip-on-board applications, i.e., place the order to the IC vendor and receive the KGD from it. It should be noted that this is more desirable than the bumping and procurement flow used with the vendor for chip A. Even though the IC vendor was willing to accept responsibility for bump quality from its preferred bumper, in case A, there were communication problems with the bump vendor and extra shipping and handling requirements. Furthermore, that arrangement also has the disadvantage that people other than the IC vendor saw our wafers.

Wafer bumping by electroplating. As mentioned earlier, the vendor for chip D agreed to bump its wafers using its own electroplating process, and the IC wafers from all of the other vendors were bumped either by their preferred bumper or by a third-party bumper. Fortunately, in this study, all the wafers of the three vendors were bumped by the same third party with eutectic or near-eutectic solders using the electroplating process, which is completed in the eight steps shown in Fig. 6.20a.

6.2.7.5 Assembly. Twenty of the modules were assembled and four were fully functional at turn on (Fig. 6.71). The modules were assembled and then functionally tested using an electrical test fixture designed for the module. Modules passing the functional test were then soldered to the main board. The process flow for the initial assembly is shown in Fig. 6.72. Placement of the ICs onto the module was accomplished using a modified universal tape-automated bonding (TAB) machine. The solder balls were reflowed and the underfill was hand-dispensed and cured in a small lab oven. A cross section of the MCM-L is shown in Fig. 6.73.

Placement of the module on the main board was accomplished manually, and reflow of the BGA interconnect between the module and main board was accomplished in the same reflow oven at a lower overall temperature profile to avoid additional reflowing of the flip chip solder balls.

No solder joint–related defects were discovered in the postbuild analysis, which included X-rays of the joint areas. This was true for both the solder joints formed between the IC and module and the sec-

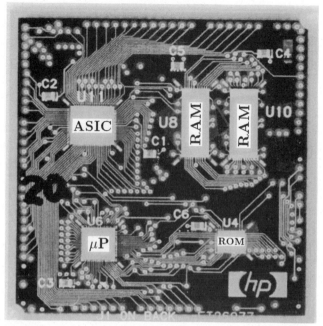

Figure 6.71   The assembled MCM-L.[42]

ond level interconnect between the module and main board. The stand-off distance between the ICs and the module was found to have a large amount of variation due to the pad size and solder volume differences between the IC chips. The solder joint reliability of the solder-bumped flip chip on the MCM-L was demonstrated.[37–39,43,46–47,127]

6.2.7.6   Summary and recommendations.   The design, sourcing, bumping, and assembly of the MCM-L using solder-bumped flip chip technology was successfully demonstrated. The assembly was found to be relatively straightforward; however, there are still many technical challenges to overcome, and these should not be underestimated. It has become clear, even in this early feasibility stage, that the most difficult challenges facing this technology are associated with the existing infrastructure for wire bonding applications.

IC vendors need to develop the design technology which permits an area array of bond pads. Conversion of existing ICs designed for wire bonding will be severely limited by the small pad sizes and pitches found in many designs. This investment by the IC vendors will be critical to the long-term success of flip chip technology.

Standardization of bond pad layouts will permit multivendor sourcing for the same IC functionality. This is critical for the movement of

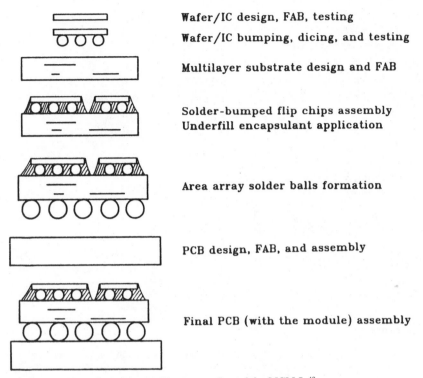

Wafer/IC design, FAB, testing

Wafer/IC bumping, dicing, and testing

Multilayer substrate design and FAB

Solder-bumped flip chips assembly
Underfill encapsulant application

Area array solder balls formation

PCB design, FAB, and assembly

Final PCB (with the module) assembly

Figure 6.72   The complete assembly process flow of the MCM-L.[42]

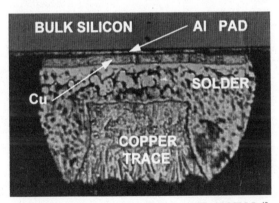

Figure 6.73   A cross section of the assembled MCM-L.[42]

flip chip technology into high-volume consumer electronics applications, especially for commodity products such as memory devices.

Solder-bumped KGD needs to become a packaging alternative provided by the IC vendors. Ideally this would be accommodated by in-house bumping capability. From the end user's standpoint, the next most desirable alternative would be to have an IC vendor's preferred bumper act much like a wire bond packaging house. The IC vendor would assume the responsibility for the delivered bump quality. Any direct ownership for bump quality by the end user of the ICs is not a preferred long-term solution for solder-bumped flip chips on boards.

Bare die burn-in for certain classes of ICs is required. This technology is available for very specific applications—but at a cost premium. Test strategies that permit low-cost bare die burn-in or alternatives to burn-in need to be developed.

Acknowledgments

The principal author (Lau) would like to thank Bill Wun of Hewlett-Packard (HP) for his contribution to Ref. 41; Matt Heydinger, Judy Glazer, and Den Uno of HP for their contributions to Ref. 40; Mike Kelly of HP for his contribution to Ref. 42; Eric Schneider and Tom Baker of HP for their contributions to Ref. 127. It has been a privilege to work with them. He also wants to thank the management of HP for its strong support on the flip chip program.

References

1. Davis, E., W. Harding, R. Schwartz, and J. Corning, "Solid Logic Technology: Versatile, High Performance Microelectronics," *IBM Journal of Research and Development,* April 1964, pp. 102–114.
2. Suryanarayana, D., and D. S. Farquhar, "Underfill Encapsulation for Flip Chip Applications," *Chip On Board Technologies for Multichip Modules,* J. H. Lau, ed., Van Nostrand Reinhold, New York, 1994, pp. 504–531.
3. Totta, P. A., and R. P. Sopher, "SLT Device Metallurgy and Its Monolithic Extension," *IBM Journal of Research and Development,* May 1969, pp. 226–238.
4. Goldmann, L. S., and P. A. Totta, "Chip Level Interconnect: Solder Bumped Flip Chip," *Chip On Board Technologies for Multichip Modules,* J. H. Lau, ed., Van Nostrand Reinhold, New York, 1994, pp. 228–250.
5. Goldmann, L. S., R. J. Herdzik, N. G. Koopman, and V. C. Marcotte, "Lead Indium for Controlled Collapse Chip Joining," *Proceedings of the IEEE Electronic Components Conference,* 1977, pp. 25–29.
6. Totta, P., "Flip Chip Solder Terminals," *Proceedings of the IEEE Electronic Components Conference,* 1971, pp. 275–284.
7. Goldmann, L. S., "Geometric Optimization of Controlled Collapse Interconnections," *IBM Journal of Research and Development,* May 1969, pp. 251–265.
8. Goldmann, L. S., "Optimizing Cycle Fatigue Life of Controlled Collapse Chip Joints," *Proceedings of the 19th IEEE Electronic Components and Technology Conference,* 1969, pp. 404–423.
9. Goldmann, L. S., "Self Alignment Capability of Controlled Collapse Chip Joining," *Proceedings of the 22nd IEEE Electronic Components and Technology Conference,* 1972, pp. 332–339.
10. Shad, H. J., and J. H. Kelly, "Effect of Dwell Time on Thermal Cycling of the Flip Chip Joint," *Proceedings of the ISHM,* 1970, pp. 3.4.1–3.4.6.
11. Hymes, I., R. Sopher, and P. Totta, "Terminals for Microminiaturized Devices and Methods of Connecting Same to Circuit Panels," U.S. Patent 3,303,393, 1967.
12. Karan, C., J. Langdon, R. Pecararo, and P. Totta, "Vapor Depositing Solder," U.S. Patent 3,401,055, 1968.

13. Seraphim, D. P., and J. Feinberg, "Electronic Packaging Evolution," *IBM Journal of Research and Development,* May 1981, pp. 617–629.
14. Tsukada, Y., Y. Mashimoto, and N. Watanuki, "A Novel Chip Replacement Method for Encapsulated Flip Chip Bonding," *Proceedings of the 43rd IEEE/EIA Electronic Components & Technology Conference,* June 1993, pp. 199–204.
15. Tsukada, Y., Y. Maeda, and K. Yamanaka, "A Novel Solution for MCM-L Utilizing Surface Laminar Circuit and Flip Chip Attach Technology," *Proceedings of the 2nd International Conference and Exhibition on Multichip Modules,* April 1993, pp. 252–259.
16. Tsukada, Y., S. Tsuchida, and Y. Mashimoto, "Surface Laminar Circuit Packaging," *Proceedings of the 42nd IEEE Electronic Components and Technology Conference,* May 1992, pp. 22–27.
17. Miller, L. F., "Controlled Collapse Reflow Chip Joining," *IBM Journal of Research and Development,* May 1969, pp. 239–250.
18. Miller, L. F., "A Survey of Chip Joining Techniques," *Proceedings of the 19th IEEE Electronic Components and Technology Conference,* 1969, pp. 60–76.
19. Miller, L. F., "Joining Semiconductor Devices with Ductile Pads," *Proceedings of ISHM,* 1968, pp. 333–342.
20. Norris, K. C., and A. H. Landzberg, "Reliability of Controlled Collapse Interconnections," *IBM Journal of Research and Development,* May 1969, pp. 266–271.
21. Oktay, S., "Parametric Study of Temperature Profiles in Chips Joined by Controlled Collapse Technique," *IBM Journal of Research and Development,* May 1969, pp. 272–285.
22. Bendz, D. J., R. W. Gedney, and J. Rasile, "Cost/Performance Single Chip Module," *IBM Journal of Research and Development,* 1982, pp. 278–285.
23. Blodgett, A. J., Jr., "A Multilayer Ceramic Multichip Module," *IEEE Transactions on Components, Hybrids, and Manufacturing Technology,* 1980, pp. 634–637.
24. Fried, L. J., J. Havas, J. Lechaton, J. Logan, G. Paal, and P. Totta, "A VLSI Bipolar Metallization Design with Three-Level Wiring and Area Array Solder Connections," *IBM Journal of Research and Development,* 1982, pp. 362–371.
25. Clark, B. T., and Y. M. Hill, "IBM Multichip Multilayer Ceramic Modules for LSI Chips—Designed for Performance Density," *IEEE Transactions on Components, Hybrids, and Manufacturing Technology,* 1980, pp. 89–93.
26. Blodgett, A. J., and D. R. Barbour, "Thermal Conduction Module: A High-Performance Multilayer Ceramic Package," *IBM Journal of Research and Development,* 1982, pp. 30–36.
27. Dansky, A. H., "Bipolar Circuit Design for a 5000-Circuit VLSI Gate Array," *IBM Journal of Research and Development,* 1981, pp. 116–125.
28. Oktay, S., and H. C. Kammer, "A Conduction-Cooled Module for High-Performance LSI Devices," *IBM Journal of Research and Development,* 1982, pp. 55–66.
29. Chu, R. C., U. P. Hwang, and R. E. Simons, "Conduction Cooling for an LSI Package: A One-Dimensional Approach," *IBM Journal of Research and Development,* 1982, pp. 45–54.
30. Howard, R. T., "Packaging Reliability and How to Define and Measure It," *Proceedings of the IEEE Electronic Components Conference,* 1982, pp. 376–384.
31. Howard, R. T., "Optimization of Indium-Lead Alloys for Controlled Collapse Chip Connection Application," *IBM Journal of Research and Development,* 1982, pp. 372–389.
32. Satoh, R., K. Arakawa, M. Harada, and K. Matsui, "Thermal Fatigue Life of Pb-Sn Alloy Interconnections," *IEEE Transactions on Components, Hybrids, and Manufacturing Technology,* 14(1):224–232, March 1991.
33. Logsdon, W. A., P. K. Liaw, and M. A. Burke, "Fracture Behavior of 63Sn-37Pb," *Engineering Fracture Mechanics,* 36(2):183–218, 1990.
34. Lau, J. H., *Flip Chip Technologies,* McGraw-Hill, New York, 1996.
35. Lau, J. H., *Ball Grid Array Technology,* McGraw-Hill, New York, 1995.
36. Lau, J. H., "Thermal Fatigue Life Prediction of Flip Chip Solder Joints by Fracture Mechanics Method," *International Journal of Engineering Fracture Mechanics,* 45:643–654, July 1993.

37. Lau, J. H., "Thermomechanical Characterization of Flip Chip Solder Bumps for Multichip Module Applications," *Proceeding of IEEE International Electronics Manufacturing Technology Symposium,* September 1992, pp. 293–299.
38. Lau, J. H., "Thermal Fatigue Life Prediction of Encapsulated Flip Chip Solder Joints for Surface Laminar Circuit Packaging," *ASME Paper No. 92W/EEP-34,* 1992 ASME Winter Annual Meeting.
39. Lau, J. H., and D. Rice, "Thermal Fatigue Life Prediction of Flip Chip Solder Joints by Fracture Mechanics Methods," *Proceedings of the 1st ASME/JSME Electronic Packaging Conference,* April 1992, pp. 385–392.
40. Lau, J. H., M. Heydinger, J. Glazer, and D. Uno, "Design and Procurement of Eutectic Sn/Pb Solder-Bumped Flip Chip Test Die and Organic Substrates," *Circuit World,* **21:**20–24, March 1995.
41. Wun, B., and J. H. Lau, "Characterization and Evaluation of the Underfill Encapsulants for Flip Chip Assembly," *Circuit World,* **21:**25–32, March 1995.
42. Kelly, M., and J. H. Lau, "Low Cost Solder Bumped Flip Chip MCM-L Demonstration," *Circuit World,* **21:**14–17, July 1995.
43. Lau, J. H., "A Brief Introduction to Flip Chip Technologies for Multichip Module Applications," in *Flip Chip Technologies,* J. H. Lau, ed., McGraw-Hill, New York, 1996, pp. 1–82.
44. Lau, J. H., *Solder Joint Reliability: Theory and Applications,* Van Nostrand Reinhold, New York, 1991.
45. Lau, J. H., and D. W. Rice, "Solder Joint Fatigue in Surface Mount Technology: State of the Art," *Solid State Technology,* October 1985, pp. 91–104.
46. Lau, J. H., T. Krulevitch, W. Schar, M. Heydinger, S. Erasmus, and J. Gleason, "Experimental and Analytical Studies of Encapsulated Flip Chip Solder Bumps on Surface Laminar Circuit Boards," *Circuit World,* **19**(3):18–24, March 1993.
47. Lau, J. H., "Solder Joint Reliability of Flip Chip and Plastic Ball Grid Array Assemblies Under Thermal, Mechanical, and Vibration Conditions," *Proceedings of IEEE Japan IEMT Symposium,* December 1995, pp. 13–19.
48. Lau, J. H., *Chip On Board Technologies for Multichip Modules,* Van Nostrand Reinhold, New York, 1994.
49. Wong, C. P., J. M. Segelken, and C. N. Robinson, "Chip On Board Encapsulation," *Chip On Board Technologies for Multichip Modules,* J. H. Lau, ed., Van Nostrand Reinhold, New York, 1994.
50. Rai, A., Y. Dotta, T. Tsukamoto, T. Fujiwara, H. Ishii, T. Nukii, and H. Matsui, "COB (Chip On Board) Technology: Flip Chip Bonding Onto Ceramic Substrates and PWB (Printed Wiring Boards)," *ISHM Proceedings,* 1990, pp. 474–481.
51. Rai, A., Y. Dotta, T. Nukii, and T. Ohnishi, "Flip Chip COB Technology on PWB," *Proceedings of IMC,* June 1992, pp. 144–149.
52. Lowe, H., "No-Clean Flip Chip Attach Process," *International TAB/Advance Packaging and Flip Chip Proceedings,* February 1994, pp. 17–24.
53. Giesler, J., S. Machuga, G. O'Malley, and M. Williams, "Reliability of Flip Chip on Board Assemblies," *International TAB/Advance Packaging and Flip Chip Proceedings,* February 1994, pp. 127–135.
54. Tsukada, Y., "Solder Bumped Flip Chip Attach on SLC Board and Multichip Module," *Chip On Board Technologies for Multichip Modules,* J. H., Lau, ed., Van Nostrand Reinhold, New York, 1994, pp. 410–443.
55. Hu, K., C. Yeh, R. Doot, A. Skipor, and K. Wyatt, "Die Cracking in Flip Chip-on-Board Assembly," *Proceedings of IEEE Electronic Components & Technology Conference,* May 1995, pp. 293–299.
56. Sweet, J., D. Peterson, J. Emerson, and R. Mitchell, "Liquid Encapsulant Stress Variations as Measured with the ATC04 Assembly Test Chip," *Proceedings of IEEE Electronic Components & Technology Conference,* May 1995, pp. 300–304.
57. Greer, S., "An Extended Eutectic Solder Bump for FCOB," *Proceedings of IEEE Electronic Components & Technology Conference,* May 1996, pp. 546–551.
58. McLaren, T., S. Kang, W. Zhang, D. Hellman, T. Ju, and Y. Lee, "Thermosonic Flip Chip Bonding for an 8 × 8 VCSEL Array," *Proceedings of IEEE Electronic Components & Technology Conference,* May 1995, pp. 393–400.

59. Nishimori, T., H. Yanagihara, K. Murayama, Y. Kama, and M. Nakamura, "Characteristics and Potential Application of Polyimide-Core-Bump to Flip Chip," *Proceedings of IEEE Electronic Components & Technology Conference,* May 1995, pp. 515–519.

60. Suryanarayana, D., J. Varcoe, and J. Ellerson, "Repairability of Underfill Encapsulated Flip Chip Packages," *Proceedings of IEEE Electronic Components & Technology Conference,* May 1995, pp. 524–528.

61. Shaukatullah, H., B. Hansen, W. Storr, and F. Andros, "Thermal Enhancement of Flip Chip Packages with Radial-Finger-Contact Spring," *Proceedings of IEEE Electronic Components & Technology Conference,* May 1995, pp. 865–871.

62. Kim, D., H. Han, S. Park, G. Joo, M. Song, N. Hwang, S. Kang, H. Lee, and H. Park, "Application of the Flip Chip Bonding Technique to the 10 Gbps Laser Diode Module," *Proceedings of IEEE Electronic Components & Technology Conference,* May 1995, pp. 872–875.

63. Perfecto, E., R. Shields, R. Master, S. Purushothaman, and C. Prasad, "A Low Cost MCM-D Process for Flip Chip and Wirebonding Applications," *Proceedings of IEEE Electronic Components & Technology Conference,* May 1995, pp. 1081–1086.

64. Howell, W., D. Brouillette, J. Korejwa, E. Sprogis, S. Yankee, and J. Wursthorn, "Area Array Solder Interconnection Technology for the Three-Dimensional Silicon Cube," *Proceedings of IEEE Electronic Components & Technology Conference,* May 1995, pp. 1174–1178.

65. Chen, W., J. Gentile, and L. Higgins, "FCOB Reliability Evaluation Simulating Multiple Rework/Reflow Processes," *Proceedings of IEEE Electronic Components & Technology Conference,* May 1996, pp. 1184–1195.

66. Clementi, J., G. Dearing, J. Zimmerman, and C. Bergeron, "Reliability and Analytical Evaluations of No-Clean Flip Chip Assembly," *Proceedings of IEEE Electronic Components & Technology Conference,* May 1995, pp. 1191–1196.

67. Ogashiwa, T., T. Arikawa, H. Murai, A. Inoue, and T. Masumoto, "Reflowable Sn-Pb Bump Formation on Al Pad by a Solder Bumping Method," *Proceedings of IEEE Electronic Components & Technology Conference,* May 1995, pp. 1203–1208.

68. Eldring, J., K. Koeffers, H. Richter, A. Baumgartner, and H. Reichl, "Flip Chip Attachment of Silicon Devices Using Substrate Ball Bumping and the Technology Evaluation on Test Assemblies for 20 Gbit/s Transmission," *Proceedings of IEEE Electronic Components & Technology Conference,* May 1995, pp. 1209–1216.

69. Okuno, A., N. Oyama, K. Nagai, and T. Hashimoto, "Flip Chip Packaging Using PES (Printing Encapsulation Systems) and PES Underfill Epoxy Resin," *Proceedings of IEEE Electronic Components & Technology Conference,* May 1995, pp. 1240–1243.

70. Goldstein, J., D. Tuckerman, P. Kim, and B. Fernandez, "A Novel Flip Chip Process," *Proceedings of SMI Conference,* August 1995, pp. 59–71.

71. Magill, P., and G. Rinne, "Implementation of Flip Chip Technology," *Proceedings of SMI Conference,* August 1995, pp. 72–79.

72. Lowe, H., and R. Lyn, "Real World Flip Chip Assembly: A Manufacture's Experience," *Proceedings of SMI Conference,* August 1995, pp. 80–87.

73. Aschenbrenner, R., E. Zakel, G. Azdasht, A. Kloeser, and H. Reichl, "Fluxless Flip Chip Bonding on Flexible Substrates: A Comparison Between Adhesive Bonding and Soldering," *Proceedings of SMI Conference,* August 1995, pp. 91–101.

74. Liu, J., K. Boustedt, and Z. Lai, "Development of Flip Chip Joining Technology on Flexible Circuitry Using Anisotropically Conductive Adhesives and Eutectic Solder," *Proceedings of SMI Conference,* August 1995, pp. 102–109.

75. Patterson, T., "A Practical Versatile Approach to Flip Chip on Flex," *Proceedings of SMI Conference,* August 1995, pp. 110–114.

76. Koh, W., and M. Edwards, "Performance Enhancement of Newly Developed Chip on Board (COB) Encapsulants," *Proceedings of SMI Conference,* August 1995, pp. 138–145.

77. Schiesser, T., E. Menard, T. Smith, and J. Akin, "Microdynamic Solder Pump vs Alternatives Comparative Review of Solder Bumping Techniques for Flip Chip Attach," *Proceedings of SMI Conference,* August 1995, pp. 171–178.

78. McDermott, B., "Impact of Direct Chip Attach on the Printed Circuit Board," *Proceedings of SMI Conference,* August 1995, pp. 1026–1030.
79. Schrand, J., "Using Die in PC Card Applications," *Proceedings of SMI Conference,* August 1995, pp. 1031–1038.
80. Koopman, N., and S. Nangalia, "Fluxless Flip Chip Solder Joining," in *Flip Chip Technologies,* J. H. Lau, ed., McGraw-Hill, New York, 1996, pp. 83–121.
81. Sharma, R., and R. Subrahmanyan, "Solder Bumped Flip Chip Interconnect Technologies: Materials, Processes, Performance and Reliability," in *Flip Chip Technologies,* J. H. Lau, ed., McGraw-Hill, New York, 1996, pp. 123–153.
82. Chung, T., T. Dolbear, and Dick Nelson, "Large High I/O Solder Bumped Flip Chip Technology," in *Flip Chip Technologies,* J. H. Lau, ed., McGraw-Hill, New York, 1996, pp. 155–179.
83. Degani, Y., T. D. Dudderar, R. C. Frye, K. L. Tai, M. Lau, and B. Han, "Micro-Interconnect Technology: The Large Volume Fabrication of Cost Effective Flip Chip MCMs," in *Flip Chip Technologies,* J. H. Lau, ed., McGraw-Hill, New York, 1996, pp. 181–221.
84. Estes, R., and F. Kulesza, "Conductive Adhesive Polymer Materials in Flip Chip Applications," in *Flip Chip Technologies,* J. H. Lau, ed., McGraw-Hill, New York, 1996, pp. 223–267.
85. Breen, M., D. Duane, R. German, K. Keswick, and R. Nolan, "Compliant Bumps for Adhesive Flip Chip Assembly," in *Flip Chip Technologies,* J. H. Lau, ed., McGraw-Hill, New York, 1996, pp. 269–287.
86. Date, H., Y. Hozumi, H. Tokuhira, M. Usui, E. Horikoshi, and T. Sato, "Anisotropic Conductive Adhesive for Fine Pitch Flip Chip Interconnections," in *Flip Chip Technologies,* J. H. Lau, ed., McGraw-Hill, New York, 1996, pp. 289–300.
87. Watanabe, I., N. Shiozawa, K. Takemura, and T. Ohta, "Flip Chip Interconnection Technology Using Anisotropic Conductive Adhesive Films," in *Flip Chip Technologies,* J. H. Lau, ed., McGraw-Hill, New York, 1996, pp. 301–315.
88. Lee, C. H., "Anisotropic Conductive Flip Chip-on-Glass Technology," in *Flip Chip Technologies,* J. H. Lau, ed., McGraw-Hill, New York, 1996, pp. 317–339.
89. Baba, S., and W. Carlomagno, "Wirebonding Flip Chip Technology for Multichip Modules," in *Flip Chip Technologies,* J. H. Lau, ed., McGraw-Hill, New York, 1996, pp. 341–356.
90. Tsunoi, K., T. Kusagaya, and H. Kira, "Flip Chip Mounting Using Stud-Bumps and Adhesive for Encapsulation," in *Flip Chip Technologies,* J. H. Lau, ed., McGraw-Hill, New York, 1996, pp. 357–366.
91. Moresco, L., D. Love, W. Chow, and V. Holalkere, *Wire Interconnect Technology: An Ultra High Density Flip Chip to Substrate Connection Method,* J. H. Lau, ed., McGraw-Hill, New York, 1996, pp. 367–386.
92. DiStefano, T., and J. Fjelstad, "A Compliant Chip Size Packaging Technology," in *Flip Chip Technologies,* J. H. Lau, ed., McGraw-Hill, New York, 1996, pp. 387–413.
93. Zakel, E., and H. Reichl, "Flip Chip Assembly Using the Gold, Gold-Tin and Nickel-Gold Metallurgy," in *Flip Chip Technologies,* J. H. Lau, ed., McGraw-Hill, New York, 1996, pp. 415–490.
94. Gilg, L., "Assurance Technologies for Known Good Die," in *Flip Chip Technologies,* J. H. Lau, ed., McGraw-Hill, New York, 1996, pp. 491–521.
95. Rinne, G., "Burn-In and Test Substrate for Flip Chip ICs," in *Flip Chip Technologies,* J. H. Lau, ed., McGraw-Hill, New York, 1996, pp. 523–538.
96. Doi, K., N. Hirano, T. Okada, Y. Hiruta, T. Sudo, and M. Mukai, "Prediction of Thermal Fatigue Life for Encapsulated Flip Chip Interconnection," *Proceedings of International Symposium on Microelectronics,* October 1995, pp. 247–252.
97. Zoba, D., and M. Edwards, "Review of Underfill Encapsulant Development and Performance of Flip Chip Applications," *Proceedings of International Symposium on Microelectronics,* October 1995, pp. 354–358.
98. Bessho, Y., Y. Tomura, T. Shiraishi, M. Ono, T. Ishida, and K. Omoya, "Advanced Flip Chip Bonding Technique to Organic Substrates," *Proceedings of International Symposium on Microelectronics,* October 1995, pp. 359–364.

99. Yatsuda, H., and T. Eimura, "Flip Chip Assembly Technique for SAW Devices," *Proceedings of International Symposium on Microelectronics,* October 1995, pp. 365–370.
100. Gupta, D., "Evaluation of Alternative Processes for Au-Sn-Au Flip Chip Bonding of Power Devices," *Proceedings of International Symposium on Microelectronics,* October 1995, pp. 371–377.
101. Chrusciel, R., P. Delivorias, and K. Rispoli, "Flip Chip Interconnect, A Versatile Known Good Die Technology," *Proceedings of International Symposium on Microelectronics,* October 1995, pp. 384–389.
102. Amano, T., M. Kohno, and Y. Obara, "Solder Bumping Through Super Solder," *Proceedings of IEEE International Electronics Manufacturing Technology Symposium,* October 1995, pp. 1–4.
103. Schwiebert, M., and W. Leong, "Underfill Flow as Viscous Flow Between Parallel Plastes Driven by Capillary Action," *Proceedings of IEEE International Electronics Manufacturing Technology Symposium,* October 1995, pp. 8–13.
104. Goenka, L., and A. Achari, "Void Formation in Flip Chip Solder Bumps—Part I," *Proceedings of IEEE International Electronics Manufacturing Technology Symposium,* October 1995, pp. 14–19.
105. Kallmayer, C., D. Lin, J. Kloeser, H. Oppermann, E. Zakel, and H. Reichl, "Fluxless Flip Chip Attachment Techniques Using the Au/Sn Metallurgy," *Proceedings of IEEE International Electronics Manufacturing Technology Symposium,* October 1995, pp. 20–28.
106. Koopman, N., G. Adema, S. Nangalia, M. Schneider, and V. Saba, "Flip Chip Process Development Techniques Using A Modified Laboratory Aligner Bonder," *Proceedings of IEEE International Electronics Manufacturing Technology Symposium,* October 1995, pp. 29–35.
107. Imler, W., T. Hildebrandt, S. Paolini, K. Scholz, M. Cobarruviaz, and V. Nagesh, "Design and Fabrication of 600-dpi Light-Emitting Diode Printheads Using Precision Flip Chip Solder Bump Technology," *Proceedings of IEEE International Electronics Manufacturing Technology Symposium,* September 1994, pp. 154–159.
108. Aintila, A., A. Bjorklof, E. Jarvinen, and S. Lalu, "Electroless Ni/Au Bumps for Flip chip-on-Flex and TAB Applications," *Proceedings of IEEE International Electronics Manufacturing Technology Symposium,* September 1994, pp. 160–164.
109. Zakel, E., J. Gwiasda, J. Kloeser, J. Eldring, G. Engelmann, and H. Reichl, "Fluxless Flip Chip Assembly on Rigid and Flexible Polymer Substrates Using the Au-Sn Metallurgy," *Proceedings of IEEE International Electronics Manufacturing Technology Symposium,* September 1994, pp. 177–184.
110. Moresco, L., D. Love, V. Holalkeri, P. Boucher, W. Chou, C. Grilletto, and C. Wong, "Wire Interconnect Technology, A New Flip Chip Technique," *Proceedings of NEPCON West,* February 1996, pp. 947–955.
111. Lee, C. H., and K. Loh, "Fine Pitch COG Interconnections Using Anisotropically Conductive Adhesives," *Proceedings of NEPCON West,* February 1996, pp. 956–967.
112. Degani, Y., T. D. Dudderar, and K. Tai, "High Density MCM Assembly Using Printed Solder Paste Process," *Proceedings of NEPCON West,* February 1996, pp. 968–972.
113. Vardaman, E. J., and T. Goodman, "Worldwide Trends in Flip Chip Developments," *Proceedings of NEPCON West,* February 1996, pp. 973–975.
114. Shock, G., "Flip Chip Assembly in High Volume, High Mix Products," *Proceedings of NEPCON West,* February 1996, pp. 1433–1438.
115. Slesinger, K., "Flip Chips on Laminates: High-Volume Assembly," *Proceedings of NEPCON West,* February 1996, pp. 1439–1447.
116. Han, B., Y. Guo, T. Chung, and D. Liu, "Reliability Assessment of Flip Chip Package with Encapsulation," *Proceedings of NEPCON West,* February 1995, pp. 600–602.
117. Gutentag, C., "Methods for Handling Bare Die for C4 Assembly/Placement," *Proceedings of NEPCON West,* February 1995, pp. 603–609.
118. Zakel, E., R. Aschenbrenner, J. Gwiasda, G. Azdasht, A. Ostmann, J. Eldring, and H. Reichl, "Fluxless Flip Chip Bonding on Flexible Substrates," *Proceedings of NEPCON West,* February 1995, pp. 909–918.

119. Degani, Y., T. D. Dudderar, R. C. Frye, and K. L. Tai, "A Cost Effective MCM Manufacturing Platform Achieved by Combining Surface Mount and Silicon Technologies," *Proceedings of NEPCON West,* February 1995, pp. 932–937.
120. Chanchani, R., K. Treece, and P. Dressendorfer, "Mini Ball Grid Array (mBGA) Technology," *Proceedings of NEPCON West,* February 1995, pp. 938–945.
121. Vardaman, E. J., "International Activities in Flip Chip on Board Technologies," *Proceedings of NEPCON West,* February 1995, pp. 946–950.
122. Jimarez, M., A. Smith, and J. Zdimal, "Development of a Rework Process for Unencapsulated Flip Chips on Organic Substrates," *Proceedings of NEPCON West,* February 1995, pp. 1646–1669.
123. Simon, J., J. Gwiasda, I. Kuhls, K. Werner, and H. Reichl, "Development of Display Module Using Flip Chip on Flex," *Proceedings of IEEE Japan IEMT Symposium,* December 1995, pp. 48–51.
124. Tsunetsugu, H., T. Hayashi, and K. Katsura, "Micro-alignment Technique using 26-μm Diameter Microsolder Bumps and Its Shear Strength," *Proceedings of IEEE Japan IEMT Symposium,* December 1995, pp. 52–55.
125. Kloeser, J., A. Ostmann, R. Aschenbrenner, E. Zakel, and H. Reichl, "Approaches to Flip Chip Technology Using Electroless Nickel-Gold Bumps," *Proceedings of IEEE Japan IEMT Symposium,* December 1995, pp. 60–66.
126. Jung, E., R. Aschenbrenner, A. Ostmann, E. Zakel, and H. Reichl, "Flip Chip Soldering and Adhesive Bonding on Organic Substrates," *Proceedings of IEEE Japan IEMT Symposium,* December 1995, pp. 67–71.
127. Lau, J. H., E. Schneider, and T. Baker, "Shock and Vibration of Solder Bumped Flip Chip on Organic Coated Copper Boards," *ASME Transactions, Journal of Electronic Packaging,* June 1996, pp. 101–104.
128. Fujiuchi, S., and K. Toriyama, "Collective Screen Printing for Carrier Bump and SMT Pads," *Proceedings of IEEE Japan IEMT Symposium,* December 1995, pp. 109–112.
129. Honma, S., K. Tateyama, H. Yamada, and M. Saito, "Evaluation of Barrier Metal of Solder Bumps for Flip Chip Interconnection," *Proceedings of IEEE Japan IEMT Symposium,* December 1995, pp. 113–116.
130. Kato, Y., Y. Ueoka, E. Kono, and E. Hagimoto, "Solder Bump Forming Using Micro Punching Technology," *Proceedings of IEEE Japan IEMT Symposium,* December 1995, pp. 117–120.
131. Yanada, H., and M. Saito, "A Fine Pitch and High Aspect Ratio Bump Fabrication Process for Flip Chip Interconnection," *Proceedings of IEEE Japan IEMT Symposium,* December 1995, pp. 121–124.
132. Ohunki, Y., H. Shibuya, J. Utsunomiya, and S. Iida, "Development of Low Cost Solder Bump on LSI," *Proceedings of IEEE Japan IEMT Symposium,* December 1995, pp. 125–128.
133. Homma, Y., T. Fujiki, K. Kobayashi, Y. Shirai, and K. Akazawa, "Fast-Cure Liquid Encapsulant for ICs," *Proceedings of IEEE Japan IEMT Symposium,* December 1995, pp. 449–452.
134. DeHaven, K., and J. Dietz, "Controlled Collapse Chip Connection (C4)—An Enabling Technology," *Proceedings of IEEE Electronic Components & Technology Conference,* May 1994, pp. 1–6.
135. Puttlitz, K., and W. Shutler, "C4/BGA Comparison with Other MLC Single Chip Package Alternatives," *Proceedings of IEEE Electronic Components & Technology Conference,* May 1994, pp. 16–21.
136. Kromann, G., D. Gerke, and W. Huang, "A Hi-Density C4/CBGA Interconnect Technology for a CMOS Microprocessor," *Proceedings of IEEE Electronic Components & Technology Conference,* May 1994, pp. 22–28.
137. Switky, A., V. Sajja, J. Darnauer, and W. Dai, "A 1024-Pin Plastic Ball Grid Array for Flip Chips Die," *Proceedings of IEEE Electronic Components & Technology Conference,* May 1994, pp. 32–38.
138. O'Malley, G., J. Giesler, and S. Machuga, "The Importance of Material Selection for Flip Chip on Board Assembly," *Proceedings of IEEE Electronic Components & Technology Conference,* May 1994, pp. 387–394.

139. Machuga, S., S. Lindsey, K. Moore, and A. Skipor, "Encapsulation of Flip Chip Structures," *Proceedings of IEEE International Electronics Manufacturing Technology Symposium,* September 1992, pp. 53–58.
140. Chu, D., C. Reber, and D. Palmer, "Screening ICs on the Bare Chip Level: Temporary Packaging," *Proceedings of IEEE International Electronics Manufacturing Technology Symposium,* September 1992, pp. 223–226.
141. Yamada, H., Y. Kondoh, and M. Saito, "A Fine Pitch and High Aspect Ratio Bump Array for Flip Chip Interconnection," *Proceedings of IEEE International Electronics Manufacturing Technology Symposium,* September 1992, pp. 288–292.
142. Kromann, G., "Thermal Modeling and Experimental Characterization of the C4/Surface-Mount-Array Interconnect Technologies," *Proceedings of IEEE Electronic Components & Technology Conference,* May 1994, pp. 395–402.
143. Gasparini, N., and B. Bhattacharyya, "A Method of Designing a Group of Bumps for C4 packages to Maximize the Number of Bumps and Minimize the Number of Package Layers," *Proceedings of IEEE Electronic Components & Technology Conference,* May 1994, pp. 695–699.
144. Thompson, P., M. Begay, S. Lindsey, D. Vanoverloop, B. Vasquez, S. Walker, and B. Williams, "Mechanical and Electrical Evaluation of a Bumped-Substrate Die-Level Burn-In Carrier," *Proceedings of IEEE Electronic Components & Technology Conference,* May 1994, pp. 700–703.
145. Metzfer, D., U. Beutler, J. Eldring, and H. Reichl, "Laser Bumping for Flip Chip and TAB Applications," *Proceedings of IEEE Electronic Components & Technology Conference,* May 1994, pp. 910–916.
146. Gupta, D., "A Novel Active Area Bumped Flip Chip Technology for Convergent Heat Transfer from Gallium Arsenide Power Devices," *Proceedings of IEEE Electronic Components & Technology Conference,* May 1994, pp. 917–921.
147. Itoh, M., J. Sasaki, A. Uda, and I. Yoneda, "Use of AuSn Solder Bumps in Three-dimensional Passive Aligned Packaging," *Proceedings of IEEE Electronic Components & Technology Conference,* May 1996, pp. 1–7.
148. Hunziker, W., W. Vogt, H. Melchior, R. Germann, and C. Harder, "Low Cost Packaging of Semiconductor Laser Arrays using Passive Self-Aligned Flip-Chip Technique on Si Motherboard," *Proceedings of IEEE Electronic Components & Technology Conference,* May 1996, pp. 8–12.
149. Hayashi, T., and H. Tsunetsugu, "Optical Module with MU Connector Interface Using Self-Alignment Technique by Solder-Bump Chip Bonding," *Proceedings of IEEE Electronic Components & Technology Conference,* May 1996, pp. 13–19.
150. Nakasuga, Y., T. Hashimoto, Y. Yamada, H. Terui, M. Yanagisawa, K. Moriwaki, Y. Akahori, Y. Tohmori, K. Kato, S. Sekine, and M. Horiguchi, *Proceedings of IEEE Electronic Components & Technology Conference,* May 1996, pp. 20–25.
151. Tan, Q., and Y. Lee, "Soldering Technology for Optoelectronic Packaging," *Proceedings of IEEE Electronic Components & Technology Conference,* May 1996, pp. 26–36.
152. Han, S., K. Wang, and S. Cho, "Experimental and Analytical Study on the Flow of Encapsulant During Underfill Encapsulation of Flip-Chips," *Proceedings of IEEE Electronic Components & Technology Conference,* May 1996, pp. 327–334.
153. Wong, C., J. Segelken, K. Tai, and C. Wong, "Non-Hermetic Plastic Packaging of High Voltage Electronic Switches Utilizing a Low-Stress Glob Coating for 95/5 Pb/Sn Solder Joints of Flip-Chip Bonded Multi-chip Module High Voltage Devices," *Proceedings of IEEE Electronic Components & Technology Conference,* May 1996, pp. 347–353.
154. Le Gall, C., J. Qu, and D. McDowell, "Delamination Cracking in Encapsulated Flip Chips," *Proceedings of IEEE Electronic Components & Technology Conference,* May 1996, pp. 430–434.
155. Gektin, V., A. Bar-Cohen, and S. Witzman, "Thermo-Structural Behavior of Underfilled Flip-Chip," *Proceedings of IEEE Electronic Components & Technology Conference,* May 1996, pp. 440–447.
156. Wu, T., Y. Tsukada, and W. Chen, "Materials and Mechanics Issues in Flip-Chip Organic Packaging," *Proceedings of IEEE Electronic Components & Technology Conference,* May 1996, pp. 524–534.

157. Doot, R., "Motorola's First DCA Product: The Gold Line Pen Pager," *Proceedings of IEEE Electronic Components & Technology Conference,* May 1996, pp. 535–539.
158. Wun, K., and G. Margaritis, "The Evaluation of Fast-Flow, Fast-Cure Underfills for Flip Chip on Organic Substrates," *Proceedings of IEEE Electronic Components & Technology Conference,* May 1996, pp. 540–545.
159. Koopman, N., S. Nangalia, and V. Rogers, "Fluxless No-Clean Assembly of Solder Bumped Flip Chips," *Proceedings of IEEE Electronic Components & Technology Conference,* May 1996, pp. 552–558.
160. Puttlitz, K., K. Lidestri, and P. Totta, "Solder Transfer Technique for Flip Chip and Electronic Assembly Applications," *Proceedings of IEEE Electronic Components & Technology Conference,* May 1996, pp. 559–564.
161. Max, D., and J. Carbin, "Advances in Flip-Chip Underfill Flow and Cure Rates and Their Enhancement of Manufacturing Processes and Component Reliability," *Proceedings of IEEE Electronic Components & Technology Conference,* May 1996, pp. 1025–1031.
162. Lassen, C., "Global Technical and Commercial Developments with Flip Chip Technology," *Proceedings of IEEE Electronic Components & Technology Conference,* May 1996, pp. 1056–1058.
163. Smith, D., and A. Alimonda, "A New Flip-Chip Technology for High-Density Packaging," *Proceedings of IEEE Electronic Components & Technology Conference,* May 1996, pp. 1069–1073.
164. Kuhmann, J., H. Hensel, D. Pech, P. Harde, and H. Bach, "Self-Aligned, Fluxless Flip-Chip Bonding Technology for Photonic Devices," *Proceedings of IEEE Electronic Components & Technology Conference,* May 1996, pp. 1088–1092.
165. Yang, H., P. Deane, P. Magill, and K. Murty, "Creep Deformation of 96.5Sn-3.5Ag Solder Joints in a Flip Chip Package," *Proceedings of IEEE Electronic Components & Technology Conference,* May 1996, pp. 1136–1142.
166. Baldwin, D., R. Deshmukh, and C. Hau, "Gallium Alloy Interconnects for Flip-Chip Assembly Applications," *Proceedings of IEEE Electronic Components & Technology Conference,* May 1996, pp. 1143–1150.
167. Kawai, M., M. Harada, A. Andou, O. Yamada, R. Satoh, and T. Netsu, "Highly Accurate Design of Thermal Fatigue Life for Flipchip Joint," *Proceedings of IEEE Electronic Components & Technology Conference,* May 1996, pp. 1196–1201.
168. Lall, P., G. Gold, B. Miles, K. Banerji, P. Thompson, C. Koehler, and I. Adhihetty, "Reliability Characterization of the SLICC Package," *Proceedings of IEEE Electronic Components & Technology Conference,* May 1996, pp. 1202–1210.
169. Lau, J. H., L. Powers, J. Baker, D. Rice, and W. Shaw, "Solder Joint Reliability of Fine Pitch Surface Mount Technology Assemblies," *IEEE Transaction on Components, Hybrid, and Manufacturing Technology,* 13:534–544, September 1990.
170. Lau, J. H., T. Marcotte, J. Severine, A. Lee, S. Erasmus, T. Baker, J. Moldaschel, M. Sporer, and G. Burward-Hoy, "Solder Joint Reliability of Surface Mount Connectors," *Journal of Electronic Packaging, Transactions of ASME,* 115:180–188, June 1993.

Solder Joint Reliability of CSP Assemblies

The definition of *chip-scale package* (CSP) is that the ratio of the chip area to the package area is greater than 80 percent.[1] The advantages of CSP are that it is easy to handle, test, and standardize (Table 1.2). These advantages are the toughest technical and infrastructural issues to deal with for directly attaching bare chips on the substrate. In this section the solder joint reliability of several CSP assemblies is discussed.

7.1 Solder Joint Reliability of Mitsubishi's CSP Assembly

Figure 7.1 shows the CSPs under consideration.[2,3] The package body size and chip size were $6.35 \times 15.24 \times 0.65$ mm and $5.95 \times 14.84 \times 0.4$ mm, respectively. For the 60-pin CSP, the 63wt%Sn/37wt%Pb solder bump diameter was 0.68 mm and the bump pitch was 1 mm. For the 96-pin CSP the eutectic solder bump diameter was 0.55 mm and the bump pitch was 0.8 mm. For the 32-pin CSP the bump diameter and pitch were the same as the 96-pin CSP. A cross section of the solder bump is shown in Fig. 7.2. These CSPs were assembled on ceramic and FR-4 glass-epoxy substrates. There was no underfill epoxy in either case.

Figure 7.3 shows the stress-strain curve of the 63wt%Sn/37wt%Pb solder with the temperature of 125°C and at the frequency of 1/3600 Hz. This curve was obtained from the hysteresis loops of the solder in a tensile and compression cycle test by the incremental step method.[2] The other material properties of the CSP assemblies are shown in Table 7.1. The boundary condition was −40 to 125°C at one cycle per hour.

TABLE 7.1   Material Properties for Mitsubishi's CSP Simulation[2,3]

	Young's modulus (kg/mm²)	Poisson's ratio	TCE ($\times 10^{-6}$/°C)
Chip (Si)	17,000	0.28	3.5
Encapsulation resin	1,280	0.35	19.0
Inner bump (Cu)	12,000	0.35	16.8
Bump (63Sn/37Pb)*	1,850	0.36	24.7
Glass-epoxy sub.	2,320	0.18	15.8
Al$_2$O$_3$ sub.	26,000	0.30	6.5

* At 125°C, 1/3600 Hz.

TABLE 7.2   Simulation Results on Average Solder Joint Thermal Fatigue Life, Modified from Refs. 2, 3

	Substrates			
	Al$_2$O$_3$		FR-4	
CSP	$\Delta\gamma_p$(%)	N_f*(cycles)	$\Delta\gamma_p$ (%)	N_f* (cycles)
60-pin	1.478	1,247	1.628	1,032
96-pin	1.492	1,224	1.832	818
32-pin	1.666	986	3.310	256

* $N_f = 1.29(\Delta\gamma_p)^{-1.96}/SF$, where $SF = 4$ has been applied and $\Delta\gamma_p =$ equivalent plastic strain range.

Figure 7.4 shows a typical finite element result of the equivalent plastic strain range distribution in a deformed corner solder bump of the 60-pin CSP on a FR-4 glass-epoxy PCB. It can be seen that the maximum equivalent plastic strain occurred at the four corners of the solder bump (shear deformation) due to the thermal expansion mismatch of the chip and the FR-4 PCB.

Table 7.2 summarizes the nonlinear finite element results of these three CSPs on ceramic and FR-4 substrates. It can be seen that in all the cases, the plastic strain ranges are smaller in the ceramic assemblies than in the FR-4 assemblies. This is because the thermal expansion mismatch is larger in the FR-4 assemblies than in the ceramic assemblies. The number of cycles to failure (N) was calculated based on Solomon's fatigue data.[4,5]

Recently, Mitsubishi extended its CSP to very high pin counts[3] (Fig. 7.5 and Table 7.3). A cutout and cross section of the new package is shown in Figs. 7.6 and 7.7. It can be seen that this new CSP has an additional transferred copper land. The package assembly process is shown in Fig. 7.8a and b.

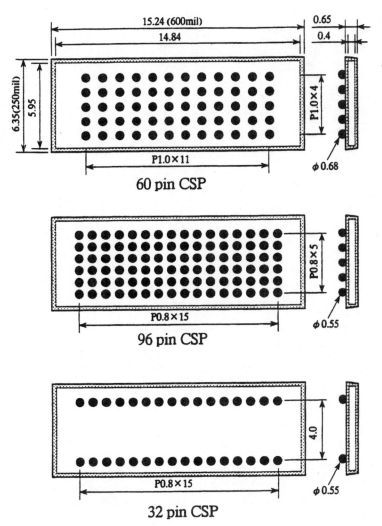

Figure 7.1 Mitsubishi's CSP family.

Figure 7.9 shows the thermal cycling test results of a rectangular outline CSP, $6.35 \times 15.24 \times 0.65$ mm in body size and $5.95 \times 14.84 \times 0.4$ mm in chip size. There were 96 (6×16) solder bumps 0.55 mm in diameter and with 0.8 mm pitch. The test condition was -40 to $125°C$ at one cycle per hour. It can be seen from Fig. 7.9 that: (1) the solder joint reliability of the double stacked solder bump structure of CSP on board is better than that of the solder-bumped bare flip chip-on-board, and (2) the solder joint of the underfilled CSP on board is reliable.

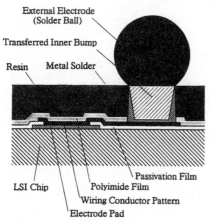

Figure 7.2 Cross-sectional view of Mitsubishi's electrode bump.[2]

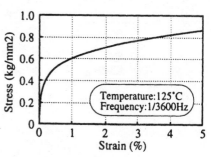

Figure 7.3 Stress-strain curve of Mitsubishi's 63Sn/37Pb solder.[2]

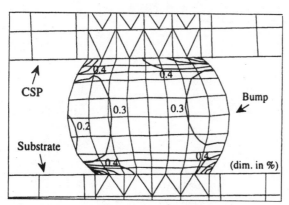

Figure 7.4 Equivalent plastic strain distribution in Mitsubishi's deformed bump.[2]

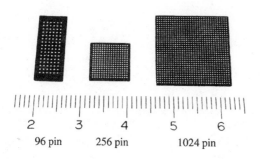

Figure 7.5 Mitsubishi's CSPs for high pin count applications.[3]

TABLE 7.3   Specifications of Mitsubishi's CSP[3]

Package body size	Pin count	Bump pitch
6.35 × 15.24 mm	96	0.8 mm
9.0 × 9.0 mm	256	0.5 mm
16.6 × 16.6 mm	1,024	0.5 mm

7.2   Solder Joint Reliability of Motorola's CSP Assembly

Figure 7.10 shows a cross section of Motorola's CSP assembly.[6–9] They called it *slightly larger than IC carrier* (SLICC) *package assembly*. It can be seen that they used high-temperature solder-bumped flip chips (C4-type) on the FR-4 or BT epoxy substrates which are coated with 60wt%Sn/40wt%Pb solder. The solder bump diameter is 0.1 mm and pitch is 0.25 mm. The solder-bumped flip chip is underfilled with epoxy encapsulant. The BGA-type's 62wt%Sn/36wt%Pb/2wt%Ag solder balls (Motorola calls them C5 solder joints) are attached at the bottom of the substrate. The diameter of the balls is 0.56 mm and pitch is 0.8 mm. The standoff height of the SLICC from the FR-4 PCB is 0.3 mm.

7.2.1   Mechanical bending of Motorola's CSP assembly

The SLICC assembly was subjected to three-point bending (Fig. 7.11) with various PCB thicknesses: 32 mils (0.81 mm), 49 mils (1.24 mm),

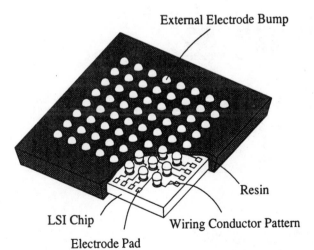

External Electrode Bump

Resin

LSI Chip

Wiring Conductor Pattern

Electrode Pad

Figure 7.6   Perspective view of Mitsubishi's molded CSP.

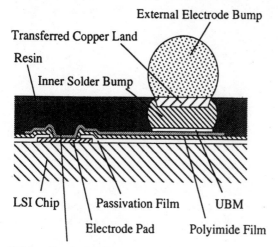

External Electrode Bump

Transferred Copper Land

Resin

Inner Solder Bump

LSI Chip   Passivation Film   UBM

Electrode Pad   Polyimide Film

Wiring Conductor Pattern

Figure 7.7 Cross-sectional view of Mitsubishi's new electrode bump.[3]

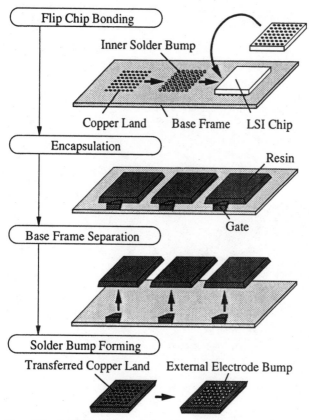

Flip Chip Bonding

Inner Solder Bump

Copper Land   Base Frame   LSI Chip

Encapsulation

Resin

Gate

Base Frame Separation

Solder Bump Forming

Transferred Copper Land   External Electrode Bump

Figure 7.8 (*a*) Mitsubishi's packaging process. (*b*) Mitsubishi's solder-bumping process.[3]

Structure		Result of Test (Failures/Sample Size)			
		0cyc	100cyc	200cyc	500cyc
Bare Die		0/30	10/30	25/30	30/30
CSP		0/20	0/20	0/20	8/20
CSP with underfilling		0/3	0/3	0/3	0/3

Figure 7.9 Mitsubishi's CSP solder joint reliability.[3]

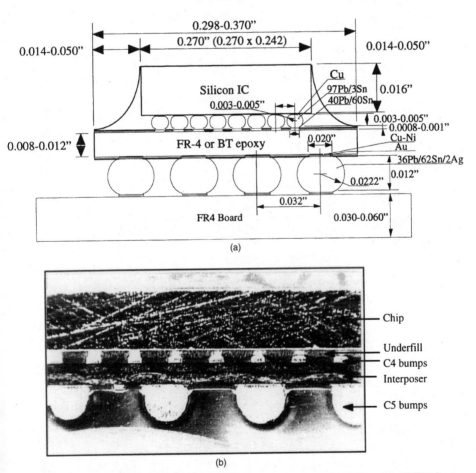

(a)

(b)

Figure 7.10 (a) Schematic of Motorola's SLICC. (b) Cross section of Motorola's SLICC.[6]

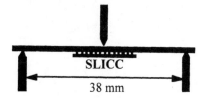

Figure 7.11 Motorola's SLICC under three-point bending.[6]

56 mils (1.42 mm; double-sided), and 55 mils (1.4 mm; six-layer). The maximum load-deflection curves are shown in Fig. 7.12.[6,7] As expected, for the same SLICC, the thicker the PCB, the higher the stiffness (slope).

The finite element model of the SLICC assembly is shown in Fig. 7.13 (11,872 3D elements, 14,384 nodes, and 40,629 degrees of freedom). The constitutive behavior of the solders was modeled using a sinh-viscoplastic law given by Brown et al.[6] The Young's modulus of the underfill material is a function of temperature and is given in Fig. 7.14. It can be seen that there are two different cure conditions: 30 minutes at 150°C, and 15 minutes wait plus 60 minutes at 150°C. The underfill's TCE does not change with cure condition and stays at 20 ppm/°C up to the glass transition temperature. Figure 7.15 shows the deformed shape (solid lines) of the SLICC on the 55-mil (1.4-mm)-thick PCB. The predicted load-deflection curve is plotted as the dashed line

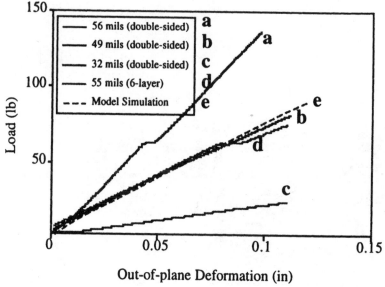

Figure 7.12 Load-deflection curves for Motorola's SLICC on different PCBs.[6]

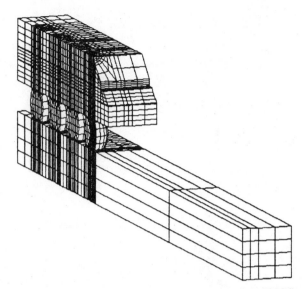

Figure 7.13 3D finite element model of Motorola's SLICC assembly.[6]

in Fig. 7.12 with legend *e* which matched the solid line with legend *d* very well. It is worthwhile to point out that the SLICC assembly can be subjected to bending resulting in a radius of curvature of 1.94 in (49 mm) without incurring any C4 and C5 solder joint failure.[6] (C5 solder joints are those connecting the substrate and the FR-4 board.)

7.2.2 Thermal fatigue of Motorola's CSP assembly

A Motorola SLICC assembly (55-mil- or 1.4-mm-thick PCB with a TCE of 15 ppm/°C) was subjected to a liquid-to-liquid thermal shock test (−55

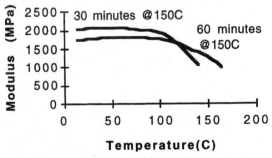

Figure 7.14 Elastic modulus of underfill versus temperature.[6]

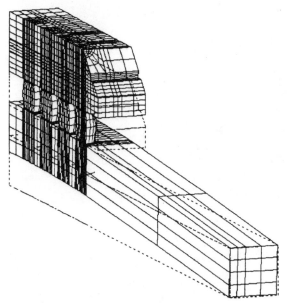

Figure 7.15   Deformation of Motorola's SLICC assembly.[6]

to 125°C). It was observed that the waiting time during underfill cure had little or no effect on the thermal fatigue life of the chip-underfill interface.[6] It was also found that the C4 solder joints were very reliable (>13,000 liquid-to-liquid thermal shock cycles) unless there were delaminations at the chip-underfill interface. Typically the delamination initiation site is at the edge of the chip at the chip-underfill interface, which is also the site of maximum von Mises stress concentration, both at the cold and the hot end of the shock test. Figures 7.16 and 7.17 show the von Mises stress distribution in a chip-underfill interface (30 minutes at 150°C cure condition) at –55 and 125°C, respectively.

The plastic work (a combination of stresses and strains) in the C5 solder joints is shown in Figs. 7.18 and 7.19, respectively, at –55 and 125°C. It can be seen that, in the primary region (top right) the plastic work per cycle is in the neighborhood of 12 psi-in/in but is 2.03 psi-in/in in the secondary region (bottom left). Based on these values, Motorola predicted that it would take 655 cycles to attain crack initiation in the primary region and 3930 cycles in the secondary region.[6]

The failure mode in the primary region typically is that the crack initiates at the outside periphery of the joint and moves inward along the joint-chip interface. In the secondary region, the crack initiates at the inside periphery of the joint and travels outward along the joint-board interface.

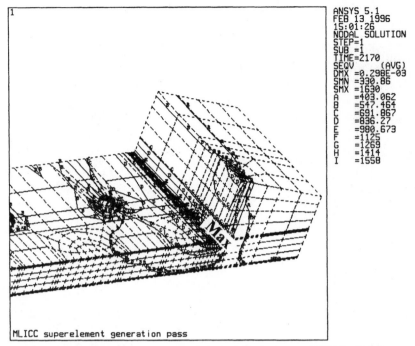

Figure 7.16 von Mises stress at the chip-underfill interface at 125°C for 30 minutes at a 150°C cure condition.[6]

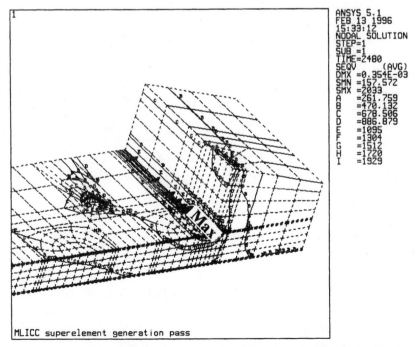

Figure 7.17 von Mises stress at the chip-underfill interface at –55°C for 30 minutes at a 150°C cure condition.[6]

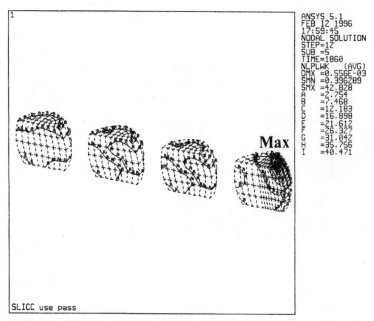

Figure 7.18   Plastic work in the eutectic solder balls during thermal shock—hot end at 125°C.[6]

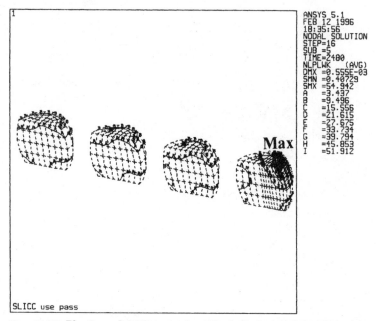

Figure 7.19   Plastic work in the eutectic solder balls during thermal shock—cold end at –55°C.[6]

As mentioned previously, most of the thermal fatigue life of ductile materials such as solders is spent in propagating the crack. In this case, Motorola predicted that it will take at least 20,000 cycles to completely fail the C5 solder joints. This prediction is supported by its data because no C5 solder joint failures have been observed in accelerated tests so far.[6]

7.3 Solder Joint Reliability of Tessera's CSP Assembly

A daisy-chained test chip (12.7 × 12.7 mm) for Tessera's CSP (also called µBGA) solder joint reliability study is shown in Fig. 7.20.[10] A close-up look at the elastomer-compliant µBGA is shown in Figs. 7.21 and 7.22. This µBGA is attached to a FR-4 PCB (181 × 130 × 1.6 mm) (Fig. 7.23) with 63wt%Sn/37wt%Pb RMA 400 mesh solder. Four different ratios of the diameter of the copper land to mask resist openings were studied: 0.3/0.3, 0.3/0.25, 0.3/0.2, and 0.2/0.3 mm.

The thermal shock, temperature cycling, high-temperature storage, biased HAST (highly accelerated stress test) test conditions, and the passing criteria for the solder joint reliability tests of Tessera's µBGA are given in Table 7.4.[10] The test results are shown in Table 7.5 and Fig. 7.24. It can be seen that for all four land/resist openings, the solder joints passed all of the above tests.

7.4 Solder Joint Reliability of NEC's CSP Assembly

Figure 7.25*a* through *c* show the cross sections of NEC's fine-pitch BGA (FPBGA), molded FPBGA, and molded and pad-extended FPBGA. NEC

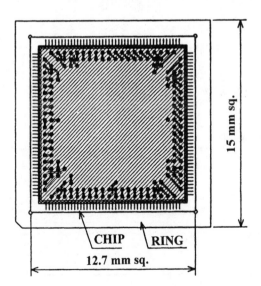

Figure 7.20 Daisy-chained µBGA for SMT evaluation.[10]

TABLE 7.4   μBGA's PCB Assembly Test Requirement[10]

Test	Test condition	Passing criteria
Thermal shock	−55°C, 5 min/+125°C, 5 min 100, 200, 300 cycles	Open check (resistance (room temp.) max. +5%), cross section
Temperature cycling	−55°C, 30 min/+125°C, 30 min 100, 300, 500, 1000 cycles	Open check (resistance (−55,125°C) max. +5%), cross section
Biased HAST	+130°C/85%RH, +5V DC BIAS 50, 100, 150, 200 h	Max. leakage current 10^{-10} Amp at 5V DC, cross section
High-temperature storage	+125°C/in the air 100, 200, 300, 500, 1000 h	Open check (resistance (room temp.) max. +5%), cross section

used tape as the interposer for redistribution. The solder balls (0.15 to 0.3 mm in diameter) were arrayed with 0.5-mm pitch and made with NEC's micropunching solder bump formation process.[11] Even though NEC normally uses eutectic solder, other solders can be used.

Because of the lamination (with adhesive) of the interposer to the chip, the shear stresses and strains at the solder joints (due to the thermal expansion mismatch between the chip and the substrate) in NEC's CSP assemblies are reduced. Some of the finite element results for FPBGA as well as for direct chip attach (DCA) on FR-4 PCB are shown in Table 7.6.[11] It can be seen that the maximum stresses in the FPBGA solder joint are less than that in the DCA. Also, the thermal cycling and HAST results show that NEC's solder joints are reliable[11] (Table 7.7).

7.5   Solder Joint Reliability of Nitto Denko's CSP Assembly

Nitto Denko's chip/carrier tape interface encapsulation CSP, and all chip sides encapsulation (molded) CSP (MCSP) are shown in Fig. 7.26a

TABLE 7.5   μBGA's PCB Assembly Test Results[10]

Structures Land/Resist opening (mm/mm)	Test								
	Temperature cycling			Thermal shock			High-temperature storage		
	100 cycles	200 cycles	300 cycles	200 cycles	300 cycles	700 cycles	100 hours	200 hours	300 hours
0.300/0.300	0/6	0/6	0/6	0/6	0/6	0/6	0/6	0/6	0/6
0.300/0.250	0/6	0/6	0/6	0/6	0/6	0/6	0/6	0/6	0/6
0.300/0.200	0/6	0/6	0/6	0/6	0/6	0/6	0/6	0/6	0/6
0.200/0.300	0/6	0/6	0/6	0/6	1/5	0/5	0/6	0/6	0/6

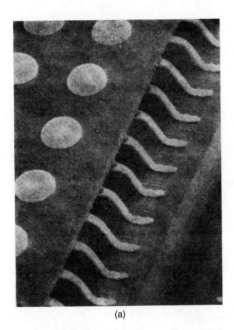

(a)

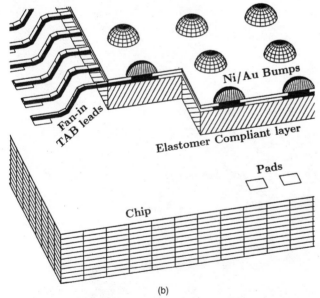

(b)

Figure 7.21   (a) μBGA's gold beam lead. (b) μBGA's elastomer.[9,10]

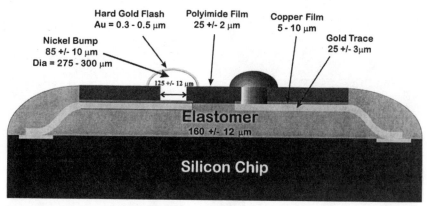

Figure 7.22   Cross section of µBGA.[9]

through *c.*[12] The high-viscosity (10^4–10^7 poise at 350°C) and low-elastic modulus (120 kg/mm²) thermo-adhesive polyimide (TAPI) was used as the encapsulant for the interface.

Finite element analysis has been applied to Nitto Denko's CSP on FR-4 PCB. The chip size was 15 × 15 mm. The CSP's eutectic solder bump was on a 1-mm pitch and with a 0.2-mm diameter. For comparison purposes, a DCA model with the same dimensions for the chip and

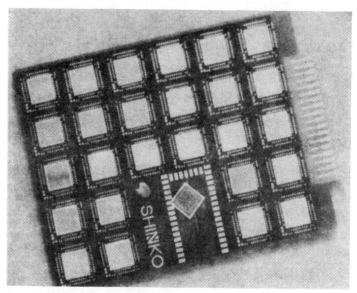

Figure 7.23   µBGAs on a FR-4 PCB.[10]

TABLE 7.6 NEC's Finite Element Results[11]

Portion	Flip chip max. stress (kgf/mm²)	FPBGA max. stress (kgf/mm²)
Solder bump	15.6	7.1
Outer pad	62.1	13.4
Substrate pad	8.6	5.4

TABLE 7.7 NEC's CSP Thermal Cycling and HAST Results

Tests	Failures/Test samples
Thermal cycling, 200 cycles	0/5
HAST, 288 hours	0/5

TABLE 7.8 Nitto Denko's CSP Finite Element Results[12]

	Thermal stress (kg/mm²)
MCSP	10.6
DCA	19.3

outer bumps was constructed. Figure 7.27 shows the schematic model of the MCSP assembly for finite element analysis and the results are shown in Table 7.8 for a temperature change of 220°C. It can be seen that the thermal stress in the MCSP solder joint is smaller than that in the DCA solder joint. This is due to the TAPI carrier tape which absorbs and redistributes the thermal stress in the solder joints.

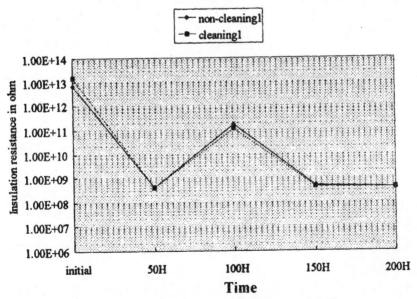

Figure 7.24 High-temperature test results for μBGA PCB assembly.[10]

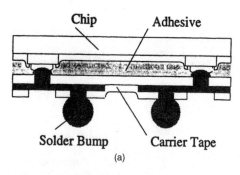

Chip   Adhesive

Solder Bump   Carrier Tape

(a)

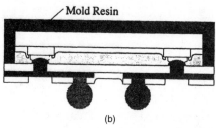

Mold Resin

(b)

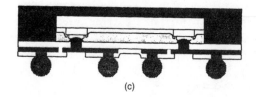

(c)

Figure 7.25 NEC's CSPs: (*a*) FPBGA; (*b*) molded FPBGA; (*c*) molded and pad-extended FPBGA.[11]

7.6 Solder Joint Reliability of a Low-Cost CSP Assembly

Figure 7.28 shows a CSP package assembly.[1] It can be seen that the chip was attached to the ceramic interposer with gold studs and underfill epoxy, and then soldered to the PCB. The chip size was 7 × 7 × 0.41 mm with 100 peripherally distributed gold bumps on a 0.25-mm pitch. The dimensions of the gold bump were 0.1 × 0.1 × 0.025 mm (Fig. 7.29). The ceramic interposer dimensions were 7.45 × 7.45 × 0.25 mm (Fig. 7.30). There were 100 arrayed eutectic solder bumps (on a 0.65-mm pitch) at the bottom of the ceramic substrate. After the assembly of the CSP on the PCB, the bump height was about 0.2 mm. The diameter of the pads on the ceramic and PCB substrates was about 0.3 mm.

Figure 7.31 shows one-half of the assembly along the diagonal direction with the TCE (or CTE) of its key elements, and Fig. 7.32 shows the finite element model for the analysis (8896 2D eight-node plane-strain elements and 27,217 nodes).

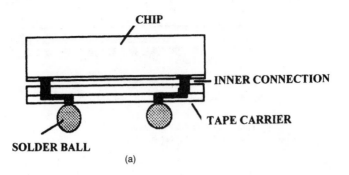

(a)

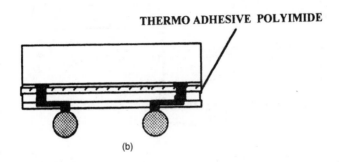

(b)

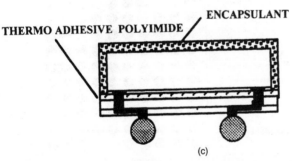

(c)

Figure 7.26 Nitto Denko products: (a) molded CSP; (b) chip/tape encapsulated CSP; (c) all chip sides encapsulated CSP.[12]

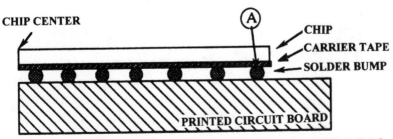

Figure 7.27 Schematic model of Nitto Denko's CSP. (The corner solder ball A for detailed finite element analysis.)[12]

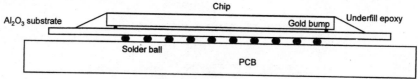

Figure 7.28   Schematic cross section of a low-cost CSP.[1]

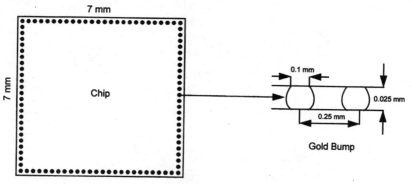

Figure 7.29   Chip size and gold bump pitch for the CSP.[1]

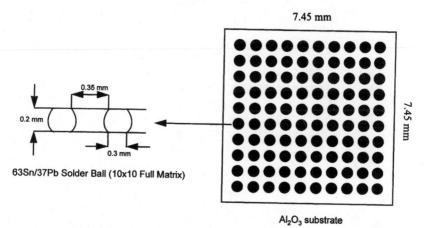

Figure 7.30   Ceramic substrate and eutectic solder balls.[1]

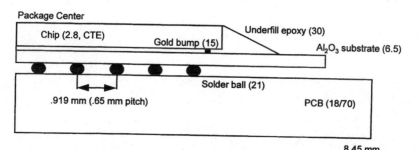

Figure 7.31   Schematic cross section in the diagonal direction for finite element modeling.[1]

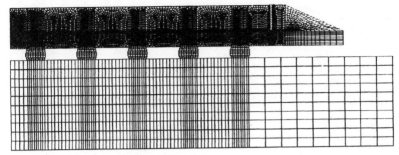

Figure 7.32   Finite element model.[1]

7.6.1 Temperature-dependent elastoplastic analysis

The physical and mechanical properties of the key elements are shown in Table 7.9.[1] It can be seen that the material properties of the silicon, gold, ceramic, FR-4, and underfill epoxy were assumed to be constant except for those of the eutectic solder, which was temperature-dependent. Figure 7.33*a* and *b* shows the stress-strain curves and Fig. 7.34 shows the Young's modulus of the eutectic solder.

The temperature imposed on the CSP assembly is shown in Fig. 7.35. Since this is an elastoplastic analysis, the temperature applied to the model is from –40 to 125°C. The whole-field deformation is shown in Figs. 7.36 and 7.37. It can be seen that the maximum deflection (moving downward in Fig. 7.36) is at the center of the PCB and equal to 0.0144 mm (Fig. 7.37). This is due to the thermal expansion mismatch between the ceramic substrate and FR-4 PCB and a temperature range of 165°C.

The maximum solder joint von Mises stress and equivalent plastic strain range distributions along the diagonal distance from the package center are shown in Figs. 7.38 and 7.39, respectively. The maximum values (von Mises stress = 49 MPa and equivalent plastic strain = 0.0066) occurred at the corner solder joint because it is the farthest (4.136 mm) from the package center. It is interesting to point

TABLE 7.9   Material Properties of a Low-Cost CSP Assembly[1]

	Young's modulus (GPa)	Poisson's ratio	TCE 10^{-6}(m/m–°C)
Silicon	131	0.3	2.8
Solder	Temperature-dependent	0.4	21
Gold	78	0.3	15
Al_2O_3	300	0.22	6.5
FR-4	11	0.28	18
Underfill epoxy	6	0.35	30

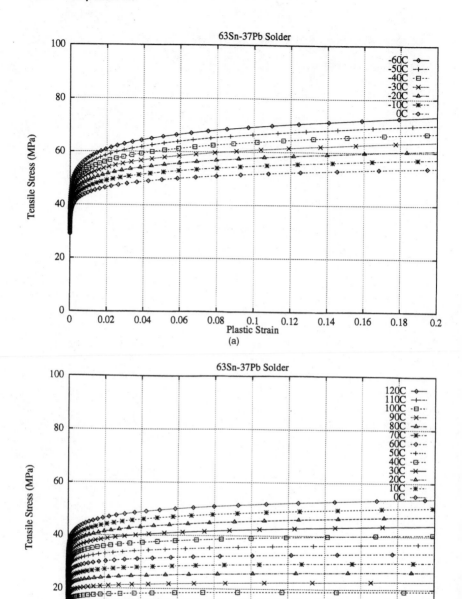

Figure 7.33 (*a*) Temperature-dependent stress-strain curves (−65 to 0°C) of eutectic solder. (*b*) Temperature-dependent stress-strain curves (5 to 125°C) of eutectic solder.[1]

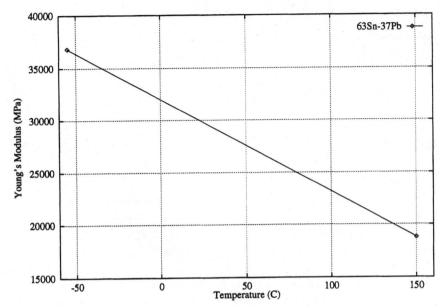

Figure 7.34   Temperature-dependent Young's modulus of eutectic solder.[1]

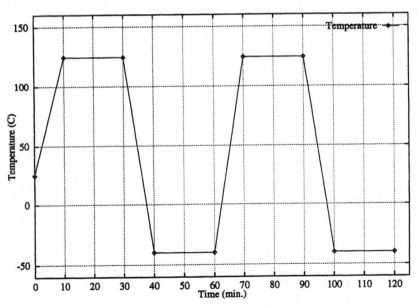

Figure 7.35   Thermal cycling temperature profile (boundary condition).[1]

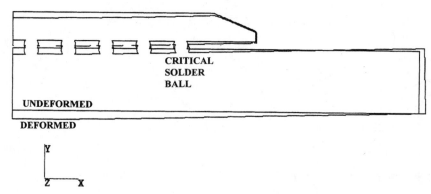

CRITICAL
SOLDER
BALL

UNDEFORMED

DEFORMED

Figure 7.36 Deformation of the CSP assembly, −40 to 125°C (elastoplastic solution).[1]

out that, based on the concept of DNP, the calculated shear strain, $4.136(165)(18 - 6.5)/0.2 = 0.0392$ is approximately six times that calculated by the nonlinear finite element method. This is due, in part, to the fact that the DNP concept neglects the resistant contribution of all the interior solder bumps and the bending of the entire assembly.

The detailed von Mises stress and equivalent plastic strain range distributions in the corner solder bump are shown in Figs. 7.40 and 7.41. It can be seen that the location of the maximum von Mises stress and equivalent plastic strain range occurred at the upper right-hand corner of the solder joint. The second-highest location of the von Mises stress and equivalent plastic strain occurred at the lower left-hand cor-

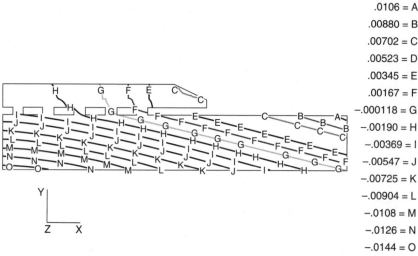

```
.0106 = A
.00880 = B
.00702 = C
.00523 = D
.00345 = E
.00167 = F
−.000118 = G
−.00190 = H
−.00369 = I
−.00547 = J
−.00725 = K
−.00904 = L
−.0108 = M
−.0126 = N
−.0144 = O
```

Figure 7.37 Deformation contours of the CSP assembly, −40 to 125°C (elastoplastic solution).[1]

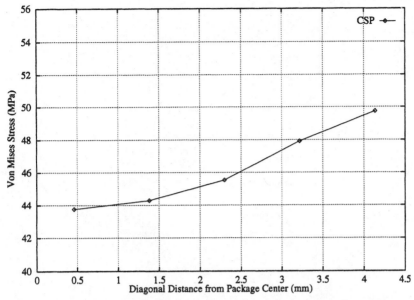

Figure 7.38 von Mises stress distribution along the diagonal direction of the CSP assembly, −40 to 125°C (elastoplastic solution).[1]

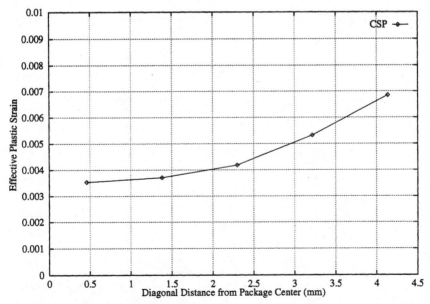

Figure 7.39 Equivalent plastic strain distribution along the diagonal direction of the CSP assembly, −40 to 125°C (elastoplastic solution).[1]

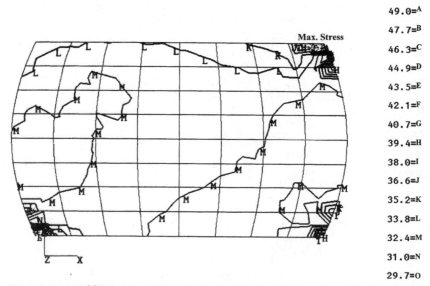

Figure 7.40   von Mises stress (MPa) distribution in the critical solder ball, −40 to 125°C (elastoplastic solution).[1]

ner of the solder joint. Thus, when the solder joint cracks, the initiation begins at the upper right-hand corner and propagates near the interface between the chip and the solder. A second crack could initiate (later) at the lower left-hand corner and propagate near the interface between the PCB and the solder joint.

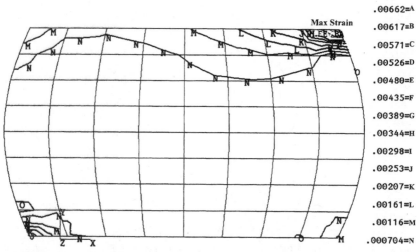

Figure 7.41   Equivalent plastic strain distribution in the critical solder ball, −40 to 125°C (elastoplastic solution).[1]

Once the equivalent plastic strain range is determined, thermal fatigue life can be estimated by the Coffin-Manson relation. For eutectic solders, Solomon fit his data at –50, 35, and 125°C into the Coffin-Manson equation and obtained the average life as $N_f = 1.29(\Delta\gamma)^{-1.96}$, where $\Delta\gamma$ is the plastic shear strain range (in our case, it is 0.00662). Thus, the average thermal fatigue life of this CSP assembly solder joints is about 21,000 cycles or 5400 cycles with a safety factor of four.[4,5]

7.6.2 Temperature-dependent elasto-plastic-creep analysis

The Norton's creep relation, Eq. (4.1), described in Sec. 4.1.3, has been used for the elasto-plastic-creep analysis of the CSP solder joints. For eutectic solder the material constants of the Norton equation are: $B^* = 0.205$ $1/\text{MPa}^{5.25}$ sec, $H = 0.49$ eV, and $n = 5.25$ (Table 4.2). The temperature-dependent stress-strain curves and Young's modulus for the solder are given in Figs. 7.33 and 7.34.

The temperature imposed on the CSP assembly is shown in Fig. 7.35. It can be seen that for each cycle (60 minutes) the temperature condition was between –40 and 125°C with 10 minutes ramp and 20 minutes hold. Two complete cycles were executed.

Again, the location of the maximum stress and strain hysteresis responses was at the corner solder joint (Fig. 7.42).[1] The hysteresis loops of the shear stress and shear strain at various locations (NODES A, B, C, D, E, F, G, H, and I on the upper and lower interfaces) in the corner solder joint are shown in Figs. 7.43 (upper interface between ceramic and solder) and 7.44 (lower interface between solder and PCB). It can be seen that, just as in the elastoplastic case, the maximum shear stress and shear strain range location occurs at the upper right-hand corner (NODE I on interface 1) of the solder joint.

7.7 Solder Joint Reliability of LG Semicon's CSP Assemblies

Figure 7.45 shows a PCB assembly with an ultrathin and crackfree bottom leaded plastic (BLP) package designed by LG Semicon (Glod-Star).[13] It is an LOC package (Table 1.1, Fig. 7.45a). Its size is smaller than the small outline package with J-lead (SOJ) and thin small outline package (TSOP) (Fig. 7.45b). In the BLP package, the leadframe sustains the chip rigidly during the wire bonding and epoxy molding processes and provides a path for the electrical power or signals between the chip and the PCB (Fig. 7.46a). A detailed view of the leadframe is shown in Fig. 7.46b. Notches a and b improve the leadframe down set and trimming processes. Notch c improves the hermeticity of the package by detouring the interfacial path with epoxy molding compound. The BLP package is designed to house memory IC devices, e.g., the 4 MB DRAM chip with 20 I/O pins.

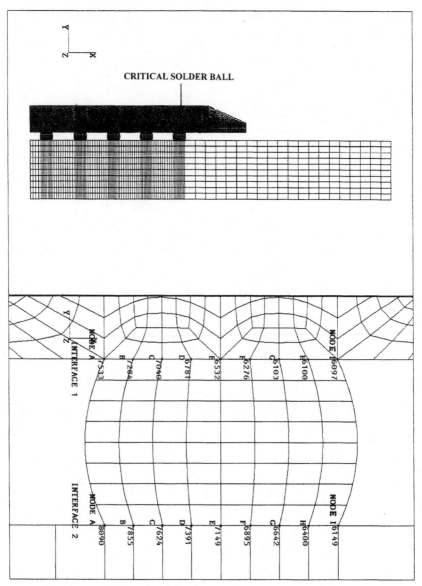

Figure 7.42 Critical solder ball location for creep analysis.[1]

Table 7.10 shows the current design of the BLP package as well as the proposed design for better solder joint reliability. It can be seen that the new design uses copper (instead of alloy-42) leadframe which has a better thermal expansion mismatch with the solder and a larger compliance. Also, the solder joint height for the new design is 40 per-

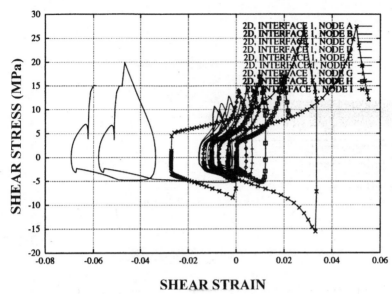

Figure 7.43 Hysteresis loops of the shear stress and shear strain at various locations in the CSP's corner solder joint (upper interface between ceramic and solder).[1]

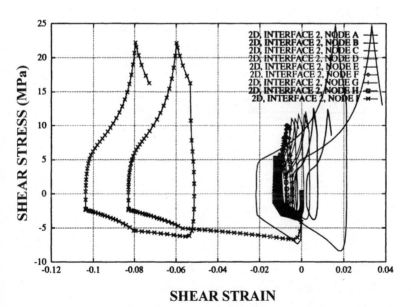

Figure 7.44 Hysteresis loops of the shear stress and shear strain at various locations in the CSP's corner solder joint (lower interface between solder and PCB).[1]

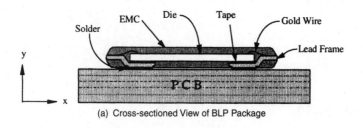

(a) Cross-sectioned View of BLP Package

| Height : 3.76 mm | 0.48~0.82 mm | 1.27 mm |
| Width : 8.50 mm | 5.21 mm | 9.22 mm |

SOJ   BLP   TSOP

(b) Comparison of Package Dimensions

Figure 7.45 Hysteresis loops of the shear stress and shear strain at various locations in the CSP's corner solder joint (lower interface between solder and PCB).[13]

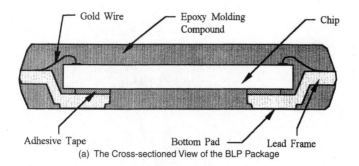

(a) The Cross-sectioned View of the BLP Package

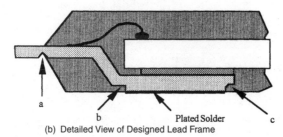

(b) Detailed View of Designed Lead Frame

Figure 7.46 Cross-sectional views of LG Semicon's BLP package.[13]

TABLE 7.10   LG Semicon's Current and Proposed Design Parameters

Design parameters	Current design	Proposed design
Solder joint height	100 μm	140 μm
PCB thickness	1.27 mm	0.4 mm
Die thickness	280 μm	200 μm
Lead frame	Alloy 42	Copper
CTE of EMC	11 ppm/°C	16 ppm/°C

cent taller. Furthermore, the thickness of the die and PCB for the new design is thinner. All these parameters contribute to better solder joint reliability for the new design.

Figure 7.47 shows the finite element model for the solder joint thermal strain analysis of the current and proposed BLP packages and PCB assemblies. It was found that the average shear strain of the current design (0.0128) was reduced to 0.0076 for the new design (i.e., the proposed BLP package should have a longer solder joint thermal fatigue life).[13] The proposed BLP package on PCB assemblies is being tested under thermal cycling (–30 to 85°C). After 1200 cycles there are no electrical failures observed.

Acknowledgments

The authors would like to thank Dr. Wen-Je Jung of Ford Motor Company for his great contribution for Ref. 1. It has been a great pleasure and fruitful experience to work with him.

Figure 7.47   LG Semicon's FEM Model for BLP Package and Board Assembly.[13]

References

1. Lau, J. H., W. Jung, and Y.-H. Pao, "Temperature-Dependent Elasto-Plastic-Creep Analysis of a Chip Scale Package (CSP)'s Solder Joints," *ASME Winter Annual Meeting Paper,* November 1996.
2. Yasunaga, M., S. Baba, M. Matsuo, H. Matsushima, S. Nakao, and T. Tachikawa, "Chip Scale Package (CSP): A Lightly Dressed LSI Chip," *Proceedings of the IEEE International Electronics Manufacturing Technology Symposium,* 1994, pp. 169–176.
3. Baba, S., Y. Tomita, M. Matsuo, H. Matsushima, N. Ueda, and O. Nakagawa, "Molded Chip Scale Package for High Pin Count," *Proceedings of IEEE Electronic Components & Technology Conferences,* May 1996, pp. 1251–1257.
4. Solomon, H. D., "Fatigue of 60/40 Solder," *IEEE Transactions on Components, Hybrids, and Manufacturing Technology,* December 1986, pp. 423–432.
5. Lau, J. H., *Solder Joint Reliability,* Van Nostrand Reinhold, New York, 1991.
6. Lall, P., G. Gold, B. Miles, K. Banerji, P. Thompson, C. Koehler, and I. Adhihetty, "Reliability Characterization of the SLICC Package," *Proceedings of IEEE Electronic Components & Technology Conferences,* May 1996, pp. 1202–1210.
7. Banerji, K., and P. Lall, "Development of the SLICC (Slightly Larger Than IC Carrier) Package," *Proceedings of NEPCON East,* June 1995, pp. 441–451.
8. Banerji, K., "Development of the Slightly Larger Than IC Carrier (SLICC)," *Proceedings of NEPCON West,* February 1994, pp. 1249–1256.
9. Lau, J. H., *Ball Grid Array Technology,* McGraw-Hill, New York, 1995.
10. Koyama, T., K. Abe, N. Sakaguchi, and S. Wakabayashi, "Reliability of mBGA Mounted On a Printed Circuit Board," *Proceedings of the SMI Conference,* August 1995, pp. 43–55.
11. Matsuda, S., K. Kata, and E. Hagimoto, "Simple-Structure, Generally Applicable Chip-Scale Package," *Proceedings of IEEE Electronic Components & Technology Conferences,* May 1995, pp. 218–223.
12. Tanigawa, S., K. Igarashi, M. Nagasawa, and N. Yoshio, "The Resin Molded Chip Size Package (MCSP)," *Proceedings of the IEEE International Electronics Manufacturing Technology Symposium,* October 1995, pp. 410–415.
13. Kim, Y., B. Han, S. Choi, and M. Kim, "Bottom Leaded Plastic (BLP) Package: A New Design with Enhanced Solder Joint Reliability," *Proceedings of IEEE Electronic Components & Technology Conferences,* May 1996, pp. 448–452.

Solder Joint Reliability of Fine Pitch SMT Assemblies

When the lead pitch of a surface-mount component (SMC) is equal to or less than 0.65 mm, then it is called a fine-pitch SMC.[1-64] In this chapter the solder joint reliability of the two most commonly used packages, TSOP (thin small outline package) and PQFP (plastic quad flat pack), under thermal, mechanical, and vibrational conditions are presented.

8.1 Thermal Reliability of TSOP Assemblies

TSOP is used for packaging the memory devices, especially for SRAM (static random access memory) and DRAM (dynamic RAM). Figure 8.1 shows a typical cross section of the package along with a PQFP. It can be seen that TSOP is a very low-profile package normally used for miniature and lightweight products such as PCMCIA cards and portable electronics.

8.1.1 Reliability of a 28-pin TSOP assembly

Figure 8.2 shows a schematic of a TSOP surface-mount assembly. It consists of four major parts: the package body (Fig. 8.3), the alloy-42 leads (Fig. 8.4), the 63wt%Sn-37wt%Pb solder joints, and the FR-4 epoxy/glass PCB. Because of the large ratio of the silicon to plastic (i.e., leading to a large global thermal expansion mismatch between the package body and the PCB, Figs. 8.2 and 8.3 and Table 8.1), the use of alloy-42 lead frame (i.e., large local thermal expansion mismatch between the lead frame and solder joint, Table 8.1), the low-profile package (i.e., less compliance, Fig. 8.4), and the concerns associated

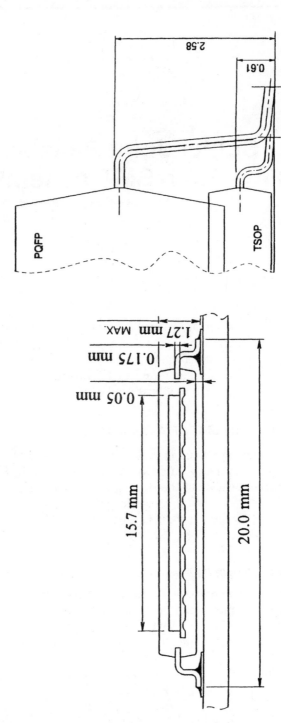

Figure 8.1 Low-profile TSOP package.

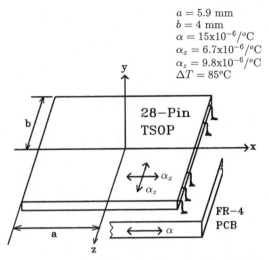

$$a = 5.9 \text{ mm}$$
$$b = 4 \text{ mm}$$
$$\alpha = 15 \times 10^{-6}/^{\circ}\text{C}$$
$$\alpha_x = 6.7 \times 10^{-6}/^{\circ}\text{C}$$
$$\alpha_z = 9.8 \times 10^{-6}/^{\circ}\text{C}$$
$$\Delta T = 85^{\circ}\text{C}$$

Figure 8.2 Three major sources to produce thermal stress in the 28-pin TSOP solder joint: (1) global thermal coefficient of expansion (TCE) mismatch between the TSOP and PCB; (2) local TCE mismatch between the lead and solder; and (3) lead stiffness.[10]

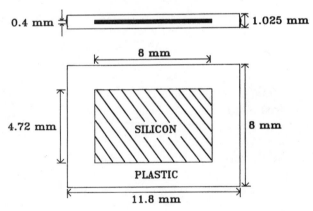

Figure 8.3 Top and side views of the 28-pin TSOP (leads not shown).

with fine-pitch technology, solder joint reliability of TSOPs on FR-4 PCB is under scrutiny.

The fine-pitch component used in the test was a 28-lead, 0.5-mm-pitch, Type-I TSOP (with live memory devices)[10] (Figs. 8.2 through 8.4). The dimensions of this TSOP were $11.8 \times 8 \times 1.025$ mm. The ratio of the volume of silicon to plastic was very large. The thermal coefficient of expansion of the silicon (2.9 ppm/°C) and the plastic (22 ppm/°C) dif-

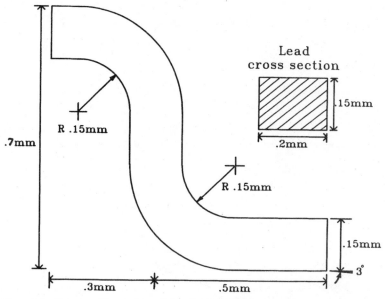

Figure 8.4 Cross-sectional view of the 28-pin TSOP lead.

fered greatly (Table 8.1). Consequently, the effective thermal coefficient of expansion of the whole package body was of interest and was measured by the 943 Thermomechanical Analyzer and the Thermal Analyst 2000 provided by Du Pont Instruments.

Figure 8.5 shows the effective thermal coefficient of expansion of the 28-pin TSOP body in the long or x direction (α_x) and the short or z direction (α_z). For temperatures less than 150°C, $\alpha_x = 6.7$ ppm/°C and $\alpha_z = 9.8$ ppm/°C. The reasons for $\alpha_z > \alpha_x$ are: (1) the ratio of silicon to plastic in the x direction (68 percent) is larger than that in the z direction (59 percent) (Fig. 8.3), (2) the thermal coefficient of expansion of silicon is smaller than that of plastic, and (3) the leads are oriented in the x direction. Comparing the thermal coefficient of expansion of the TSOP

TABLE 8.1   Material Properties of the TSOP Assembly

	Young's modulus (GPa)	Poisson's ratio	Thermal coefficient of linear expansion 10^{-6}(m/m-°C)
Solder	10	0.40	21
Alloy-42	148	0.30	5
Silicon	131	0.30	2.9
Plastic	14	0.30	22
Copper	121	0.35	17
FR-4	11	0.28	15

body α_x and α_z and the PCB ($\alpha = 15$ ppm/°C), the difference (global thermal expansion mismatch) is very large. Furthermore, the lead frame material of the TSOP was made of alloy-42 with a 300-microinch-thick (7.62 microns) 90wt%Sn-10wt%Pb-plated surface. The thermal coefficient of linear expansion of alloy-42 was 5 ppm/°C. Thus, the local thermal expansion mismatch between the gull-wing lead and the Sn/Pb solder joint (21 ppm/°C) is also large.

Finally, the critical dimensions of the gull-wing-shaped leads, the foot length, width, and thickness, were 0.28, 0.2, and 0.15 mm, respectively (Fig. 8.4). The standoff of the gull-wing lead was short compared with the other surface-mountable fine-pitch components, such as PQFP. Thus, the TSOP lead was stiffer, i.e., had less compliance. Consequently, when the TSOP assembly was subjected to a temperature change, the leads and solder joints experienced large stresses and strains due to the large global and local thermal expansion mismatches and the low profile of the TSOP.

Because of the small size of the test board and for cost savings, a multiple-board manufacturing method was adopted. The dimensions of the large panel were $406.4 \times 304.8 \times 1.57$ mm. It consisted of 40 small test boards, each with dimensions of $50.8 \times 25.4 \times 1.57$ mm. These test boards were connected by tiny connecting tabs. The test board material was made of single-layer FR-4 glass fiber–filled epoxy.

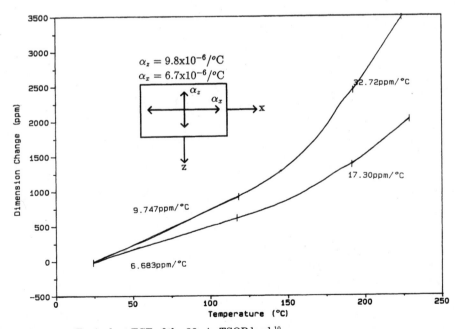

Figure 8.5   Equivalent TCE of the 28-pin TSOP lead.[10]

The solder paste contained 50 percent by volume (88 percent by weight) solder balls in an RMA (rosin, mildly activated) flux. The solder ball composition was the eutectic 63wt%Sn-37wt%Pb with particle size (−325, +500), i.e., the diameter was no greater than 0.0432 mm or less than 0.0254 mm. The viscosity was 1000 + 102 kcps. This solder paste was printed through a metal stencil with a squeegee. The stencil had an opening 0.61 × 0.25 mm and the wet paste thickness was 0.15 mm. The TSOPs were then placed in position on the panel using a pick and place machine.

The solder joints for all the TSOPs were mass-reflowed by infrared soldering with a peak temperature of 240°C. Immediately after the solder was reflowed, the assembled panel was solvent-cleaned and inspected. Finally, the panel was depanelized into 40 small test boards, each with a 28-pin TSOP. For electrical measurement purposes, the test board was soldered to a 28-pin header for socketing. It should be pointed out that all the TSOP solder joints in this book were carefully made under special attention. Only those solder joints with very good heel, toe, side fillets, and sufficient standoff were tested.

The test boards were tested for full electrical functionality including an opens/shorts test routine using a Fast Access Tester for memory components before and during intermittent readout of the temperature cycling. The units that showed an electrical open were retested on the curve tracer to verify the opens. Solder joint opens were confirmed by pressing slightly on the open pin, which resulted in full continuity and full functionality of all the failed units. Units showing an electrical open were counted as failures and were removed from the population during further testing. The number of cycles to failure (N) was defined as the first solder joint failure in a 28-pin TSOP assembly. The solder joints were inspected under SEM to examine the quality of the joints as well as to document the failure modes.

The interposer test boards were tested in an air-to-air thermal cycling Ransco chamber. They were subjected to two temperature cycling (T/C) profiles (shown in Fig. 8.6):

$$\text{T/C B: } -55 \text{ to } 125°C \tag{8.1}$$

$$\text{T/C C: } -65 \text{ to } 150°C \tag{8.2}$$

For both conditions, the ramp time, dwell time, and period (time to complete one cycle) were 1, 12, and 26 minutes, respectively. At each condition, 22 interposer test boards were tested. Before testing, all solder joints were inspected.

Table 8.2 shows the test results under T/C B ($\Delta T = 180°C$) and C ($\Delta T = 215°C$) conditions. In this chapter, reliability of the solder joints was modeled by Weibull's CDF and analyzed by Johnson's statistical method

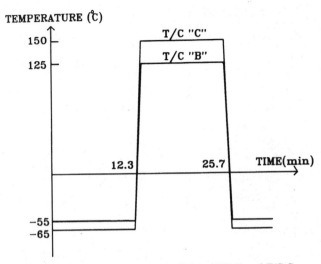

TEMPERATURE (℃)

Figure 8.6   Temperature cycling conditions: T/C B and T/C C.

with ranking (Chap. 2). Furthermore, an acceleration model was developed to bridge the temperature and frequency gaps between the testing and operating conditions.

Applying the principles of least squares and ranking to the temperature cycling test results (Table 8.2), the best fit Weibull parameters (for the median rank) are: $\beta = 5.14$ and $\theta = 6363$ for T/C B ($\Delta T = 180°C$), and $\beta = 3.48$ and $\theta = 4340$ for T/C C ($\Delta T = 215°C$). With these values, Eq. (2.39) is called the life distribution of the 28-pin TSOP solder joint at test conditions and is plotted in Fig. 8.7.

The temperature cycling test frequency was at 1 cycle/26 min ($f_t = 1$ cycle/26 min = 55.38 cycle/day) and the temperature ranges were $\Delta T = 180°C$ (T/C B) and $\Delta T = 215°C$ (T/C C). The operating conditions were assumed to be: $f_o = 1$ cycle/day and $\Delta T_o = 85°C$ (f_o is the cycling frequency at operating condition, ΔT_o is the temperature range at operating condition). Consequently, an acceleration factor is needed to transfer the life distribution, reliability function, and failure rate at the higher testing conditions to the lower operating condition.

Assuming a linear acceleration like Eq. (2.55), in the present study, we proposed the following acceleration factor:

$$AF = \left(\frac{f_o}{f_t}\right)^{1/3}\left(\frac{\Delta T}{215}\right)^{-2.38} \tag{8.3}$$

For $f_o = 1$ cycle/day, $f_t = 55.38$ cycle/day, $\Delta T = \Delta T_o = 85°C$, Eq. (8.3) becomes:

$$AF = \left(\frac{1}{55.38}\right)^{1/3}\left(\frac{85}{215}\right)^{-2.38} = 2.39$$

TABLE 8.2   28-Pin TSOP Test Results for Statistical Analysis[10]

Failure no.	Cycle to failure (1 cycle = 26 min) Temperature range (°C) 180	215	Median (50%) rank
1	3,750	1,250	0.0315
2	4,250	2,750	0.0761
3	4,250	3,250	0.1207
4	4,250	3,250	0.1653
5	4,250	3,250	0.2099
6	4,750	3,250	0.2545
7	4,750	3,250	0.2992
8	5,750	4,250	0.3438
9	5,750	4,250	0.3884
10	5,750	4,250	0.4330
11		4,250	0.4776
12		4,250	0.5223
13		4,250	0.5669
14		4,250	0.6115
15		4,250	0.6561
16		4,250	0.7007
17		4,250	0.7454
18		4,250	0.7900
19		4,250	0.8346
20		4,250	0.8792
21		4,250	0.9238
22		4,250	0.9684

In Eq. (8.3), the acceleration factor is a function of the product of the cycling frequency effect and the cycling temperature effect. The cycling frequency term $(f_o/f_t)^{1/3}$ is exactly the same as in Eqs. (2.67) through (2.69). The cycling temperature term $(\Delta T/215°C)^{-2.38}$ was generated as follows:

1. In Fig. 8.7, the number of cycles to failure at the 50 percent failure point from both test conditions was recorded, i.e.,

$$N_{50\%} (\Delta T = 180°C) = 5800 \quad \text{and} \quad N_{50\%} (\Delta T = 215°C) = 3800 \quad (8.4)$$

2. In Fig. 8.8 (temperature range versus $N_{50\%}$), the points (180,5800) and (215,3800) were plotted, and a straight line was constructed through these two points. Then we have

$$N_{50\% \text{ at } \Delta T} = (\Delta T/215°C)^{-2.38}(3800) \quad (8.5)$$

Once the acceleration factor AF = 2.39 is determined, the life distribution, reliability function, and failure rate under operating conditions can be determined by Eqs. (2.71) through (2.73), respectively, with $\beta = 3.48$ and $\theta = 4340$.

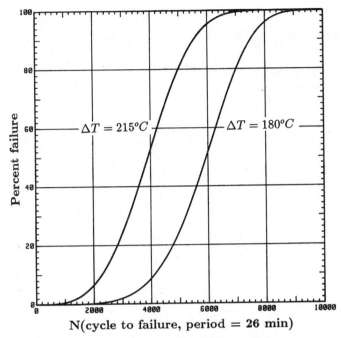

Figure 8.7 Life distribution (Weibull) of the 28-pin TSOP solder joints at test conditions.[10]

$$F_o(x_o) = 1 - e^{-(x_o/10373)^{3.48}}$$

$$R_o(x_o) = e^{-(x_o/10373)^{3.48}} \tag{8.6}$$

$$h_o(x_o) = 0.0000387(x_o/4340)^{2.48}$$

Figure 8.9 shows the life distribution $F(x)$ of the 28-pin TSOP solder joints under operating conditions. The 1 percent cumulative failure point occurred at approximately 66,500 hours or 7.5 years, and the 50 percent cumulative failure point occurred at about 224,000 hours or 25.5 years. These values are beyond a typical design lifetime. The failure rate $h(x)$ of the TSOP solder joints operating at 1 cycle/day and with a temperature range of 85°C is shown in Fig. 8.10.

8.1.2 Reliability of a 32-pin TSOP assembly

The 32-pin TSOP for the present study is shown in Fig. 8.11a and b.[11] The equivalent thermal coefficient of expansion of the 32-pin TSOP body in the long or x direction (α_x) and the short or z direction (α_z) is shown in Fig. 8.12. The test board, paste, printing and pick and place, solder material, reflow, cleaning, and thermal cycling test setup are

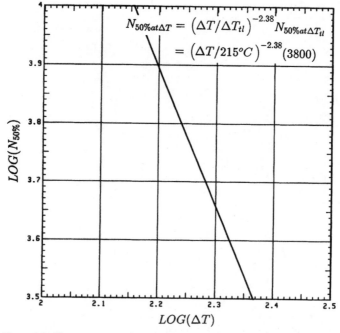

Figure 8.8   Temperature range versus cycle to failure at 50 percent failure point.[10]

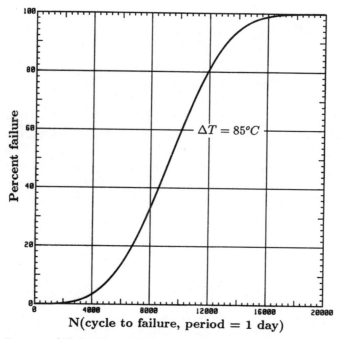

Figure 8.9   Life distribution (Weibull) of the 28-pin TSOP solder joints at operating condition.[10]

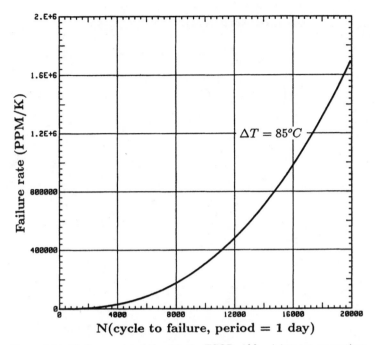

Figure 8.10 Failure rate of the 28-pin TSOP solder joints at operating condition.[10]

exactly the same as for the 28-pin TSOP assembly. The only difference is that the package size and the 32-pin TSOP are larger. Also, this time we only ran the test at one temperature condition, which is shown in Fig. 8.13 with the test results shown in Table 8.3.

In this case, the Weibull slope is $\beta = 4.44$ and the characteristic life $\theta = 3710$. The life distribution of the 32-pin TSOP solder joint is shown in Fig. 8.14. Unlike the test for the 28-pin TSOP, this time we have only one set of test data and use Eq. (2.67) to determine the acceleration factor AF. We assume the same operating conditions: one cycle/day and a temperature range of 85°C. Then we have

$$AF = (f_o/f_t)^{1/3}(\Delta T_t/\Delta T_o)^2(\Phi_o/\Phi_t)$$

$$= (1/55.38)^{1/3}(180/85)^2(1.7) \qquad (8.7)$$

$$= 2$$

Substituting this value into Eqs. (2.60) through (2.62), respectively, we have the life distribution (Fig. 8.15), reliability function (Fig. 8.16), and failure rate (Fig. 8.17) of the 32-pin TSOP solder joint at operating conditions.

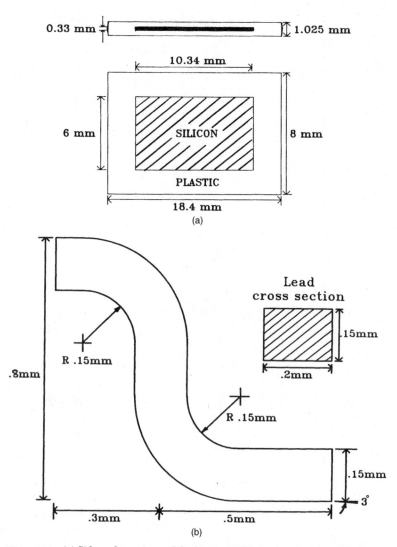

Figure 8.11 (a) Side and top views of the 32-pin TSOP (leads not shown). (b) Cross-sectional view of the 32-pin TSOP lead.[11]

Figure 8.15 shows the life distribution $F(x)$ of the 32-pin Type-I TSOP solder joints under the operating condition (1 cycle/day, ΔT_o = 85°C). It can be seen that the 1 percent cumulative failure point occurred at approximately 2636 cycles, i.e., 63,300 hours or 7.2 years, and the 50 percent cumulative failure point occurred at about 6830 cycles, i.e., 163,900 hours or 18.7 years. These values are beyond the design lifetime. Figure 8.15 also shows the life distribution of the 28-pin TSOP solder joints. It can be seen that for the first 3000 days of

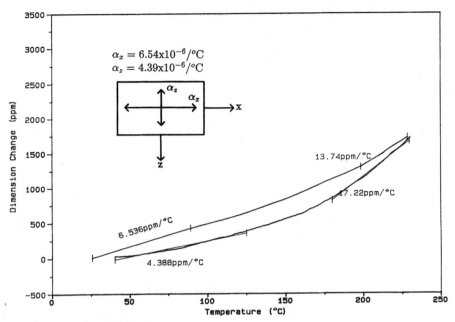

Figure 8.12   Equivalent TCE of the 32-pin TSOP.[11]

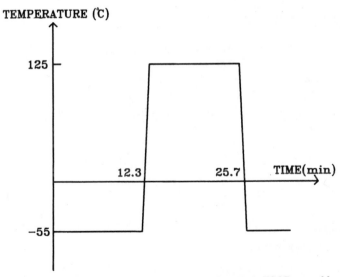

Figure 8.13   Temperature cycling condition for the 32-pin TSOP assembly.

TABLE 8.3   32-Pin TSOP Test Results for Statistical Analysis[11]

Failure no.	Cycle to failure (1 cycle = 26 min) Temperature range (°C) 180	Median (50%) rank
1	2,250	0.0315
2	2,250	0.0761
3	2,250	0.1207
4	2,250	0.1653
5	2,750	0.2099
6	2,750	0.2545
7	2,750	0.2992
8	2,750	0.3438
9	2,750	0.3884
10	3,250	0.4330
11	3,250	0.4776
12	3,750	0.5223
13	3,750	0.5669
14	3,750	0.6115
15	3,750	0.6561
16		0.7007
17		0.7454
18		0.7900
19		0.8346
20		0.8792
21		0.9238
22		0.9684

operation, the 28- and 32-pin TSOP solder joints have about the same level of reliability. However, after 3000 days of operation, the 28-pin TSOP solder joint shows a higher level of reliability. Figure 8.16 shows the reliability of the 28- and 32-pin TSOP solder joints.

Figure 8.17 shows the failure rate of the 32-pin TSOP solder joints operating at 1 cycle/day and with a temperature range of 85°C. Figure 8.17 also shows the failure rate of the 28-pin TSOP solder joints. It can be seen that, for the first 2000 days of operation, the failure rate of the 28- and 32-pin solder joints are about the same. However, after 2000 days of operation, the 32-pin TSOP solder joints have a much higher failure rate than the 28-pin TSOP solder joints.

8.1.3 Type-I or Type-II TSOPs?

There are two types of TSOP, Type I and Type II. Figure 8.18 schematically shows a 32-pin Type-I TSOP and a 32-pin Type-II TSOP. Type-I TSOP has leads on the shorter edges, while Type-II TSOP has leads on the longer edges of the TSOPs. Also, the dimensions of Type-II TSOPs are slightly larger than those of Type-I TSOPs. However, the lead width of Type-II TSOPs (0.4 mm) is twice that of Type-I TSOP (0.2 mm); there-

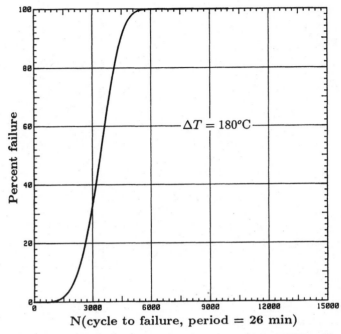

Figure 8.14 Life distribution (Weibull) of the 32-pin TSOP solder joints under test conditions.

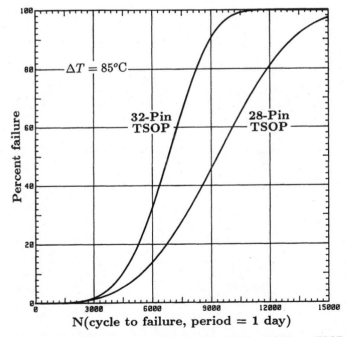

Figure 8.15 Life distribution (Weibull) of the 32-pin and 28-pin TSOP solder joints under operating condition.

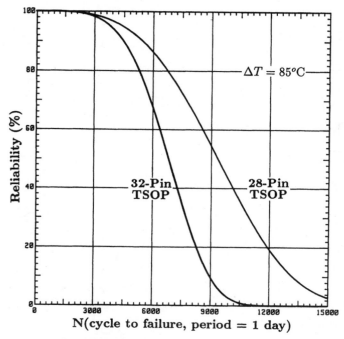

Figure 8.16 Reliability function (Weibull) of the 32-pin and 28-pin TSOP solder joints under operating condition.

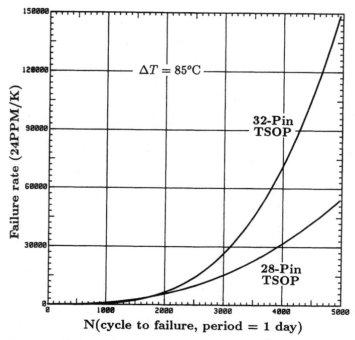

Figure 8.17 Failure rate (Weibull) of the 32-pin and 28-pin TSOP solder joints under operating condition.

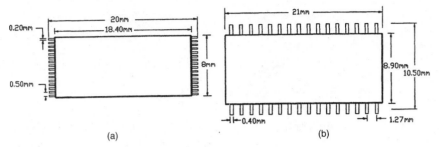

Figure 8.18   Types of TSOPs: (*a*) Type-I; (*b*) Type-II.

fore, the lead stiffness of Type-II TSOPs is much larger than that of Type-I TSOPs. Which type of TSOP solder joint has a better thermal reliability? A quantitative comparison is given by the following nonlinear finite element analyses.

Figure 8.19 shows the finite element model (dashed lines) for the analysis of the corner lead and solder joint of the 32-pin Type-I TSOP assembly. It consisted of 1024 3D solid elements (544 for the alloy-42 lead and 480 for the Sn/Pb solder joint). Each element had 20 nodes, and each node had 3 degrees of freedom. The boundary conditions were: (1) the TSOP assembly was subjected to a temperature change

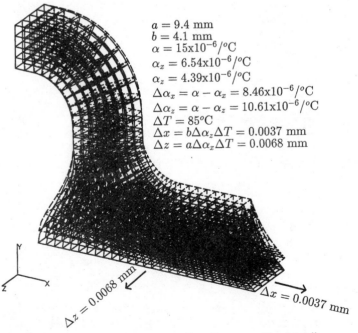

$a = 9.4$ mm
$b = 4.1$ mm
$\alpha = 15 \times 10^{-6}/°C$
$\alpha_x = 6.54 \times 10^{-6}/°C$
$\alpha_z = 4.39 \times 10^{-6}/°C$
$\Delta\alpha_x = \alpha - \alpha_x = 8.46 \times 10^{-6}/°C$
$\Delta\alpha_z = \alpha - \alpha_z = 10.61 \times 10^{-6}/°C$
$\Delta T = 85°C$
$\Delta x = b\Delta\alpha_z\Delta T = 0.0037$ mm
$\Delta z = a\Delta\alpha_x\Delta T = 0.0068$ mm

Figure 8.19   Deformation of the 32-pin Type-I corner solder joint.[11]

from 0 to 80°C, (2) at the upper end of the lead, all points cannot move in the x and z directions, and (3) at the bottom of the solder joint, all points moved 0.0066 mm in the x direction and 0.0036 mm in the z direction due to the global thermal expansion mismatch between the TSOP and PCB.

The Sn/Pb solder is an elastoplastic material with low yield strength and high ductility. The Young's modulus of the solder was 1.5 psi (10,000 MN/m) and the Poisson's ratio was 0.4. It should be emphasized that the mechanical properties of solders were strongly temperature-, frequency-, rate-, and time-dependent as discussed in Chaps. 3 through 7 of the book. These variables must be included in the analysis to have an accurate estimation of the strains and, consequently, the number of cycles to failure of the solder joint. However, for the sake of simplicity and due to the objective of this study (for comparison purposes), all the material properties were assumed to be temperature-, frequency-, rate-, and time-independent. Furthermore, the stress-strain curve at 120°C (yield stress = 1200 psi or 8.3 MN/m^2, yield strain = 0.0008 and the strain-hardening parameter = 0.1) was used for the Sn/Pb solder joint.

Figure 8.19 shows the whole-field displacements (solid lines) of the TSOP assembly during heating. It can be seen that the gull-wing lead and solder joint not only expanded in the x direction but also expanded in the z direction and resulted in a very complex state of stresses and strains. This was due to: (1) the horizontal thermal expansion mismatch between the alloy-42 lead and the Sn/Pb solder joint, and (2) the especially low standoff of the lead.

The von Mises stress distribution in the Type-I TSOP solder joint is shown in Fig. 8.20. The maximum von Mises stress occurred at the edge (in the positive z direction) near the middle portion of the solder joint. This maximum stress was concentrated in a very small area and declined rapidly away from this location. Thus, any solder joint damage (cracks) should initiate at this location and propagate in the solder joint along the interface between the gull-wing lead and the solder joint and through the width of the solder joint.

The accumulated plastic strain components in all directions acting in the solder joint are not presented here. However, the accumulated effective plastic strain is shown in Fig. 8.21. The strain contours in the solder joint were not uniform. The maximum strain (0.0227) was concentrated in a very small area at the edge in the positive z direction near the center portion of the solder joint. This maximum value decreased rapidly to a much smaller value a short distance away from the center portion.

Figure 8.22 shows the finite element model (dashed lines) for the analysis of the corner lead and solder joint of the 32-pin Type-II TSOP assembly. It consisted of 1446 3D solid elements (770 for the alloy-42 lead and 676 for the Sn/Pb solder joint). The element type was exactly

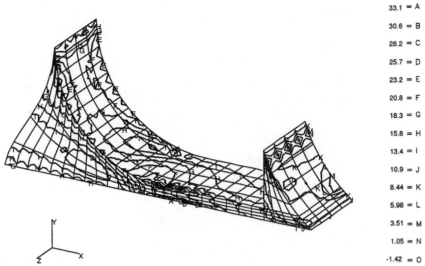

33.1 = A

30.6 = B

28.2 = C

25.7 = D

23.2 = E

20.8 = F

18.3 = G

15.8 = H

13.4 = I

10.9 = J

8.44 = K

5.98 = L

3.51 = M

1.05 = N

-1.42 = O

Figure 8.20   von Mises stress (MPa) in the 32-pin Type-I corner solder joint.

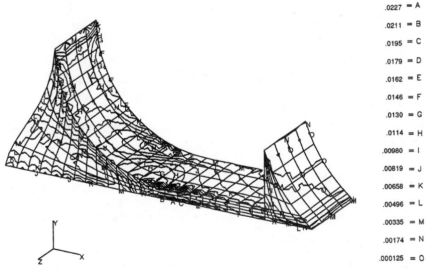

.0227 = A

.0211 = B

.0195 = C

.0179 = D

.0162 = E

.0146 = F

.0130 = G

.0114 = H

.00980 = I

.00819 = J

.00658 = K

.00496 = L

.00335 = M

.00174 = N

.000125 = O

Figure 8.21   Equivalent plastic strain in the 32-pin Type-I corner solder joint.

the same as those for the Type-I TSOP lead and solder joint. The boundary conditions were also the same except for the displacement boundary condition at the bottom of the solder joint where all points moved 0.0036 mm in the x direction and 0.0066 mm in the z direction.

The whole-field displacements (solid lines) of the Type-II TSOP corner lead and joint under heating were very similar to those of the Type-I. Figures 8.23 and 8.24 show, respectively, the effective stress and accu-

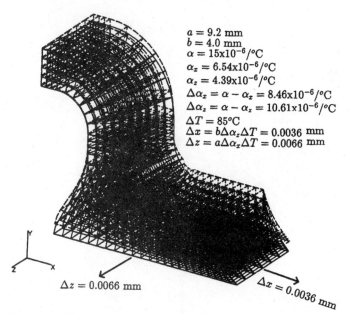

$a = 9.2$ mm
$b = 4.0$ mm
$\alpha = 15\text{x}10^{-6}/°C$
$\alpha_x = 6.54\text{x}10^{-6}/°C$
$\alpha_z = 4.39\text{x}10^{-6}/°C$
$\Delta\alpha_x = \alpha - \alpha_x = 8.46\text{x}10^{-6}/°C$
$\Delta\alpha_z = \alpha - \alpha_z = 10.61\text{x}10^{-6}/°C$
$\Delta T = 85°C$
$\Delta x = b\Delta\alpha_z\Delta T = 0.0036$ mm
$\Delta z = a\Delta\alpha_x\Delta T = 0.0066$ mm

$\Delta z = 0.0066$ mm

$\Delta x = 0.0036$ mm

Figure 8.22   Deformation of the 32-pin Type-II corner solder joint.[11]

mulated effective plastic strain in the Type-II TSOP solder joint. These values are much larger than those of the Type-I TSOP solder joint. This is due to the larger stiffness of the Type-II TSOP lead than the Type-I TSOP lead. One way for Type-II TSOP solder joints to have the same level of thermal reliability as Type-I TSOPs is to reduce the Type-II TSOP lead width by a factor of 2, i.e., 0.2 mm. The other way is to reduce the Type-II TSOPs body size.

8.1.4   Copper or alloy-42 leads for TSOPs?

In this section, the advantage of TSOP with copper leads is shown by comparing: (1) the calculated stress and strain in the 28-pin TSOP solder joints with that of the TSOP with alloy-42 lead frames, and (2) the experimental thermal life distribution of the 32-pin TSOP solder joints with that of the TSOP with alloy-42 leads. The disadvantage of TSOPs with copper leads is shown by considering the technology limitations and manufacturing constraints.

8.1.4.1   Plastic strain in the 28-pin TSOP solder joints with copper leads.
For copper lead frames, the Young's modulus is 121,000 MN/m and the TCE is 17 ppm/°C. Thus, the local thermal expansion mismatch between the copper lead and the solder is smaller than that between the alloy-42 lead and the solder. Also, the stiffness of the copper lead is 18 percent smaller than that of the alloy-42 lead (assuming they have

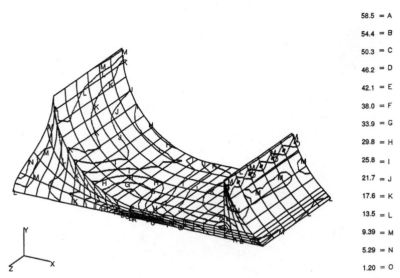

58.5	= A
54.4	= B
50.3	= C
46.2	= D
42.1	= E
38.0	= F
33.9	= G
29.8	= H
25.8	= I
21.7	= J
17.6	= K
13.5	= L
9.39	= M
5.29	= N
1.20	= O

Figure 8.23   von Mises stress (MPa) in the 32-pin Type-II corner solder joint.[11]

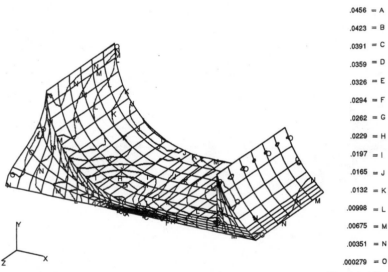

.0456	= A
.0423	= B
.0391	= C
.0359	= D
.0326	= E
.0294	= F
.0262	= G
.0229	= H
.0197	= I
.0165	= J
.0132	= K
.00998	= L
.00675	= M
.00351	= N
.000279	= O

Figure 8.24   Equivalent plastic strain in the 32-pin Type-II corner solder joint.[11]

the same geometry). Furthermore, the global thermal expansion mismatch between the TSOP body with copper lead frames and the FR-4 PCB is smaller than that between the TSOP body with alloy-42 lead frames and the FR-4 PCB.

Figures 8.25 and 8.26 show the whole-field displacements (deflections) of the TSOP corner lead and joint under heating. The solid lines

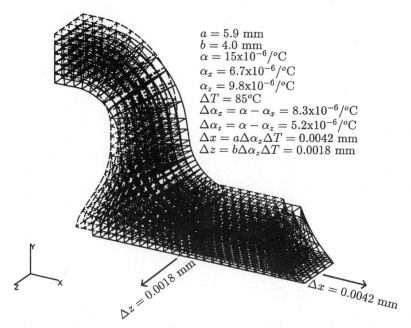

$$a = 5.9 \text{ mm}$$
$$b = 4.0 \text{ mm}$$
$$\alpha = 15\text{x}10^{-6}/°\text{C}$$
$$\alpha_x = 6.7\text{x}10^{-6}/°\text{C}$$
$$\alpha_z = 9.8\text{x}10^{-6}/°\text{C}$$
$$\Delta T = 85°\text{C}$$
$$\Delta\alpha_x = \alpha - \alpha_x = 8.3\text{x}10^{-6}/°\text{C}$$
$$\Delta\alpha_z = \alpha - \alpha_z = 5.2\text{x}10^{-6}/°\text{C}$$
$$\Delta x = a\Delta\alpha_x\Delta T = 0.0042 \text{ mm}$$
$$\Delta z = b\Delta\alpha_z\Delta T = 0.0018 \text{ mm}$$

Figure 8.25 Boundary-value problem of the 28-pin TSOP (with Cu-lead) corner solder joint.[12]

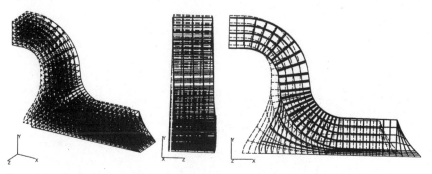

Figure 8.26 Deformation of the 28-pin TSOP (with Cu-lead) corner solder joint.[12]

are for the displaced mesh. It can be seen that the gull-wing lead and solder joint not only expand in the z direction but also expand in the x direction resulting in a very complex state of stresses and strains. This deformed shape is very similar to that of the TSOP with alloy-42 leads. However, the magnitudes of the cumulative effective plastic strain were smaller than those of the TSOP with alloy-42 leads.[12] It was shown that the maximum cumulative effective plastic strain in the TSOP solder joint with copper leads was 0.0115.[12] This value was 20 percent less than that (0.0143) of the TSOP solder joint with alloy-42 leads. Applying Solomon's isothermal fatigue data,[24] the average predicted thermal fatigue life of the TSOP solder joint increases by 53 percent if the lead frame material is changed from alloy-42 to copper. This is due to the smaller thermal expansion mismatch and the greater compliance of the copper leads.

8.1.4.2   Thermal cycling of the 32-pin TSOP solder joints with copper leads. The test board assembly, setup, and test conditions of the 32-pin TSOP with copper leads were exactly the same as those for the 32-pin TSOP with alloy-42 leads. The test result of the solder joints is shown in Table 8.4 and is represented by the Weibull distribution (Fig. 8.27), where $\beta = 2.21$ and $\theta = 5048$. For comparative purposes, the thermal cycling result of the alloy-42-leaded TSOP solder joints is also shown in Fig. 8.27. It can be seen that at 50 percent failure ($N_f = 4277$ cycles for TSOP solder joints with copper leads and $N_f = 3416$ cycles for TSOP solder joints with alloy-42 leads), the thermal fatigue life of the TSOP solder joints with copper leads increased by 25 percent. Also, the characteristic value (N_f at 63.2 percent failure) of the solder joints at test condition increased by 36 percent when the lead frame material

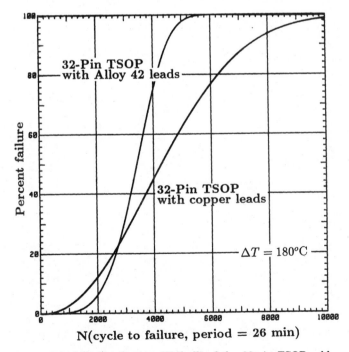

Figure 8.27   Life distribution (Weibull) of the 32-pin TSOP solder joints with alloy-42 leads and copper leads.[12]

TABLE 8.4   32-Pin TSOP (with Cu-Lead) Test Results
for Statistical Analysis[12]

Failure no.	Cycle to failure (1 cycle = 26 min) Temperature range (°C) 180	Median (50%) rank
1	1,250	0.0315
2	1,750	0.0761
3	1,750	0.1207
4	2,250	0.1653
5	2,250	0.2099
6	2,250	0.2545
7	3,250	0.2992
8	4,250	0.3438
9	4,250	0.3884
10	4,250	0.4330
11	4,750	0.4776
12	4,750	0.5223
13	4,750	0.5669
14	4,750	0.6115
15	5,250	0.6561
16	5,250	0.7007
17	5,250	0.7454
18	5,750	0.7900
19	7,250	0.8346
20	7,250	0.8792
21	7,250	0.9238
22	7,250	0.9684

changes from alloy-42 to copper ($\theta = 3710$ cycles for TSOP solder joints with alloy-42 leads).

8.1.4.3   TSOP solder joint reliability.   Solder joint reliability is a function of the following variables[3]: IC packaging geometry and material; lead geometry, material, and finishing composition; PCB material, thickness, surface composition, and system (displacement) constraint; single/double-sided assembly; paste/flux material; reflow temperature profile; cleaning technique or no-clean at all; solder joint geometry; mechanical and physical properties of bulk solders and joints; operating conditions (cycling frequency, temperature range, dwell time at high temperature, etc.); and the expected life. Thus, the incremental gains from copper must be carefully weighed against the benefits of alloy-42 and the trade-offs with all other variables.

Package profile.   The fundamental issue for TSOPs *is not* alloy-42 versus copper, but the low profile of the package itself. By virtue of their low profiles, TSOPs with either alloy-42 or copper leads will always have a significantly higher stiffness compared to other conventional SMT packages. Typically, TSOPs are 40 to 60 times less compliant than PLCCs or PQFPs. As a result, TSOPs with either copper or alloy-42 leads fall short of the general expectation that low-profile packages, such as TSOP, should perform exactly like the other SMT tall packages on the board without altering anything at the FR-4 board/system level.[25,26]

Lead geometry. It is obvious that a larger cross-section area of the solder joint would reduce stress and strain concentrations that, in effect, would improve reliability. Larger cross sections, however, require larger lead width. Increased lead width in turn increases lead stiffness. *Dog bone–shaped* lead was proposed as a solution, but it is very difficult to manufacture, since the leads can break off in the narrow regions during IC assembly, testing, and handling.

Reducing the lead frame thickness (from 0.152 to 0.127 mm) definitely yields higher benefits in terms of increasing the lead compliance. The resulting loss of rigidity due to reduced thickness, however, increases handling difficulties, warped lead frames, bent leads, and causes serious difficulty in controlling coplanarity. For the same thickness, the rigidity of an alloy-42 lead is an advantage.

SMT process. The importance of *good* fine-pitch SMT processes cannot be emphasized enough. Fine-pitch component solder joints have less tolerance for defects such as large voids that may undermine reliability. An ideal solder joint should have good heel, toe, and side fillets, and sufficient standoff.

Excess solder paste can cause solder to wick up the shoulder of the lead, increasing the thickness and also the stiffness of the lead significantly. This increase in stiffness applies to both alloy-42 and copper leads.

It is noted that the quantitative increase in lead stiffness by the SMT process is not yet reported in the literature. The effects, however, are observed as reduced life in test data. In fact, noncorner leads have been reported to fail before the corner leads. This discrepancy can only be explained by statistical variation of the solder joints in terms of local defects or increased local stiffness from solder-coated leads. All current models predict the corner solder joints to fail first because they are furthest away from the neutral point.

8.1.4.4 Disadvantages of TSOP with copper

Package thermal mismatch. As mentioned earlier, the TCE of copper is closer to that of epoxy FR-4 PCB and solder. Although this helps to reduce the stresses and strains in the solder joint, it creates a large internal mismatch in the TSOP body between silicon and copper. It is often proposed that this can be overcome by compliant die attach material. This is feasible in thick packages like PLCCs or PQFPs. In the case of thin packages, the thickness of all materials is reduced to

achieve the thin profile (Fig. 8.28). The thinner die attach cannot overcome the mismatch problems because of interfacial delamination,
warping, and higher stresses on the die surface. These effects are exac-
erbated during the surface-mount reflow process. Alloy-42 and silicon
have very little thermal mismatch and thus a better package integrity.

Package moisture effects. The majority of current TSOP suppliers opt for
alloy-42 because of the manufacturing hardships involved with copper-
leaded TSOPs. This makes it difficult to find hard data on copper-leaded
TSOPs. As a result, the following discussion is based on information
available on other SMT packages.

Adhesion of molding compounds to copper is inferior to that of adhesion to alloy-42 because of the oxidation characteristics of the two
materials. As a result, copper-leaded TSOPs have significantly lower
moisture resistance compared to alloy-42-leaded TSOPs. As a matter of
fact, copper-leaded TSOPs may be categorized as highly moisture sensitive and some suppliers recommend a special baking step before SMT

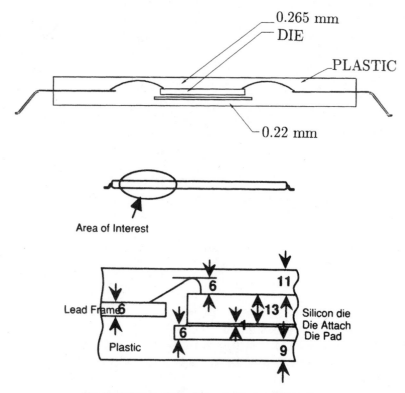

Nominal Dimensions Rounded to nearest mils

Figure 8.28   Cross section of TSOP (not to scale).

reflow in order to avoid popcorn cracking,[19,20,23] which is a special burden on an SMT manufacturing line. Alloy-42, on the other hand, shows good adhesion and is less susceptible to delamination. Alloy-42 TSOPs very often do not require dessicant moisture barrier bags, which is a welcome relief to the SMT line discipline.

Kim[21] reported that the adhesive strength of the molding compound (Novolac epoxy) to copper decreases monotonically with increasing moisture content. This has been supported by the thermodynamic calculations of energy of adhesion in the presence and absence of moisture. The energy of adhesion substantially decreases in the presence of moisture.

Adhesion-strength degradation or delamination (as a result of exposure to 85°C/85% RH followed by three vapor-phase reflows), as measured by scanning acoustic microscopy (SAM), has been found to increase with decreasing peel strength due to absorbed moisture.[21] The results demonstrate that interface adhesion plays a vital role in minimizing delamination during SMT operation. Note that die pad delamination is often a precursor to such popcorn cracking and has been observed in C-SAM.[22] The delamination dome allows excess moisture buildup. Thus the stress generated by the steam pressure of this excess moisture at SMT reflow temperatures of 220°C, exceeds the sum total of the flexural strength and the adhesive strength of the molding compound, resulting in popcorn cracking. Thus, the adhesion characteristics increase the susceptibility of copper to popcorn cracking.

Bhattacharya et al.[17] estimated that the gap between the molding compound and the lead frame is close to 0.32 mils (0.008 mm) on the basis of CTE mismatches alone. The doming effect seen at the bottom of the package due to the saturation pressures associated with reflow temperatures is estimated to be as high as 22 mils (0.56 mm). Since doming up to a few mils only has been reported, it is believed that a certain portion of the flashed steam is probably dissipated across the package. Thus the doming effect is not exaggerated under the saturated pressure. Under saturation pressures, a similar doming effect can be expected for TSOPs. Since the amount of plastic below the die pad is only about 0.2 mm, much thinner than conventional packages, it is more prone to cracking than thicker packages. Mechanical interlocking with features such as dimples and slits, provides additional margin but cannot compensate for lack of initial adhesion.

Package coplanarity effects. For fine-pitch devices, there is often a demand for tighter coplanarity (0.1 mm maximum and desired 0.076 mm). Such a requirement is difficult to meet with copper if burn-in and multiple socketings are involved in the test process. A possible alternative to gull-wing formation after testing is expensive as it requires deviation from current testing procedures.

8.1.4.5 Summary. The advantages of TSOPs with copper leads have been shown by determining the effective stress and cumulative effective plastic strain in the corner solder joints. It was found that their magnitudes are smaller than those of TSOPs with alloy-42 leads. The average thermal fatigue life of the copper-leaded TSOP solder joints is predicted to be 53 percent better than that of the alloy-42-leaded TSOPs.

Also, the advantages of TSOPs with copper leads have been shown by thermal cycling tests and statistical analysis. It was found that the thermal characteristic life of the copper-leaded TSOP solder joints is 36 percent better than that of the alloy-42-leaded TSOPs. Finally, the incremental gains from copper have been weighed against the trade-offs of other variables such as the package profile, lead geometry, and SMT process.

The disadvantages of TSOPs with copper leads include technology limitations and manufacturing constraints, such as the internal thermal expansion mismatch within the package, the moisture sensitivity that causes popcorn cracking, and the loss of rigidity that leads to handling difficulties, warped lead frames, bent leads, and difficulty in controlling coplanarity.

8.2   Thermal Reliability of 256-Pin PQFP Assembly

Since its introduction in 1987, PQFP (plastic quad flat pack) has been widely used for packaging the ASICs, memories, and microprocessors. It moved SMT to a higher level of density and performance by reducing the lead pitch from 1.27 to 0.65 mm and less. The solder joint reliability of 0.4-mm-pitch, 256-pin PQFP assemblies are discussed in this section.[5] Also, the water-clean and no-clean mass reflow of the PQFP is provided.[2,4]

8.2.1   The 0.4-mm-pitch fine pitch component

The surface-mount component used in this study was a 256-pin, 0.4-mm-pitch PQFP, as shown in Fig. 8.29a and b. The package met the EIAJ (Electronic Industry Association of Japan) package specification. The dimensions of this PQFP were $28 \times 28 \times 3.35$ mm and it was made of plastic. The thermal coefficient of linear expansion of the plastic was 22 ppm/°C. The Young's modulus of the plastic was 2×10^6 psi (14,000 MN/m) and the Poisson's ratio was 0.3. The lead frame material was made of copper. The thermal coefficient of linear expansion of the copper was 17 ppm/°C, the Young's modulus was 121,000 MN/m^2 and the Poisson's ratio was 0.35. The critical dimensions of

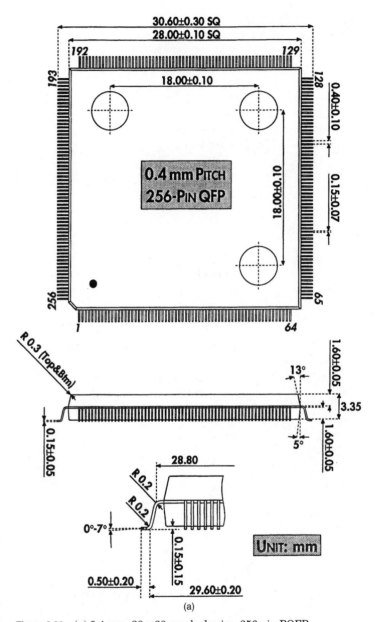

Figure 8.29   (*a*) 0.4 mm, 28 × 28 mm body size, 256-pin PQFP.

the gull-wing-shaped leads, the foot length, width, and thickness, were 0.5, 0.15, and 0.127 mm, respectively. These PQFP lead frames were interconnected inside the package in an alternating pattern so as to provide a daisy-chained connection when the PQFP is soldered to the PCB.

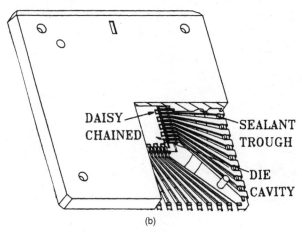

(b)

Figure 8.29 (*Continued*) (*b*) Daisy-chained design of the 256-pin PQFP for reliability testings.

8.2.2 PCB design

Figure 8.30 shows one of the PCBs with 25 sites of 256-pin PQFPs.[4,5] The dimensions of the PCB were 210 × 210 × 1.57 mm. The PCB was made of FR-4 epoxy/glass substrate with etched copper conductors and land patterns (solder pads). The thermal coefficient of linear expansion of the FR-4 material was 15 ppm/°C. The Young's modulus of the PCB was 11,000 MN/m² and the Poisson's ratio was 0.28. The finished PCB was coated with a photoimaged, low-profile, dry-film solder mask (0.058-mm-thick). The dimensions of each solder pad were 1.26 × 0.2 mm. The surface composition of the pad was Cu(0.043-mm-thick)-Sn/Pb(0.007-mm-thick). These pads were interconnected in an alternating pattern so as to provide a daisy-chained connection with the 256-pin PQFP.

8.2.3 Solder pastes

Based on the flux activity and residue cleaning methods, today's solder pastes in the electronics industry may be classified in two groups, namely: (1) the rosin-based and solvent-cleaning pastes (e.g., R-grade, RMA-grade, and RA-grade, where R = rosin, M = mildly, and A = activated), and (2) the water-cleaning (e.g., water-saponifiable and water-soluble) and no-clean pastes. The important parameters that affect the paste selection and the defect-free manufacturing are: flux, single-point viscosity, particle size (powder size), metal composition, and percent of metal.

Flux is the substance that prepares and protects the surface to be soldered by removing surface oxides; it then provides a clean metallic

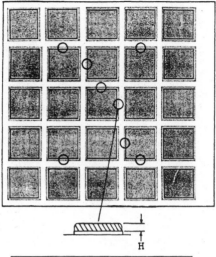

Brd. No.	Paste Height (mm)
1	0.134 – 0.207
2	0.130 – 0.214
3	0.108 – 0.168
4	0.118 – 0.179

Figure 8.30   256-pin test board (25 sites).

surface and prevents further oxidation during the soldering process. Flux also assists in the transfer of heat from the molten solder to the joint area so that the base metals reach a high enough temperature to be wetted by the solder. Furthermore, flux reduces the interfacial surface tension between the solder and the base metal, enabling the solder to flow over and metallurgically wet the solderable surface. In this study, two types of solder paste flux were used: (1) water-soluble flux (Rheomet Solderpaste Formula R-587), and (2) no-clean flux (Rheomet Solderpaste Formulas R-255 and R-255H). These two fluxes were manufactured by Kester Solder.

The paste viscosity is an important parameter in controlling the yield of stencil printing. For fine-pitch applications the viscosity needs to be high enough so that its reduction due to printing operation does not affect the printability. (Of course, the most ideal condition is to keep the paste viscosity uniform throughout the printing process.) A Brookfield RVDT viscometer was used to test each container opened.

They were tested at a stabilized 25°C temperature with a spindle rotation of 5 rpm. In this study, the viscosity used for the water-clean paste was 990 kcps (kcps = 1000 centipoises) and for the no-clean paste was 850 kcps. It is noted that the no-clean paste dried up faster than the water-clean paste.

The powder size in the solder paste was chosen to match both the stencil opening and the thickness of the wet paste being printed. In this study, for both processes, we used the particle size (−325, +500), i.e., the diameter was no greater than 0.0432 mm or less than 0.0254 mm. Generally, the powder size should be four to five times smaller than the aperture size of the stencil for good fine-line printing. Smaller powder pastes have less tendency to slump but the resulting printed deposits are usually thinner. Spherical solder particles are preferred over other shaped particles.

The metal composition for both processes was 63%wt tin and 37%wt lead with a melting point about 183°C. For fine-pitch applications, at least 90 percent metal-content paste is required. It should be noted that the metal content in solder paste determines the solder fillet size.

8.2.4 Stencil design

The desired capabilities of a fine-pitch stencil are solder paste transmissibility and continuous printability without spreading. In order to maintain a repeatable and uniform paste printing operation it is very important to control the accuracy of the stencil thickness, opening dimensions, and straight wall surfaces. The material and process selected for the stencil fabrication will determine these parameters.

Brass is a good material for maintaining stencil openings and very uniform wall finishes, and is cheaper than stainless steel. However, brass is not as durable as stainless steel. Recently, in order to improve the smoothness of the etched wall in a stainless steel stencil, a subtractive or electropolishing process has been applied. This additional process buffs the steel wall to improve paste transmissibility. One of the other durable materials available for stencil design is molybdenum. This material has better etching characteristics than stainless steel but is more expensive. In this study, the molybdenum material was used for the stencil because of the decidedly superior wall profile expected. The stencil opening to the solder pad was approximately 1:1. The stencil was 0.007 in (0.178 mm) thick.

8.2.5 Stencil printing

After optimization of paste and stencil specifications, the registration accuracy and reproducibility of the printed pattern alignment to the

PCB must be addressed. Among the large number of parameters (paste, stencil, etc.) affecting printing results, several operation parameters, such as squeegee pressure, squeegee speed, and snap-off, become increasingly critical when processing fine-pitch components. Also, a fully automatic vision printer is a must. Too high a squeegee pressure will cause the prints to smear and may damage the stencil. Too low a squeegee pressure will cause the solder to skip and be insufficient. A pressure-setting technique that has been most successful is to start with low pressure and then increase the pressure until a clean sweep is observed on the top of the stencil. In this study, the squeegee pressure was set to about 200 kPa with a squeegee hardness of 90 durometers. The squeegee speed depends on the viscosity of the paste. For the present fine-pitch applications, a speed of 30 mm/sec was applied to the no-clean paste with a viscosity of 850 kcps, and a speed of 20 mm/sec was applied to the water-clean paste with a viscosity of 990 kcps.

The *snap-off* in off-contact printing is the distance between the bottom of the stencil and the PCB. The degree of the off-contact affects the desirable setting of squeegee pressure and speed. If the snap-off is too large, most of the squeegee pressure is used to push the stencil in contact with the PCB and results in insufficient solder paste deposited. On the other hand, if the snap-off is too small, the paste will be smeared. Usually, for fine-pitch applications the snap-off distance should be between 0.127 mm to 0.254 mm. In this study, the snap-off distance was 0.2 mm.

The printer used in this study was a Fuji Model GSP-III 5000 in-line solder stencil system. This system provides automatic alignment of the stencil to the board through recognition of two global fiducial targets on the test board surface and two coinciding locations on the stencil surface. In order to reduce dry out of the paste deposited on the stencil surface, the entire work area of the Fuji system was covered and environmentally controlled.

8.2.6 Solder volume measurement

In production, only periodic sampling of the solder volume is customary for process monitoring. During process development, however, it is important to record the solder volume and consistency on each substrate sample. A Cybernetics Laser Scanning Microscope was employed to measure the height, length, and width of the deposited solder paste at eight locations on the test board (it took about five minutes). The summary of the measurements (height) made for four boards is shown in Fig. 8.30. It can be seen that the paste height measurements are quite consistent. It should be noted that, in order to avoid excessive slumping, the board must be transferred into the next assembly stage with minimum delay.

8.2.7 Pick and place

The pick and place of any fine-pitch components must be made with the assistance of a vision system capable of extreme accuracy and alignment tolerance. A fully automated pick and place system (Fuji IP-II 5000) was used in this study. This system had the combination of speed, flexibility, reliability, and vision capability for precision placement of large components with lead pitches as small as 0.3 mm. The vision system can place the 256-pin PQFP within 0.05 mm of *true position* (all leads aligned to land patterns on all test substrates). To ensure that the typical tolerance variation of the commercial FR-4 board material would not compromise the placement accuracy of any PQFPs, two 1.0-mm-diameter fiducial targets were provided for each component. Even with 100 percent fiducial scanning, the placement process for each board was completed within three minutes.

8.2.8 Mass reflow

The main purpose of the reflow process is to melt the solder particles, wet the surfaces to be joined, and then solidify the solder into a metallurgical bond. The primary areas of concern during reflow are flux spread, slumping, and solder balls. As heat is applied to solder paste, prior to solder melting, the paste's volatile solvents evaporate and the flux chemically activates. Excessive heat during the preheating and predrying stage or a fast ramp-up of temperature can create solder balling and inadequate fluxing activity during reflow. Also, exposure of the plastic PQFPs to high temperatures for relatively long periods of time can cause internal damage, e.g., plastic cracking due to the popcorn effect.[19]

A reflow profile test model was assembled prior to the processing of the primary test boards. Using thermocouples and a recording device, precise temperature levels were recorded throughout the reflow cycle. Following five or six passes of the profile test model through the process, adjustments were programmed into the controller and the desired profile curve was achieved (Fig. 8.31). The profile exposed the plastic PQFPs to temperatures between 215.0 and 225.6°C, ensuring liquefaction of the solder paste, but the dwell time at that temperature was limited to less than one minute.

In this study, the system selected for reflowing the test boards was the Conceptronics Mark IV (a forced-convection in-line conveyer system). The 10-zone machine allows for very precise thermal profile adjustment and provides for the use of nitrogen, an inert gas, to insure a very pure reflow environment.

temperature (^{o}C)

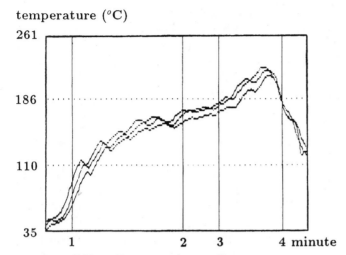

Figure 8.31   Solder reflow temperature profile.

8.2.9 Cleaning

Residues that remain around solder joints after the reflow process are a mixture of the flux vehicle of the paste and residue resulting from any interaction between the component and PCB. All of these residues should be cleaned using the appropriate cleaning agent for the paste in question. For the water-clean paste, an in-line conveyer-type aqueous cleaning system (the Westkleen FORMULA III MIL/SMT+) was used. This system is self-contained (two wash zones, one rinse zone, and two air knife zones) and provides high-quality cleaning without impacting the environment. In this study, the water temperature was set at 140–150°F, the water pressure was set at 420 kPa, and the conveyer speed was set at 2 ft/min (610 mm/min). The test boards were cleaned as soon as the solder reflow process was completed. For the no-clean paste solder reflow, of course, no cleaning step was necessary.

8.2.10 Notes on 0.4-mm-pitch and 256-pin PQFP assembly

The 0.4-mm-pitch, 28-mm-body-size, 256-pin PQFPs have been assembled on FR-4 epoxy PCBs by both water-clean and no-clean processes. The important fine-pitch technology parameters such as the PCB design, solder paste selection, stencil design and manufacture, printing technology, pick and placement, mass reflow, and cleaning have also been provided and discussed. Furthermore, the microstructure of the solder joints has been examined for a better understanding of the present processes. It was found that the solder joint microstructure was similar for both the no-clean and water-clean mass reflow processes.

The PCB flatness and copper pad surface composition are very important. The PQFP coplanarity must be small (no more than 0.0762 mm) and plating quality on the lead becomes important.

The solder paste selection and stencil design for paste transmissibility are the heart of very fine pitch technology. The stencil accuracy and pick and place accuracy play key roles. Also, the no-clean paste dried up faster than the water-clean paste.

Since this was our first attempt[2] to assemble the 0.4-mm-pitch, 28-mm-body-size, 256-pin PQFPs on FR-4 PCBs by both water-clean and no-clean processes, the parameters mentioned in this report are by no means optimal. More process developments are needed in order to have a high-yield production process. It is clear, however, that 0.4-mm pitch can be successfully assembled on conventional surface-mount lines with only minor modification to the procedures and attention to fine adjustment of the critical parameters of the process. It is our hope that the information presented herein may assist in removing roadblocks, avoiding unnecessary false starts, and accelerate process development of very fine pitch and large PQFPs.

8.2.11 Test setup and temperature profile

All test boards (four water-clean and four no-clean) were grouped and cabled together in such a way that the resistance of all the solder joints could be measured. The boards were tested in an air-to-air thermal cycling chamber. They were subjected to the temperature cycling profile shown in Fig. 8.32. It can be seen that the cycle period, ramp-up time, ramp-down time, dwell time at maximum temperature (125°C) and dwell time at minimum temperature (−40°C) were 45, 15, 6, 20, and 4 minutes, respectively.

A data acquisition system continually monitored the electrical resistance of the PQFP solder joints and logged the failure times. The tests were run nonstop for more than six months. The failure of a solder joint was defined as total opening (separation) between the lead and the PCB. The number of cycles to fail was defined as the first solder joint failure in a 256-pin PQFP component. The same sample size (93 PQFPs, or 23,808 solder joints) was applied to the water-clean and no-clean cases.

8.2.12 Statistical analysis of test results

Table 8.5 shows the results under the current test condition (−40 to 125°C and 1 cycle/45 min). In this case, reliability of the solder joints was modeled by Weibull's probability density function and analyzed by Johnson's statistical method with ranking shown in Chap. 2. It can be

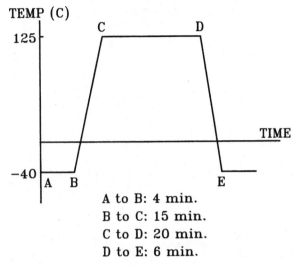

A to B: 4 min.
B to C: 15 min.
C to D: 20 min.
D to E: 6 min.

Figure 8.32   Thermal cycling temperature profile.

shown that the best fit Weibull parameters (for the median rank) are $\beta = 1.27$ and $\theta = 42{,}897$ for the water-clean solder joints, and $\beta = 2.14$ and $\theta = 12{,}114$ for the no-clean solder joints. With these values, Eq. (2.39) is called the life distribution of the 256-pin PQFP solder joints under the test condition (1 cycle/45 min, −40 to 125°C) and is shown in Fig. 8.33. It can be seen that for the first 2000 cycles the percent failure of the water-clean and no-clean solder joints is about the same. However, for higher temperature cycles the percent failure of the no-clean solder joints is higher than that of the water-clean solder joints, and the difference is a highly nonlinear function of the temperature cycle. This may be due to the thermochemical reaction of the uncleaned flux residues and contamination on the no-clean solder joints.

If one considers all the PQFP (water-clean and no-clean) solder joints as a whole population (Table 8.6) the Weibull parameters are $\beta = 1.43$ and $\theta = 26{,}520$, and the life distribution is shown in Fig. 8.34.

As mentioned earlier, the temperature cycling test frequency f_t was at 1 cycle/45 min and the temperature range was $\Delta T_t = 165°C$. For most of the computer operating conditions, $f_o = 1$ cycle/day and $\Delta T_o = 85°C$. Thus, from Eq. (2.67), we have:

$$AF = (f_o/f_t)^{1/3}(\Delta T_t/\Delta T_o)^2(\Phi_o/\Phi_t)$$

$$= (1/32)^{1/3}(165/85)^2(1.7) \qquad (8.8)$$

$$= 2.02$$

TABLE 8.5   256-Pin PQFP Test Results for Statistical Analysis

Failure no.	Cycles to failure (−40 to 125°C, 45 min.)		Median (50%) rank
	Water-clean	No-clean	
1	1,605	1,492	0.0075
2	1,610	1,529	0.0182
3	1,657	1,822	0.0289
4	3,402	3,622	0.0396
5	4,610	3,679	0.0503
6	4,703	3,862	0.0610
7	5,211	4,294	0.0717
8		4,377	0.0824
9		4,406	0.0931
10		4,565	0.1039
11		4,599	0.1146
12		4,649	0.1253
13		4,661	0.1360
14		4,798	0.1467
15		4,941	0.1574
16		4,956	0.1681
17		5,188	0.1788
18		5,199	0.1895
19		5,266	0.2002
20		5,310	0.2109

Once the acceleration factor AF = 2.02 has been determined, the life distribution, reliability function, and failure rate under operating conditions can be determined by Eqs. (2.60) through (2.62), with $\beta = 1.27$ and $\theta = 42{,}897$ for the water-clean solder joints as follows:

$$F_o(x_o) = 1 - e^{-(x_o/86652)^{1.27}} \tag{8.9}$$

$$R_o(x_o) = e^{-(x_o/86652)^{1.27}} \tag{8.10}$$

$$h_o(x_o) = 0.0000121(x_o/42897)^{0.27} \tag{8.11}$$

For the no-clean solder joints, $\beta = 2.14$, $\theta = 12{,}114$, and

$$F_o(x_o) = 1 - e^{-(x_o/24470)^{2.14}} \tag{8.12}$$

$$R_o(x_o) = e^{-(x_o/24470)^{2.14}} \tag{8.13}$$

$$h_o(x_o) = 0.0000392(x_o/12114)^{1.14} \tag{8.14}$$

Figure 8.35 shows the life distribution $F_o(x_o)$ of the 256-pin PQFP water-clean and no-clean solder joints under the operating condition ($f_o = 1$ cycle/day, $\Delta T_o = 85°C$). It can be seen that the 1 percent cumulative failure point of the no-clean solder joints occurred at approximately 2850 cycles, i.e., 68,400 hours or 7.8 years, and the 50 percent cumulative failure point occurred at about 20,620 cycles, i.e., 494,880

TABLE 8.6   256-Pin PQFP Test Results for Statistical Analysis

Failure no.	Combined water- and no-clean cycles to failure (−40 to 125°C, 45 min.)	Median (50%) rank
1	1,492	0.0038
2	1,529	0.0091
3	1,605	0.0145
4	1,610	0.0198
5	1,657	0.0252
6	1,822	0.0306
7	3,402	0.0359
8	3,622	0.0413
9	3,679	0.0467
10	3,862	0.0520
11	4,294	0.0574
12	4,377	0.0628
13	4,406	0.0681
14	4,565	0.0735
15	4,599	0.0789
16	4,610	0.0842
17	4,649	0.0896
18	4,661	0.0950
19	4,703	0.1003
20	4,798	0.1057
21	4,941	0.1111
22	4,956	0.1164
23	5,188	0.1218
24	5,199	0.1271
25	5,211	0.1325
26	5,266	0.1379
27	5,310	0.1432

hours or 56.5 years. These values are beyond most computers' design lifetimes. The life distribution of the 256-pin PQFP solder joints (both water-clean and no-clean) is shown in Fig. 8.36. The failure modes and analyses have been discussed in Chaps. 3 and 4.

8.3   Mechanical Reliability of PQFP Assembly

The mechanical responses of 304-pin, 0.5-mm pitch, 40 by 40 mm body size plastic quad flat pack (PQFP) solder joints and leads have been determined in this study. The effects of overload environmental stress factors on the mechanical responses of the leads and solder joints have been determined by bending and twisting experiments.[1]

8.3.1   The 304-pin PQFP SMC

The advent of integrated circuit technology and the requirements of high density for high-speed circuitry is driving the design of surface-mount components (SMCs) to higher pin counts. In general, higher pin

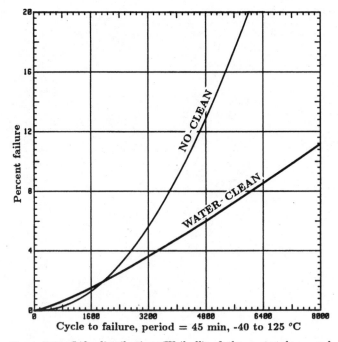

Figure 8.33 Life distribution (Weibull) of the water-clean and no-clean 256-pin PQFP solder joints at test condition.

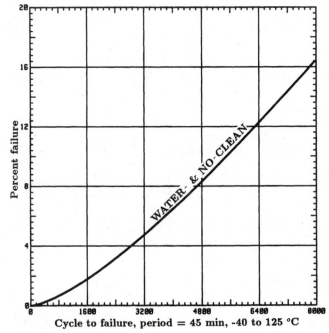

Figure 8.34 Life distribution (Weibull) of the 256-pin PQFP solder joints at test condition.

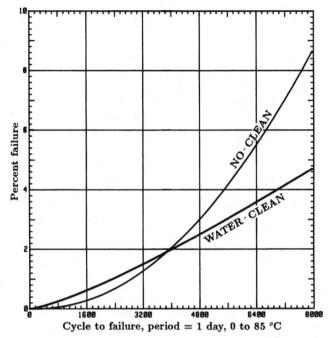

Figure 8.35 Life distribution (Weibull) of the water-clean and no-clean 256-pin PQFP solder joints at operating condition.

count is accomplished by increasing the body dimensions of the SMC and using fine-pitch technology. Because of the large body size of the SMC and the concerns associated with fine-pitch technology, the solder joint and lead reliability of large PQFPs on FR-4 printed circuit board (PCB) is under scrutiny. In this section, the fine-pitch SMC is a 0.5-mm-pitch, 304-pin PQFP (Fig. 8.37). Figure 8.38 schematically shows a 304-pin PQFP assembly. It consists of four major parts, namely the FR-4 epoxy/glass PCB, the 60wt%Sn/40wt%Pb or 63wt%Sn/37wt%Pb solder joints, the copper lead, and the PQFP (Fig. 8.38).

During manufacturing, handling, shipping, rework, etc., the PCB is subjected to bending and twisting. The effects of these (overload) environmental stress factors on PQFP's leads and solder joints will be addressed by bending and twisting experiments.

8.3.2   Test board assembly

The bending and twisting test boards were assembled by a mass reflow process. The primary areas of concern during reflow are flux spread,

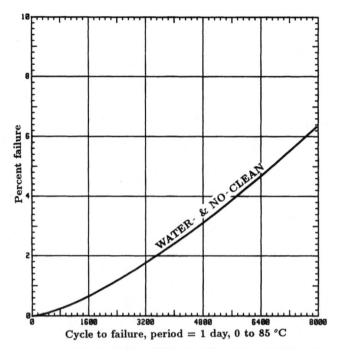

Figure 8.36 Life distribution (Weibull) of the 256-pin PQFP solder joints at operating condition.

slumping, and solder balls. As heat is applied to solder paste, prior to solder melting, the paste's volatile solvents evaporate and the flux chemically activates. Excessive heat during the preheating and predrying stage or a fast ramp-up of temperature can create solder balling and inadequate fluxing activity during reflow. Also, exposure of the plastic PQFPs to high temperatures for relatively long periods of time can cause internal damage, e.g., plastic cracking due to the popcorn effect.[19]

A reflow profile test model was assembled prior to the processing of the test boards. Using thermocouples and a recording device, precise temperature levels were recorded throughout the reflow cycle. Following a few passes of the profile test model through the process, adjustments were programmed into the controller and the desired profile curve was achieved (Fig. 8.39). The profile exposed the plastic PQFPs to temperatures between 183 and 219°C ensuring liquefaction of the solder paste, but the dwell time at that temperature interval was limited to 48 seconds. The belt speed was at 38 in/min (86.4 cm/min) and the temperature (°C) at the 10-zone forced-convection in-line conveyor system is shown in Fig. 8.39.

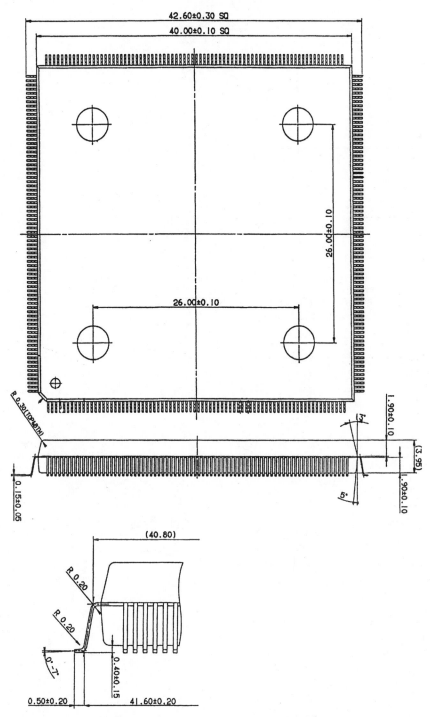

Figure 8.37   0.5-mm pitch, 40 × 40 mm body size, 304-pin PQFP.[1]

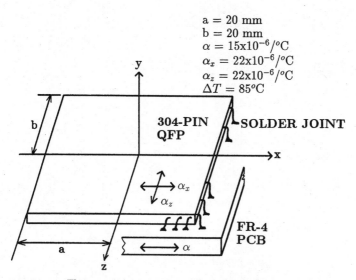

Figure 8.38   Three major sources to produce thermal stress in the 304-pin QFP solder joint: (1) global thermal coefficient of expansion (TCE) mismatch between the QFP and PCB; (2) local TCE mismatch between the lead and solder; and (3) lead stiffness.

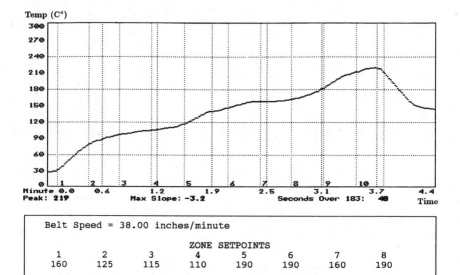

Figure 8.39   Solder reflow temperature profile.

8.3.3 Bending test of the 304-pin PQFP assembly

The setup of the three-point bending test of the 304-pin PQFP assembly is shown in Fig. 8.40. It can be seen that the test board had a clear span of 152.4 mm and was supported at both ends. The load was applied at the center of the test board with a crosshead speed equal to 3.81 mm/min. The 40 × 40 mm PQFP was placed at the center of the 200 × 76.2 × 1.58 mm test board. The planned sample size was 10; however, because of the repeatability of the test results, most of the tests stopped at the fifth specimen.

Figure 8.41 shows the load-deflection curves of the test board with and without the 304-pin PQFP. It can be seen that the PQFP increased the stiffness of the test board. The Young's modulus of the FR-4 PCB can be determined from the load-deflection curve without the PQFP component and the following equation:

$$E = \frac{L^3}{48I} \left(\frac{F}{\Delta} \right) \tag{8.15}$$

where L is the length of the test board (152.4 mm), I is the moment of inertia of the cross-sectional area of the test board ($I = \frac{1}{12}(76.2)(1.58)^3 = 25$ mm^3), and F (force) and Δ (deflection) are the pairs (in this study, we used five pairs) of values from the test results. Thus, we have $E = 22$ GPa. It should be noted that a more accurate value of the Young's modulus can be obtained by $(1 - \nu_1\nu_2)E$ where ν_1 and ν_2 are the Poisson's ratios of the PCB in the material principal directions 1 and 2, respectively.

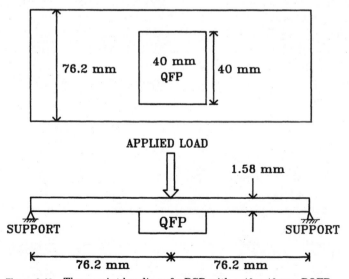

Figure 8.40   Three-point bending of a PCB with a 40 × 40 mm PQFP.

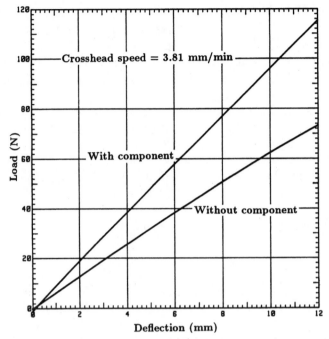

Figure 8.41   Three-point bending load-deflection curves.[1]

Figure 8.42 shows the load- and unload-deflection curves of the test board with the 304-pin QFP SMC. It can be seen that, when unloaded, the FR-4 PCB does not return to its original position. As a matter of fact, the larger the applied loads (or deflections) the greater the offset (permanent) deflections. This is because of the plastic deformation of the PCB. However, it can also be seen from Fig. 8.42 that there were no solder joint or lead failures even with a load at 120 N and a deflection at 12.7 mm. For this size board, this magnitude of deflection is beyond most of the manufacturing, shipping, handling, and rework conditions.

Another set of tests (to failure) showed that failures did not happen until the applied load reached 158.7 N and the deflection reached 19.5 mm (Fig. 8.43). For this size board, this magnitude of deflection is far beyond all the manufacturing, handling, shipping, and rework conditions. The failure model of the solder joint under three-point bending was that a few leads were broken (near the shoulder of the gull wing) on both sides (near the middle) of the PQFP in the longer direction of the test board. Also, a few leads (near the shoulder of the gull wing) and solder joints (near the interface between the gull-wing lead and the solder joint) were broken on both sides (near the corners) of the PQFP in the longer direction of the test board.

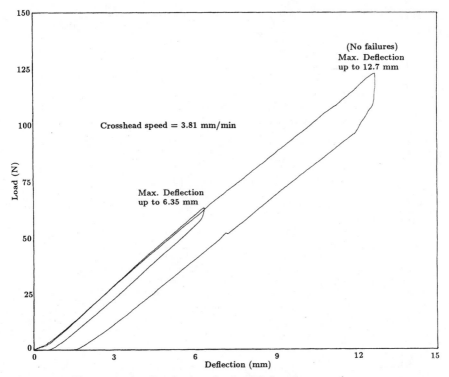

Figure 8.42   Three-point bending load- and unload-deflection curves.[1]

8.3.4   General results for three-point bending

Closed-form solutions for three-point bending are shown in this section. Figure 8.44 shows the setup of a three-point bending of a PCB with a chip (or chip carrier) attached to its middle. The governing equations are

$$EI\frac{d^2v}{dx^2} = -\frac{Px}{2} \qquad 0 \leq x \leq \frac{L}{2} - \frac{D}{2} \qquad (8.16)$$

$$E_e I_e \frac{d^2v}{dx^2} = -\frac{Px}{2} \qquad \frac{L}{2} - \frac{D}{2} \leq x \leq \frac{L}{2} \qquad (8.17)$$

where a = location of applied forces
    E = Young's modulus of PCB
    I = area moment of inertia of PCB
    v = displacement
    x = coordinate
    P = load

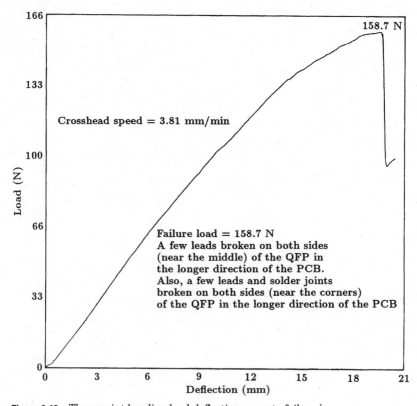

Figure 8.43   Three-point bending load-deflection curve to failure.[1]

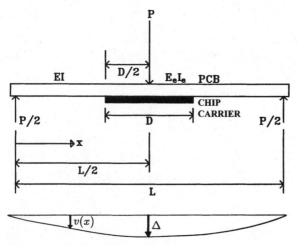

Figure 8.44   Three-point bending of a chip carrier (or bare chip) on a PCB.

L = PCB length
D = module length
E_e = effective Young's modulus of PCB and module
I_e = effective area moment of inertia of PCB and module
Δ = maximum displacement
$\alpha = D/L$
$\beta = E_e I_e / EI$

The general solutions of Eqs. (8.16) and (8.17) are

$$EI\frac{dv}{dx} = -\frac{Px^2}{4} + C_1 \qquad 0 \le x \le \frac{L}{2} - \frac{D}{2} \tag{8.18}$$

$$EIv(x) = -\frac{Px^3}{12} + C_1 x + C_2 \qquad 0 \le x \le \frac{L}{2} - \frac{D}{2} \tag{8.19}$$

$$E_e I_e \frac{dv}{dx} = -\frac{Px^2}{4} + C_3 \qquad \frac{L}{2} - \frac{D}{2} \le x \le \frac{L}{2} \tag{8.20}$$

$$E_e I_e v(x) = -\frac{Px^3}{12} + C_3 x + C_4 \qquad \frac{L}{2} - \frac{D}{2} \le x \le \frac{L}{2} \tag{8.21}$$

The boundary conditions of the present problem are

At $x = 0$, $v = 0$ in Eq. (8.19), i.e.,

$$C_2 = 0 \tag{8.22}$$

At $x = \frac{L}{2}$, $\frac{dv}{dx} = 0$ in Eq. (8.20), i.e.,

$$C_3 = \frac{PL^2}{16} \tag{8.23}$$

At $x = \frac{L}{2} - \frac{D}{2}$, the v in Eqs. (8.19) and (8.21) should be equal, i.e.,

$$\frac{1}{EI}\left[-\frac{P}{12}\left(\frac{L}{2} - \frac{D}{2}\right)^3 + C_1\left(\frac{L}{2} - \frac{D}{2}\right) + C_2\right]$$
$$= \frac{1}{E_e I_e}\left[-\frac{P}{12}\left(\frac{L}{2} - \frac{D}{2}\right)^3 + C_3\left(\frac{L}{2} - \frac{D}{2}\right) + C_4\right] \tag{8.24}$$

At $x = \frac{L}{2} - \frac{D}{2}$, the $\frac{dv}{dx}$ in Eqs. (8.18) and (8.20) should be equal, i.e.,

$$\frac{1}{EI}\left[-\frac{P}{4}\left(\frac{L}{2} - \frac{D}{2}\right)^2 + C_1\right] = \frac{1}{E_e I_e}\left[-\frac{P}{4}\left(\frac{L}{2} - \frac{D}{2}\right)^2 + C_3\right] \tag{8.25}$$

After solving Eqs. (8.22) through (8.25), we have the coefficients of the solutions for the three-point bending.

$$C_1 = \frac{EI}{E_e I_e}\left[-\frac{P}{4}\left(\frac{L}{2}-\frac{D}{2}\right)^2+\frac{PL^2}{16}\right]+\frac{P}{4}\left(\frac{L}{2}-\frac{D}{2}\right)^2 \tag{8.26}$$

$$C_2 = 0 \tag{8.27}$$

$$C_3 = \frac{PL^2}{16} \tag{8.28}$$

$$C_4 = \frac{P}{12}\left(\frac{L}{2}-\frac{D}{2}\right)^3-\frac{PL^2}{16}\left(\frac{L}{2}-\frac{D}{2}\right)$$
$$+\frac{E_e I_e}{EI}\left[\frac{P}{6}\left(\frac{L}{2}-\frac{D}{2}\right)^3\right]+\frac{PL^2}{16}\left(\frac{L}{2}-\frac{D}{2}\right)\left[\frac{2D}{L}-\left(\frac{D}{L}\right)^2\right] \tag{8.29}$$

The dimensionless results in terms of loading, structural geometry, and material properties are shown in Figs. 8.45 through 8.48 for engineering practice convenience.

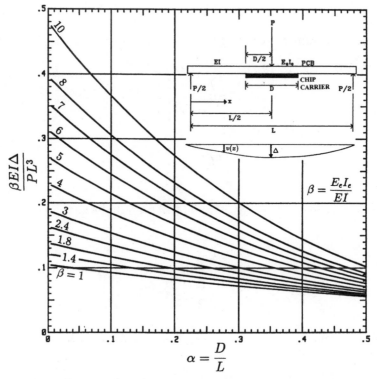

Figure 8.45   Dimensionless curves for the responses of three-point bending of a chip carrier (or bare chip) on a PCB.

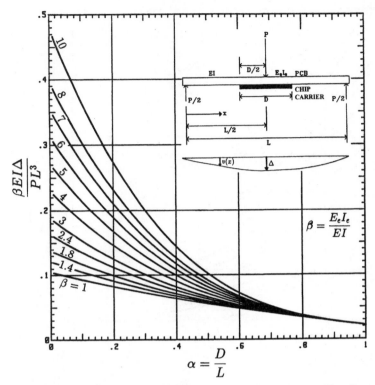

Figure 8.46   Dimensionless curves for the responses of three-point bending of a chip carrier (or bare chip) on a PCB.

8.3.5   General results for four-point bending

Closed-form solutions for four-point bending are shown in this section. Figure 8.49 shows the setup of a four-point bending of a PCB with a chip (or chip carrier) attached to its middle. The governing equations are

$$EI \frac{d^2v}{dx^2} = -\frac{Px}{2} \qquad 0 \le x \le a \tag{8.30}$$

$$EI \frac{d^2v}{dx^2} = -\frac{Pa}{2} \qquad a \le x \le \frac{L}{2} - \frac{D}{2} \tag{8.31}$$

$$E_eI_e \frac{d^2v}{dx^2} = -\frac{Pa}{2} \qquad \frac{L}{2} - \frac{D}{2} \le x \le \frac{L}{2} \tag{8.32}$$

The general solutions for Eqs. (8.30) through (8.32) are

$$EI \frac{dv}{dx} = -\frac{Px^2}{4} + C_1 \qquad 0 \le x \le a \tag{8.33}$$

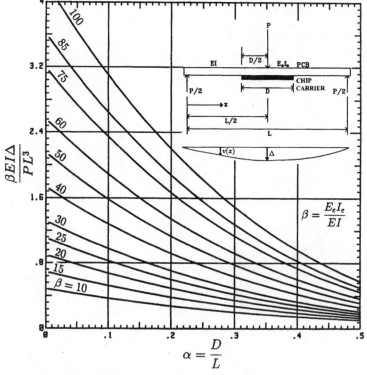

Figure 8.47 Dimensionless curves for the responses of three-point bending of a chip carrier (or bare chip) on a PCB.

$$EIv(x) = -\frac{Px^3}{12} + C_1 x + C_2 \qquad 0 \le x \le a \tag{8.34}$$

$$EI\frac{dv}{dx} = -\frac{Pa}{2}x + C_3 \qquad a \le x \le \frac{L}{2} - \frac{D}{2} \tag{8.35}$$

$$EIv(x) = -\frac{Pa}{4}x^2 + C_3 x + C_4 \qquad a \le x \le \frac{L}{2} - \frac{D}{2} \tag{8.36}$$

$$E_e I_e \frac{dv}{dx} = -\frac{Pa}{2}x + C_5 \qquad \frac{L}{2} - \frac{D}{2} \le x \le \frac{L}{2} \tag{8.37}$$

$$E_e I_e v(x) = -\frac{Pa}{4}x^2 + C_5 x + C_6 \qquad \frac{L}{2} - \frac{D}{2} \le x \le \frac{L}{2} \tag{8.38}$$

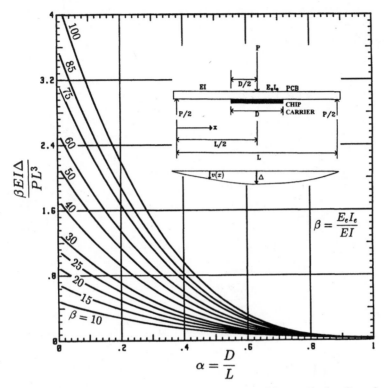

Figure 8.48 Dimensionless curves for the responses of three-point bending of a chip carrier (or bare chip) on a PCB.

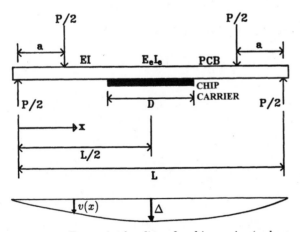

Figure 8.49 Four-point bending of a chip carrier (or bare chip) on a PCB.

The boundary conditions of the four-point bending are

At $x = 0$, $v = 0$ in Eq. (8.34), i.e.,

$$C_2 = 0 \tag{8.39}$$

At $x = \dfrac{L}{2}$, $\dfrac{dv}{dx} = 0$ in Eq. (8.37), i.e.,

$$C_5 = \frac{PLa}{4} \tag{8.40}$$

At $x = a$, the v in Eqs. (8.34) and (8.36) are equal, i.e.,

$$-\frac{Pa^3}{12} + aC_1 = -\frac{Pa^3}{4} + aC_3 + C_4 \tag{8.41}$$

At $x = a$, the $\dfrac{dv}{dx} = 0$ in Eqs. (8.33) and (8.35) are equal, i.e.,

$$-\frac{Pa^2}{4} + C_1 = -\frac{Pa^2}{2} + C_3 \tag{8.42}$$

At $x = \dfrac{L}{2} - \dfrac{D}{2}$, the v in Eqs. (8.36) and (8.38) are equal, i.e.,

$$\frac{1}{EI}\left[-\frac{Pa}{4}\left(\frac{L}{2} - \frac{D}{2}\right)^2 + C_3\left(\frac{L}{2} - \frac{D}{2}\right) + C_4 \right]$$
$$= \frac{1}{E_e I_e}\left[-\frac{Pa}{4}\left(\frac{L}{2} - \frac{D}{2}\right)^2 + C_5\left(\frac{L}{2} - \frac{D}{2}\right) + C_6 \right] \tag{8.43}$$

At $x = \dfrac{L}{2} - \dfrac{D}{2}$, the $\dfrac{dv}{dx}$ in Eqs. (8.35) and (8.37) are equal, i.e.,

$$\frac{1}{EI}\left[-\frac{Pa}{2}\left(\frac{L}{2} - \frac{D}{2}\right) + C_3 \right] = \frac{1}{E_e I_e}\left[-\frac{Pa}{2}\left(\frac{L}{2} - \frac{D}{2}\right) + C_5 \right] \tag{8.44}$$

After solving those six equations, we have the coefficients of the solutions of four-point bending of a PCB with a chip (or chip carrier) attached at its center.

$$C_1 = -\frac{Pa^2}{4} + \frac{Pa}{2}\left(\frac{L}{2} - \frac{D}{2}\right) - \frac{EI}{E_e I_e}\left[\frac{Pa}{2}\left(\frac{L}{2} - \frac{D}{2}\right) - \frac{PLa}{4} \right] \tag{8.45}$$

$$C_2 = 0 \tag{8.46}$$

$$C_3 = \frac{Pa}{2}\left(\frac{L}{2} - \frac{D}{2}\right) - \frac{EI}{E_e I_e}\left[\frac{Pa}{2}\left(\frac{L}{2} - \frac{D}{2}\right) - \frac{PLa}{4} \right] \tag{8.47}$$

$$C_4 = -\frac{Pa^3}{12} \tag{8.48}$$

$$C_5 = \frac{PLa}{4} \tag{8.49}$$

$$C_6 = -\frac{Pa}{4}\left(\frac{L}{2}-\frac{D}{2}\right)^2 - \frac{E_e I_e}{EI}\left[\frac{Pa^3}{12} - \frac{Pa}{4}\left(\frac{L}{2}-\frac{D}{2}\right)^2\right] \tag{8.50}$$

The dimensionless curves are shown in Figs. 8.50 through 8.53.

8.3.6 Twisting test of the 304-pin PQFP assembly

The setup of the four-point twisting test of the 304-pin PQFP assembly is shown in Figs. 8.54 and 8.55. It can be seen that the overall dimensions of the board were: 152.4 × 152.4 × 1.58 mm and the PQFP was located at its center. The board was loaded by two forces at its two opposite corners (the other two corners were supported). These corners were not exactly at the corners of the board but 3.2 mm from the actual corners (for the purpose of applying the forces and supports). Thus, the clear distance between the forces and supports was 146 mm. In this study, the edges of the board were not reinforced, i.e., bending and

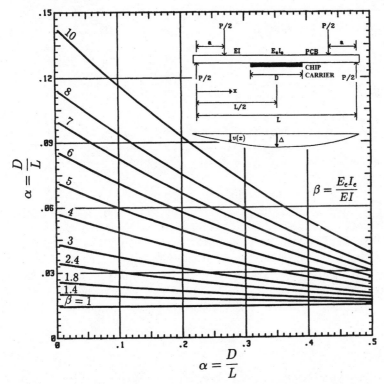

Figure 8.50 Dimensionless curves for the responses of four-point bending of a chip carrier (or bare chip) on a PCB.

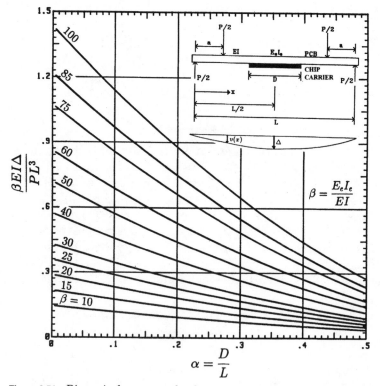

Figure 8.51 Dimensionless curves for the responses of four-point bending of a chip carrier (or bare chip) on a PCB.

shearing deformations in the edge planes were possible. The crosshead speed was 3.81 mm/min. Again, the planned sample size was 10, but because of the repeatability of the test results, all the tests stopped at the fifth specimen.

Figure 8.56 shows the load-deflection curves of the test board with and without the 304-pin PQFP. It can be seen that for both cases, most of the curves are nonlinear, except the small initial portion of the curves which are linear. This was because of the test boundary condition that allows the deformation of the edges of the board and large deflection and rotation of the board. The twisting (or torsional) stiffness of the FR-4 PCB can be determined by the test result curve (without PQFP) in Fig. 8.56. The linear twisting stiffness T can be determined by

$$T = \frac{C^2}{8} \left(\frac{F}{\Delta} \right) \tag{8.51}$$

where $C = 146$, and F and Δ can be obtained from the initial linear portion of the curve without the PQFP in Fig. 8.56 ($F/\Delta = 2.2$). Thus, we have $T = 5860$ N/mm.

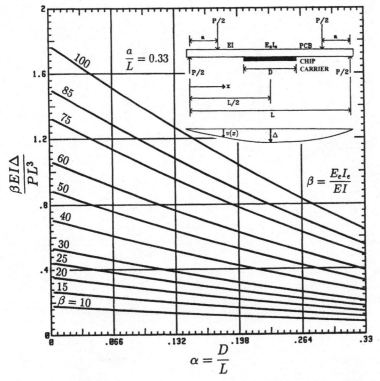

$$\alpha = \frac{D}{L}$$

Figure 8.52 Dimensionless curves for the responses of four-point bending of a chip carrier (or bare chip) on a PCB.

The load- and unload-deflection curves of the 304-pin PQFP test board are shown in Figs. 8.57 and 8.58. Because of the plastic deformation of the FR-4 PCB, there was permanent deflection of the board after unloading, similar to the bending tests. However, as far as the lead and solder joint reliability were concerned, there were no failures at a load equal to 80 N and a deflection equal to 12.7 mm (Fig. 8.57). In fact, the solder joints were not cracked until the load reached 138 N and the deflection reached 22.2 mm (Fig. 8.58). These magnitudes are far beyond most manufacturing processes. These broken solder joints were around the four corners of the PQFP. The failure mechanism of each solder joint was close to the interface between the lead and the solder joint. No lead failures were observed.

8.4 Vibrational Reliability and Solder Joint Inspection of PQFP Assemblies

PQFP's solder joint reliability under shock and vibration conditions has been reported by Lau et al.[7] In this section, a simple vibration method for detecting unsoldered leads in postsolder inspection is presented.[8]

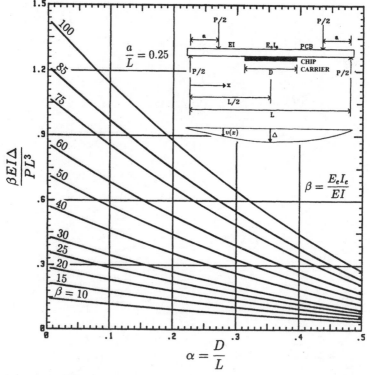

Figure 8.53 Dimensionless curves for the responses of four-point bending of a chip carrier (or bare chip) on a PCB.

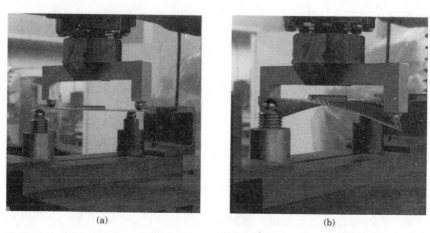

(a)   (b)

Figure 8.54 Four-point twisting of a 40×40 mm PQFP on a PCB: (a) undeformed state; (b) deformed state.[1]

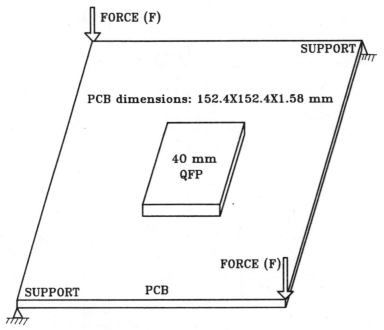

Figure 8.55   Four-point loading to produce twisting of a PCB with a 40 × 40 mm PQFP.[1]

8.4.1 Laser Doppler vibrometer (LDV)

LDV was originally developed in the late 1960s for measuring fluid flow and was known as laser Doppler anemometry. It was extended to velocity and/or vibration measurements of solid objects and hence the terms velocimetry and vibrometry, which are both commonly used.

The basic principle is as follows: a beam of laser light reflected off of a moving object is frequency-shifted by the Doppler effect in proportion to the object's velocity. By mixing that beam with a reference beam, the shift can be measured, and, hence, the velocity can be obtained. This is similar to a time-based focused beam interferometer, but the electronics of the LDV are designed to extract the vibration information from the signal, whereas the interferometer electronics extract the slowly moving or stationary position information.

In the LDV measurement system, the laser light frequency is split into two parts. One part is reflected off the object (gull-wing lead) which is assumed to be moving in a sinusoidal fashion with frequency ω_v. The frequency of the reflected light will be Doppler-shifted by an amount $\Delta\omega_v$ proportional to the velocity of the lead, and this velocity is proportional to the vibration. So the form of the light reflected can be written as

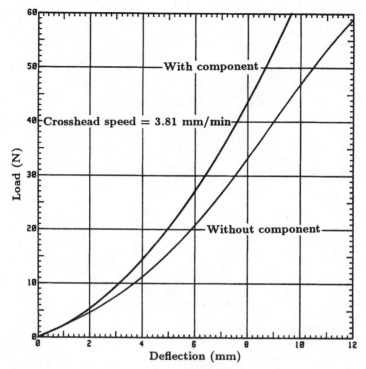

Figure 8.56 Four-point twisting load-deflection curve of a 40 × 40 mm PQFP on a PCB.[1]

$$E_v = A \cos (\omega_o + \Delta\omega_v)t \qquad (8.52)$$

The other part of the beam is frequency-shifted by a constant amount ω_b by an acoustooptic modulator, or Bragg cell. This waveform is then

$$E_b = B \cos (\omega_o + \omega_b)t \qquad (8.53)$$

These two parts are recombined and they interfere. A PIN diode detects the intensity after the recombination, and this intensity is

$$I(t) = [E_b + E_v]^2$$

$$= \frac{A^2}{2} + \frac{B^2}{2} + AB \cos (\omega_b - \Delta\omega_v)t \qquad (8.54)$$

$$+ \text{higher frequency terms}$$

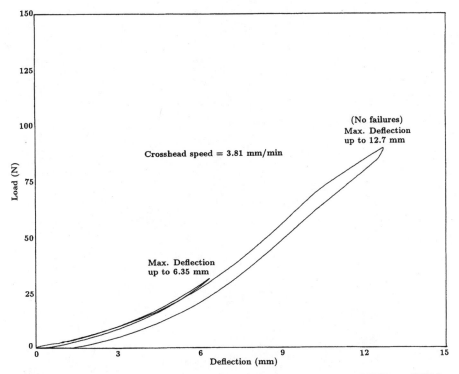

Figure 8.57   Four-point twisting load and unload curves of a 40 × 40 mm PQFP on a PCB.[1]

The first two terms in the expression are usually lost in the detector amplifier, and the high-frequency terms are not detected at all. So the vibration information carried in $\Delta\omega_v$ is recovered by frequency demodulating the signal, with the FM demodulator tuned to ω_b.

8.4.2 LDV measurement system

Figure 8.59 shows a schematic of the optical system and Fig. 8.60 shows the processing electronics. An air jet (approximately 35-kPa gage, or 5 psig) is used as an acoustic white noise source and is directed at the lead under test to stimulate natural vibration at frequency ω_v. The air jet is strong enough to force an unsoldered lead to vibrate even if at rest with the lead in contact with the pad. The measuring laser beam (frequency ω_o) is aimed at the shoulder of the lead; the shoulder is used because it reflects light back to the system better than other parts of the lead, is easy to access, and vibrates more than enough to be detected. The reflected light is collected and recombined with the reference beam of

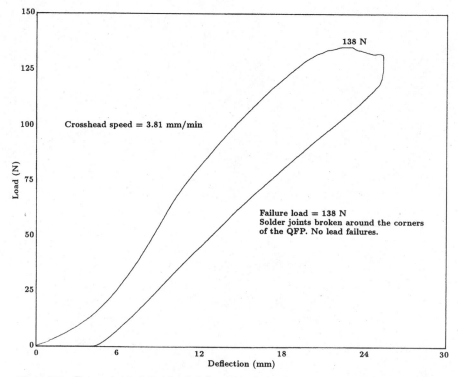

Figure 8.58   Four-point twisting load-deflection curve to failure of a 40 × 40 mm PQFP on a PCB.[1]

frequency ($\omega_o + \omega_b$), and the combined intensity is detected. The signal is demodulated, and the output of the LDV system is a signal proportional to the real-time velocity of the lead shoulder with a bandwidth up to about 200 kHz. (In practice, the usable bandwidth is limited to about 100 kHz by the noise floor level and the low signal power above this frequency, even though the demodulator can work up to 200 kHz.) From the finite element analysis,[8] the expected natural frequency (first mode, at least) for all the unsoldered leads was well below 100 kHz, so the bandwidth of the LDV system was more than enough. The velocity signal is fed to a spectrum analyzer which displays the frequency spectrum of the velocity signal. The peaks in the spectrum for each lead are the resonant frequencies of the structure ω_v.

The Doppler frequency shift and the demodulation to obtain the lead velocity occur essentially in real time, so the overall speed of the system is determined by the spectrum analyzer and/or filters that process the velocity signal to locate the frequency peaks and the mechanical system for accessing the leads. In the laboratory system built for this

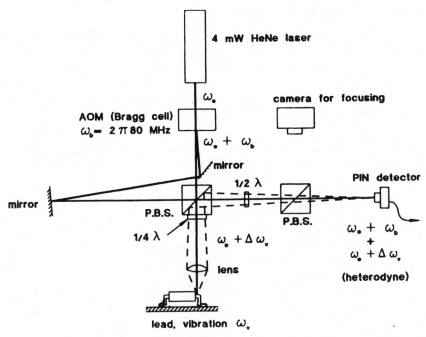

Figure 8.59 Optics for LDV. (The solid line is direct beam, dashed lines are outlines of laser light reflected off subject. B.S. = beamsplitter; P.B.S. = polarizing B.S.)[8]

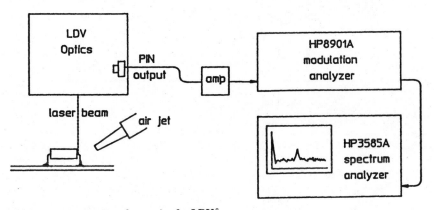

Figure 8.60 Processing electronics for LDV.[8]

study, leads were measured one at a time with no effort spent on decreasing access time. The limit on speed in the signal processing was a 1 Hz filter on the signal amplitude (video bandwidth setting) used to average the amplitude level. The spectrum analyzer resolution bandwidth setting was 1 kHz.

8.4.3   LDV experimental results and solder joint inspection

As the PQFPs rested on their backs, the leads were measured on the bottom of their feet. Because these leads had no solder on them at all, their geometry most nearly matched the leads modeled by the finite element method[8] (Fig. 8.61 without solder joint). The resulting frequency spectra for the unmounted components are not given here, but they are nearly identical to the spectra for unsoldered leads that follow.

If the vibration spectra for good and bad solder joints were different enough, they could be used as inspection criteria. To determine the feasibility of this method, sample boards containing both good and bad solder joints were measured for the PQFP leads. Figure 8.62 shows the spectra of the PQFP's lead. It contains a relatively flat spectrum obtained from a good solder joint and a spectrum from a bad joint, which has peaks at the resonant frequencies. These overlaid plots demonstrate the clear differences in the spectra of soldered leads (good solder joints) and unsoldered leads (bad solder joints), and indicate that the vibration amplitude at the resonant frequency could be used as solder joint inspection criteria for detecting unsoldered leads.

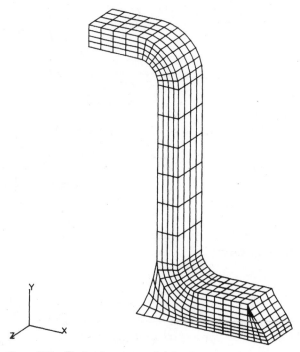

Figure 8.61   Finite element model for the vibration analysis of the PQFP lead and solder joint.

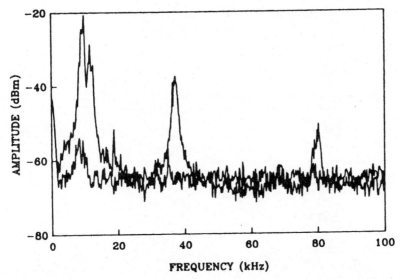

Figure 8.62 Vibration spectra for the PQFP lead (thickness is 0.15 mm). Bad solder joint spectrum has peaks; good solder joint spectrum is fairly flat.[8]

8.4.4 Comparison of analytical and experimental results

We could expect the vibration peaks to be at the resonant frequency of the free lead as predicted in the finite element model,[8] and indeed they are. The correlation between the model results and the experimental results is shown in Fig. 8.63. It shows the resonant frequency as a func-

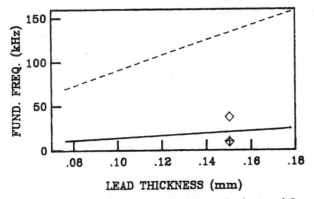

Figure 8.63 Finite element results for the fundamental frequencies as a function of lead thickness. (Dashed line = good solder joint, and solid line = unsoldered lead, or bad solder joint. Data points from LDV measurements overlaid, plus sign (+) = data point for unmounted PQFP, diamond (◇) = data point for unsoldered lead, or bad solder joint.)

tion of the lead thickness. The solid line is the finite element result with fixed-free boundary conditions corresponding to an unsoldered lead, and the dashed line is the analytical result for fixed-fixed boundary conditions corresponding to a soldered lead (Fig. 8.61). Superimposed on these plots are the data points from the LDV vibration measurements. The plus signs (+) and the diamonds (◇) are the frequency peaks obtained from the spectra of the unmounted leads and unsoldered leads (bad solder joints), respectively. It can be seen that they are in excellent agreement.

Acknowledgments

The principal author (Lau) would like to thank S. Golwalkar, R. Surratt, P. Boysan, and R. Forhringer of Intel, and S. Erasmus of Hewlett-Packard for their contributions on Refs. 10, 11, and 12, S. Erasmus for his contribution on Ref. 1, and C. Keely of Hewlett-Packard for her contribution on Ref. 8. The authors want to thank R. Govila, C. Larner, S. Twerefour, S. Dolot, and D. Gilbert of Ford Motor Company for their contributions on Refs. 2, 4, and 5. The authors learned a lot from them.

References

1. Lau, J. H., and S. Erasmus, "Experimental and Analytical Studies of 304-Pin, 0.5 mm Pitch, 40 mm Body Size Plastic Quad Flat Pack (QFP) Solder Joints and Leads Under Bending, Twisting, and Thermal Conditions," *ASME Transactions, Journal of Electronic Packaging,* 115:322–328, 1993.
2. Lau, J. H., R. Govila, C. Larner, Y.-H. Pao, S. Erasmus, S. Dolot, M. Jalilian, and M. Lancaster, "No-Clean and Solvent-Clean Mass Reflow Processes of 0.4 mm Pitch, 256-Pin Fine Pitch Quad Flat Packs (QFP)," *Circuit World,* 19(1):19–26, October 1992.
3. Lau, J. H., *Solder Joint Reliability: Theory and Applications,* Van Nostrand Reinhold, New York, 1991.
4. Lau, J. H., R. Govila, C. Larner, Y.-H. Pao, S. Erasmus, S. Dolot, and V. Solberg, "No-Clean and Water-Clean Mass Reflow Processes of 0.4 mm Pitch, 256-Pin Fine Pitch Quad Flat Packs (QFP)," *Proceedings of the 13th IEEE International Electronics Manufacturing Technology Symposium,* September 1992, pp. 305–315.
5. Lau, J. H., Y.-H. Pao, C. Larner, S. Twerefour, R. Govila, D. Gilbert, S. Erasmus, and S. Dolot, "Reliability of 0.4 mm Pitch, 256-Pin Plastic Quad Flat Pack No-Clean and Water-Clean Solder Joints," *Soldering & Surface Mount Technology,* no. 16, February 1994, pp. 42–50.
6. Lau, J. H., and G. Harkins, "Stiffness of Gull-Wing Leads and Solder Joints for a Plastic Quad Flat Pack," *IEEE Transactions on CHMT,* 13:124–130, March 1990.
7. Lau, J. H., L. Powers, J. Baker, D. Rice, and W. Shaw, "Solder Joint Reliability of Fine Pitch Surface Mount Technology Assemblies," *IEEE Transactions on CHMT,* 13:534–544, September 1990.
8. Lau, J. H., and C. A. Keely, "Dynamic Characterization of Surface Mount Components Leads for Solder Joint Inspection," *IEEE Transactions on CHMT,* 12:594–602, December 1989.
9. Lau, J. H., D. Rice, and G. Harkins, "Thermal Stress Analysis of TAB Packages and Interconnections," *IEEE Transactions on CHMT,* 13:183–188, March 1990.
10. Lau, J. H., S. Golwalkar, S. Erasmus, R. Surratt, and P. Boysan, "Experimental and Analytical Studies of 28-Pin Thin Small Outline Package (TSOP) Solder-Joint Reli-

ability," *Journal of Electronic Packaging, Transactions of ASME,* **114:**169–176, June 1992.

11. Lau, J. H., S. Golwalkar, P. Boysan, D. Surratt, R. Forhringer, and S. Erasmus, "Solder Joint Reliability of a Thin Small Outline Package (TSOP)," *Circuit World,* **20:**12–19, 1993.

12. Lau, J. H., S. Golwalkar, and S. Erasmus, "Advantages and Disadvantages of Thin Small Outline Packages (TSOP) with Copper Gull-Wing Leads," *Proceedings of ASME International Electronics Packaging Conference,* Binghamton, N.Y., September 1993, pp. 1119–1126.

13. Lau, J. H., G. Harkins, D. Rice, J. Kral, and B. Wells, "Experimental and Statistical Analyses of Surface-Mount Technology PLCC Solder-Joint Reliability," *IEEE Transactions on Reliability,* **37**(5):524–530, December 1988.

14. Lau, J. H., "Thermal Stress Analysis of SMT PQFP Packages and Interconnections," *Journal of Electronic Packaging, Transactions of ASME,* **111:**2–8, March 1989.

15. Lau, J. H., *Thermal Stress and Strain in Microelectronics Packaging,* Van Nostrand Reinhold, New York, 1993.

16. Lau, J. H., and D. W. Rice, "Solder Joint Fatigue in Surface Mount Technology: State of the Art," *Solid State Technology,* **28:**91–104, October 1985.

17. Bhattacharya, B., W. A. Huffman, W. E. Jahsman, and B. Natarajan, "Moisture Absorption and Mechanical Performance of Surface Mountable Plastic Packages," *IEEE Proceedings of the 38th Electronic Components Conference,* May 1988, pp. 49–58.

18. Emerick, A., J. Ellerson, J. McCreary, R. Noreika, C. Woychik, and P. Viswanadham, "Enhancement of TSOP Solder Joint Reliability Using Encapsulation," *IEEE Proceedings of the 43rd Electronic Components & Technology Conference,* June 1993, pp. 187–192.

19. Golwalkar, S., "Solutions to Moisture Resistance Degradation of Plastic Surface Mount Components," in *Thermal Stresses and Strains in Microelectronics Packaging,* J. H. Lau, ed., Van Nostrand Reinhold, New York, 1993.

20. Golwalkar, S., P. Boysan, R. Foehringer, and J. Jacobs, "Moisture Sensitivity of Thin Small Outline Packages," *IEEE Proceedings of the 41st Electronic Components & Technology Conference,* May 1991, pp. 745–749.

21. Kim, S., "The Role of Plastic Package Adhesion in IC Performance," *IEEE Proceedings of the 41st Electronic Components & Technology Conference,* May 1991, pp. 750–758.

22. Moore, T. M., R. McKenna, and S. J. Kelsall, "Correlation of Surface Mount Plastic Package Reliability Testing to Nondestructive Inspection by Scanning Acoustic Microscopy," *IEEE Proceedings of the 29th International Reliability Physics Symposium,* April 1991, pp. 160–166.

23. Noctor, D., E. Bader, A. Viera, P. Boysan, S. Golwalkar, and D. Foehringer, "Attachment Reliability Evaluation and Failure Analysis of Thin Small Outline Packages (TSOPs)," *IEEE Proceedings of the 43rd Electronic Components & Technology Conference,* June 1993, pp. 54–61.

24. Solomon, H. D., "Fatigue of 60/40 Solder," *IEEE Transactions on Components, Hybrids, and Manufacturing Technology,* vol. CHMT-9:423–432, December 1986.

25. Viswanadham, P., "Reliability Aspects of Fine Pitch Assembly," in *Handbook of Fine Pitch Surface Mount Technology,* J. H. Lau, ed., Van Nostrand Reinhold, New York, 1993.

26. Viswanadham, P., D. Coplin, A. Emerick, and R. Haggett, "Solder Joint Reliability of TSOPs—An Overview," *IEEE Proceedings of the 43rd Electronic Components & Technology Conference,* June 1993, pp. 863–870.

27. Lau, J. H., G. Dody, W. Chen, M. McShane, D. Rice, S. Erasmus, and W. Adamjee, "Experimental and Analytical Studies of 208-Pin Fine Pitch Quad Flat Pack Solder-Joint Reliability," *Circuit World,* **18**(2):13–19, January 1992.

28. Seraphim, D. P., R. Lasky, and C. Y. Li, *Principles of Electronic Packaging,* McGraw-Hill, New York, 1989.

29. Tummala, R. R., and E. Rymaszewski, *Microelectronics Packaging Handbook,* Van Nostrand Reinhold, New York, 1989.

30. Harper, C. A., *Handbook of Microelectronics Packaging,* McGraw-Hill, New York, 1991.

31. Pecht, M., *Handbook of Electronic Package Design,* Marcel Dekker, New York, 1991.

32. Socolowski, N., "Key Parameters to Fine Pitch Component Assemblies," *Proceedings of NEPCON West*, 1990, pp. 221–227.
33. Lea, C., *A Scientific Guide to Surface Mount Technology*, Electrochemical Publications, Scotland, 1988.
34. Wassink, R. J. K., *Soldering in Electronics*, Electrochemical Publications, Scotland, 1989.
35. Ellis, B. N., *Cleaning and Contamination of Electronics Components and Assemblies*, Electrochemical Publications, Scotland, 1986.
36. Solberg, V., *Design Guidelines for Surface Mount Technology*, TAB (McGraw-Hill) Books, New York, 1990.
37. Prasad, R. P., *Surface Mount Technology*, Van Nostrand Reinhold, New York, 1989.
38. Engel, P. A., D. V. Caletka, and M. R. Palmer, "Stiffness and Fatigue Study for Surface Mounted Module/Lead/Card Systems," *Journal of Electronic Packaging, Transactions of ASME*, **113**:129–137, June 1991.
39. Engel, P. A., and J. T. Vogelmann, "Approximate Structural Analysis of Circuit Card Systems Subjected to Torsion," *Journal of Electronic Packaging, Transactions of ASME*, **114**:203–210, June 1992.
40. Barker, D. B., I. Sharif, A. Dasgupta, and M. G. Pecht, "Effect of SMC Lead Dimensional Variabilities on Lead Compliance and Solder Joint Fatigue Life," *Journal of Electronic Packaging, Transactions of ASME*, **114**:177–184, June 1992.
41. Solomon, H. D., "Fatigue of 60/40 Solder," *IEEE Transactions on Components, Hybrids, and Manufacturing Technology*, vol. CHMT-9:423–432, December 1986.
42. Lau, J. H., *Thermal Stress and Strain in Microelectronic Packaging*, Van Nostrand Reinhold, New York, 1993.
43. Lau, J., "A Brief Introduction to Fine Pitch Technology," in *Handbook of Fine Pitch Surface Mount Technology*, J. H. Lau, ed., Van Nostrand Reinhold, New York, 1994, pp. 1–54.
44. Buschbom, M., W. Schroen, and E. Wolfe, "Fine Pitch Packaging for Surface Mount," in *Handbook of Fine Pitch Surface Mount Technology*, J. H. Lau, ed., Van Nostrand Reinhold, New York, 1994, pp. 55–80.
45. Lea, C., "Solderability of Fine Pitch Surface Mount Components," in *Handbook of Fine Pitch Surface Mount Technology*, J. H. Lau, ed., Van Nostrand Reinhold, New York, 1994, pp. 81–133.
46. Holmes, R., "Substrate Materials and Designs for Fine Pitch Technology," in *Handbook of Fine Pitch Surface Mount Technology*, J. H. Lau, ed., Van Nostrand Reinhold, New York, 1994, pp. 134–160.
47. Hwang, J., "Fine Pitch Soldering and Solder Paste," in *Handbook of Fine Pitch Surface Mount Technology*, J. H. Lau, ed., Van Nostrand Reinhold, New York, 1994, pp. 161–193.
48. Morris, J., and T. Wojcik, "Screen and Stencil Printing Technology for Fine Pitch Assembly," in *Handbook of Fine Pitch Surface Mount Technology*, J. H. Lau, ed., Van Nostrand Reinhold, New York, 1994, pp. 194–232.
49. Kou, Y., "Pick and Place Technology for Fine Pitch Assembly," in *Handbook of Fine Pitch Surface Mount Technology*, J. H. Lau, ed., Van Nostrand Reinhold, New York, 1994, pp. 233–266.
50. Racz, L., and J. Szekely, "Estimation of Solder Volume," in *Handbook of Fine Pitch Surface Mount Technology*, J. H. Lau, ed., Van Nostrand Reinhold, New York, 1994, pp. 267–307.
51. Wasielewski, J., "Vapor Phase Solder Reflow for Fine Pitch Assembly," in *Handbook of Fine Pitch Surface Mount Technology*, J. H. Lau, ed., Van Nostrand Reinhold, New York, 1994, pp. 308–332.
52. Zarrow, P., "Convection/IR and Convection Dominant Reflow Soldering of Fine Pitch Devices," in *Handbook of Fine Pitch Surface Mount Technology*, J. H. Lau, ed., Van Nostrand Reinhold, New York, 1994, pp. 333–347.
53. Dow, S., "Forced Convection Solder Reflow for Fine Pitch Assembly," in *Handbook of Fine Pitch Surface Mount Technology*, J. H. Lau, ed., Van Nostrand Reinhold, New York, 1994, pp. 348–375.

54. Adams, S., "Nitrogen Based Solder Reflow for Fine Pitch Assembly," in *Handbook of Fine Pitch Surface Mount Technology*, J. H. Lau, ed., Van Nostrand Reinhold, New York, 1994, pp. 376–410.
55. Peeples, J., "Pulsed Thermode Hotbar Bonding," in *Handbook of Fine Pitch Surface Mount Technology*, J. H. Lau, ed., Van Nostrand Reinhold, New York, 1994, pp. 411–442.
56. Fritz, H., "Aqueous and Semi-Aqueous Cleaning of Fine Pitch Assembly," in *Handbook of Fine Pitch Surface Mount Technology*, J. H. Lau, ed., Van Nostrand Reinhold, New York, 1994, pp. 443–478.
57. Melton, C., "Non-CFC Cleaning (No-Clean) of Fine Pitch Assembly," in *Handbook of Fine Pitch Surface Mount Technology*, J. H. Lau, ed., Van Nostrand Reinhold, New York, 1994, pp. 479–487.
58. Millard, D., "Inspection of Fine Pitch Assemblies," in *Handbook of Fine Pitch Surface Mount Technology*, J. H. Lau, ed., Van Nostrand Reinhold, New York, 1994, pp. 488–504.
59. Lancaster, J., "Surface Mount Repair," in *Handbook of Fine Pitch Surface Mount Technology*, J. H. Lau, ed., Van Nostrand Reinhold, New York, 1994, pp. 505–519.
60. Lancaster, M., "Total Quality Management (TQM) and Statistical Process Control (SPC) for Fine Pitch Assembly," in *Handbook of Fine Pitch Surface Mount Technology*, J. H. Lau, ed., Van Nostrand Reinhold, New York, 1994, pp. 520–549.
61. Huang, C., "Thermal Management of Fine Pitch Assembly," in *Handbook of Fine Pitch Surface Mount Technology*, J. H. Lau, ed., Van Nostrand Reinhold, New York, 1994, pp. 550–566.
62. Grasso, J., "Electrical Testing of Bare Fine Pitch Printed Circuit Boards," in *Handbook of Fine Pitch Surface Mount Technology*, J. H. Lau, ed., Van Nostrand Reinhold, New York, 1994, pp. 567–598.
63. Viswanadham, P., "Reliability Aspects of Fine Pitch Assembly," in *Handbook of Fine Pitch Surface Mount Technology*, J. H. Lau, ed., Van Nostrand Reinhold, New York, 1994, pp. 598–636.
64. Bonner, J., "Technical Management Issues of Fine Pitch Technology," in *Handbook of Fine Pitch Surface Mount Technology*, J. H. Lau, ed., Van Nostrand Reinhold, New York, 1994, pp. 637–667.

Index

(*Italics* indicate illustrations.)

ABOUT THE AUTHORS

JOHN H. LAU is the president of Express Packaging Systems (EPS, Inc.), in Palo Alto, California. His current interests cover a broad range of electronics packaging and manufacturing technology.

Prior to founding EPS in November 1995, he worked for Hewlett-Packard Company, Sandia National Laboratory, Bechtel Power Corporation, and Exxon Production and Research Company. With more than 26 years of R&D and manufacturing experience in the electronics, petroleum, nuclear, and defense industries, he is the author and coauthor of over 100 peer-reviewed technical publications, the author of twenty book chapters, and the author and editor of ten books: *Solder Joint Reliability; Handbook of Tape Automated Bonding; Thermal Stress and Strain in Microelectronics Packaging; The Mechanics of Solder Alloy Interconnects; Handbook of Fine Pitch Surface Mount Technology; Chip On Board Technologies for Multichip Modules; Ball Grid Array Technology; Flip Chip Technologies; Solder Joint Reliability of BGA, CSP, Flip Chip, and Fine Pitch SMT Assemblies,* and the forthcoming *Electronics Packaging: Design, Material, Process, and Reliability.*

John has served as one of the technical editors of the *ASME Transactions, Journal of Electronic Packaging,* and the *IEEE Transactions on Components, Packaging, and Manufacturing Technology.* He has also served as session chairman, program chairman, general chairman, and invited speaker of several IEEE, ASME, ASM, MRS, ISHM, and NEPCON conferences and symposiums. He has received a few awards from ASME and IEEE and is an IEEE Fellow.

John received his Ph.D. in Theoretical and Applied Mechanics from the University of Illinois, and M.A.Sc. in Structural Engineering from the University of British Columbia, a second M.S. in Engineering Mechanics from the University of Wisconsin, and a third M.S. in Management Science from Fairleigh Dickinson University. He also has a B.E. in Civil Engineering from National Taiwan University.

DR. YI-HSIN PAO is currently a Supervisor and Staff Technical Specialist at Ford Research Laboratory. His interests in research include advanced electronics packaging technology, design methodology, failure mode analysis, reliability prediction, solder joint fatigue, fracture mechanics, and the finite element method.

Dr. Pao joined Ford Motor Company in 1988 and has received the Ford Technological Achievement Award, the Publication Award, and the Innovation Award on his work of electronics packaging. Currently, he is serving as an associate technical editor of *ASME Transactions, Journal of Electronic Packaging* and is a board director of the Electronic Packaging Division, Society of Experimental Mechanics. He is a member of the Honor Society of Phi Kappa Phi, Phi Beta Delta Honor Society for International Scholars, American Society of Mechanical Engineers (ASME), Society of Experimental Mechanics (SEM), and Society of Automotive Engineers (SAE). He was a member of the organization committee of the First International Symposium on Microelectronic Package and PCB Technology, 1994, in Beijing, China. He was a member of the organization committee of the 1995 ASME International and Intersociety Electronic Packaging Conference (INTERpack '95) in Hawaii and a member of the Technical Program Committee, International Conference on Electronic Packaging, 1996, in Shanghai, China.

He has published two book chapters, over 40 conference and journal papers, and more than 20 technical reports in the area of electronics packaging reliability. Dr. Pao has also been invited to speak on electronics packaging at a number of universities and research organizations in the United States, Taiwan, Japan, and Hong Kong.

Dr. Pao received a Ph.D. in Applied Mechanics from Ohio State University, an M.S. in Mechanical Engineering from the University of Oklahoma, and a B.S. in Marine Engineering from the National Taiwan Ocean University.